Probability and Mathematical Statistics (Con

RAO • Linear Statistical Inference and
ROBERTSON, WRIGHT, and DYKST ... A • Order Restricted Statistical Inference
ROGERS and WILLIAMS • Diffusions, Markov Processes, and Martingales, Volume II: Îto Calculus
ROHATGI • Statistical Inference
ROSS • Stochastic Processes
RUBINSTEIN • Simulation and the Monte Carlo Method
RUZSA and SZEKELY • Algebraic Probability Theory
SCHEFFE • The Analysis of Variance
SEBER • Linear Regression Analysis
SEBER • Multivariate Observations
SEBER and WILD • Nonlinear Regression
SEN • Sequential Nonparametrics: Invariance Principles and Statistical Inference
SERFLING • Approximation Theorems of Mathematical Statistics
SHORACK and WELLNER • Empirical Processes with Applications to Statistics
STAUDTE and SHEATHER • Robust Estimation and Testing
STOYANOV • Counterexamples in Probability
STYAN • The Collected Papers of T. W. Anderson: 1943–1985
WHITTAKER • Graphical Models in Applied Multivariate Statistics
YANG • The Construction Theory of Denumerable Markov Processes

Applied Probability and Statistics

ABRAHAM and LEDOLTER • Statistical Methods for Forecasting
AGRESTI • Analysis of Ordinal Categorical Data
AGRESTI • Categorical Data Analysis
AICKIN • Linear Statistical Analysis of Discrete Data
ANDERSON and LOYNES • The Teaching of Practical Statistics
ANDERSON, AUQUIER, HAUCK, OAKES, VANDAELE, and WEISBERG • Statistical Methods for Comparative Studies
ARTHANARI and DODGE • Mathematical Programming in Statistics
ASMUSSEN • Applied Probability and Queues
*BAILEY • The Elements of Stochastic Processes with Applications to the Natural Sciences
BARNETT • Interpreting Multivariate Data
BARNETT and LEWIS • Outliers in Statistical Data, *Second Edition*
BARTHOLOMEW • Stochastic Models for Social Processes, *Third Edition*
BARTHOLOMEW and FORBES • Statistical Techniques for Manpower Planning
BATES and WATTS • Nonlinear Regression Analysis and Its Applications
BECK and ARNOLD • Parameter Estimation in Engineering and Science
BELSLEY • Conditioning Diagnostics: Collinearity and Weak Data in Regression
BELSLEY, KUH, and WELSCH • Regression Diagnostics: Identifying Influential Data and Sources of Collinearity
BHAT • Elements of Applied Stochastic Processes, *Second Edition*
BHATTACHARYA and WAYMIRE • Stochastic Processes with Applications
BIEMER, GROVES, LYBERG, MATHIOWETZ, and SUDMAN • Measurement Errors in Surveys
BLOOMFIELD • Fourier Analysis of Time Series: An Introduction
BOLLEN • Structural Equations with Latent Variables
BOX • R. A. Fisher, the Life of a Scientist
BOX and DRAPER • Empirical Model-Building and Response Surfaces
BOX and DRAPER • Evolutionary Operation: A Statistical Method for Process Improvement
BOX, HUNTER, and HUNTER • Statistics for Experimenters: An Introduction to Design, Data Analysis, and Model Building
BROWN and HOLLANDER • Statistics: A Biomedical Introduction
BUCKLEW • Large Deviation Techniques in Decision, Simulation, and Estimation
BUNKE and BUNKE • Nonlinear Regression, Functional Relations and Robust Methods: Statistical Methods of Model Building
BUNKE and BUNKE • Statistical Inference in Linear Models, Volume I
CHAMBERS • Computational Methods for Data Analysis
CHATTERJEE and HADI • Sensitivity Analysis in Linear Regression
CHATTERJEE and PRICE • Regression Analysis by Example, *Second Edition*

*Now available in a lower priced paperback edition in the Wiley Classics Library.

Applied Probability and Statistics (Continued)

CHOW · Econometric Analysis by Control Methods
CLARKE and DISNEY · Probability and Random Processes: A First Course with Applications, *Second Edition*
COCHRAN · Sampling Techniques, *Third Edition*
COCHRAN and COX · Experimental Designs, *Second Edition*
CONOVER · Practical Nonparametric Statistics, *Second Edition*
CONOVER and IMAN · Introduction to Modern Business Statistics
CORNELL · Experiments with Mixtures, Designs, Models, and the Analysis of Mixture Data, *Second Edition*
COX · A Handbook of Introductory Statistical Methods
COX · Planning of Experiments
CRESSIE · Statistics for Spatial Data
DANIEL · Applications of Statistics to Industrial Experimentation
DANIEL · Biostatistics: A Foundation for Analysis in the Health Sciences, *Fourth Edition*
DANIEL and WOOD · Fitting Equations to Data: Computer Analysis of Multifactor Data, *Second Edition*
DAVID · Order Statistics, *Second Edition*
DAVISON · Multidimensional Scaling
DEGROOT, FIENBERG, and KADANE · Statistics and the Law
*DEMING · Sample Design in Business Research
DILLON and GOLDSTEIN · Multivariate Analysis: Methods and Applications
DODGE · Analysis of Experiments with Missing Data
DODGE and ROMIG · Sampling Inspection Tables, *Second Edition*
DOWDY and WEARDEN · Statistics for Research, *Second Edition*
DRAPER and SMITH · Applied Regression Analysis, *Second Edition*
DUNN · Basic Statistics: A Primer for the Biomedical Sciences, *Second Edition*
DUNN and CLARK · Applied Statistics: Analysis of Variance and Regression, *Second Edition*
ELANDT-JOHNSON and JOHNSON · Survival Models and Data Analysis
FLEISS · The Design and Analysis of Clinical Experiments
FLEISS · Statistical Methods for Rates and Proportions, *Second Edition*
FLEMING and HARRINGTON · Counting Processes and Survival Analysis
FLURY · Common Principal Components and Related Multivariate Models
FRANKEN, KÖNIG, ARNDT, and SCHMIDT · Queues and Point Processes
GALLANT · Nonlinear Statistical Models
GIBBONS, OLKIN, and SOBEL · Selecting and Ordering Populations: A New Statistical Methodology
GNANADESIKAN · Methods for Statistical Data Analysis of Multivariate Observations
GREENBERG and WEBSTER · Advanced Econometrics: A Bridge to the Literature
GROSS and HARRIS · Fundamentals of Queueing Theory, *Second Edition*
GROVES · Survey Errors and Survey Costs
GROVES, BIEMER, LYBERG, MASSEY, NICHOLLS, and WAKSBERG · Telephone Survey Methodology
GUPTA and PANCHAPAKESAN · Multiple Decision Procedures: Theory and Methodology of Selecting and Ranking Populations
GUTTMAN, WILKS, and HUNTER · Introductory Engineering Statistics, *Third Edition*
HAHN and MEEKER · Statistical Intervals: A Guide for Practitioners
HAHN and SHAPIRO · Statistical Models in Engineering
HALD · Statistical Tables and Formulas
HALD · Statistical Theory with Engineering Applications
HAND · Discrimination and Classification
HEIBERGER · Computation for the Analysis of Designed Experiments
HELLER · MACSYMA for Statisticians
HOAGLIN, MOSTELLER, and TUKEY · Exploratory Approach to Analysis of Variance
HOAGLIN, MOSTELLER, and TUKEY · Exploring Data Tables, Trends and Shapes
HOAGLIN, MOSTELLER, and TUKEY · Understanding Robust and Exploratory Data Analysis
HOCHBERG and TAMHANE · Multiple Comparison Procedures
HOEL · Elementary Statistics, *Fourth Edition*

Continued on back end papers

*Now available in a lower priced paperback edition in the Wiley Classics Library.

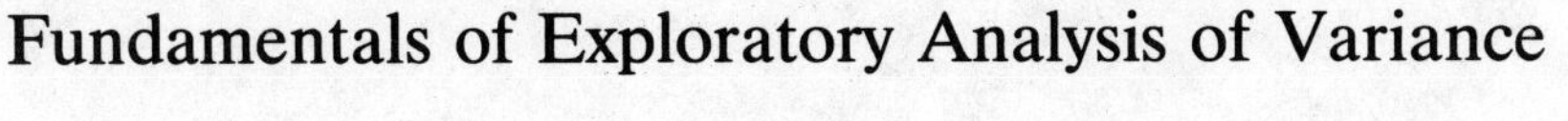

Fundamentals of Exploratory Analysis of Variance

Fundamentals of Exploratory Analysis of Variance

Edited by

DAVID C. HOAGLIN

Harvard University

FREDERICK MOSTELLER

Harvard University

JOHN W. TUKEY

Princeton University

A Wiley-Interscience Publication

JOHN WILEY & SONS, INC.

New York · Chichester · Brisbane · Toronto · Singapore

Library of Congress Cataloging in Publication Data:

Fundamentals of exploratory analysis of variance/edited by David C. Hoaglin, Frederick Mosteller, John W. Tukey.

p. cm. —(Wiley series in probability and mathematical statistics. Applied probability and statistics, ISSN 0271-6356)

Includes bibliographical references and index.

ISBN 0-471-52735-1 (cloth)

1. Analysis of variance. I. Hoaglin, David C. (David Caster), 1944– . II. Mosteller, Frederick, 1916– . III. Tukey, John Wilder, 1915– . IV. Series.

QA279.F86 1991

519.5′38—dc20 91-2407

CIP

Printed in the United States of America

10 9 8 7 6 5 4 3 2 1

To
Cuthbert Daniel

Preface

Effective data analysis often needs an exploratory component that refines the analysis and produces better understanding. The present book (the first in a multivolume monograph) approaches analysis of variance (ANOVA) from an exploratory point of view, while retaining the customary least-squares fitting methods. To focus on key basic ideas and on techniques for exploration, we have deliberately chosen to work almost exclusively with balanced data layouts. This choice avoids a considerable variety of difficulties, extending from more involved calculations to dependence of results on the order of sweeping out contributions. Subsequent volumes will include material on some frequent complications, leading to complexities that yield to modern computing power.

Because effective analysis must usually go much deeper than the ANOVA table (with its degrees of freedom, mean squares, F ratios, and P values), we take readers beyond the standard steps and considerations. Our approach emphasizes both the individual observations and the separate parts that the analysis produces. This framework allows the usual sums of squares and mean squares to take their proper place as summaries of the sizes of the parts of the decomposition. In practice, a variety of numerical and graphical displays often give insight into the structure of the data, such as sizes and spacing of individual effects, patterns of interactions, and behavior of the residuals. We provide greater flexibility and some new techniques for delving into the data.

The intended audience is broad. More specifically, the exploratory approach should offer immediate benefits to researchers who apply ANOVA to data in such fields as sociology, psychology, medicine, agriculture, and engineering. It should improve their understanding of their data by augmenting their present methods with penetrating techniques (without abandoning effective methods they already use).

For similar reasons, statisticians who use and teach ANOVA should find a variety of new techniques, practical advice, and illustrative examples.

Because the technical level of the book is not advanced, students should find it a ready introduction to ANOVA as an area of data analysis. Although brief prior exposure to ANOVA will speed the student along, the material is self-contained and thoroughly illustrated, except for occasional references to more advanced or specialized reading. As an aid to both students and professionals, nearly all chapters include exercises, and appendices give selected percentage points of the Gaussian, t, F, chi-squared, and Studentized range distributions.

The confluence of exploratory approach and classical methods has stimulated discussion of the usual approaches and development of new techniques, giving the book a number of special features.

Chapter 1 illustrates some of the kinds of findings the volume will help users discover. It familiarizes readers with the product of the work without trying to teach them how to carry out the analyses.

Chapter 2 examines a broad range of purposes for using ANOVA methods.

Chapter 3 focuses on the important stage of editing and appreciating the data before the analyst begins the main steps of analysis.

As a framework for discussing the detailed decomposition of the data, Chapters 5 and 6 emphasize overlays, each of which duplicates the dimensions of the entire data layout. Each overlay shows a separate component, and, together, they sum to the original data. For the designs we present, we use the overlay approach systematically.

People are often surprised to learn about the great variability in estimates of a population variance; Chapter 7 focuses on this obstacle.

Our exploratory approach gives a prominent role to graphical displays. Chapter 3 briefly reviews such basic EDA techniques as the boxplot and the stem-and-leaf display. Chapter 8 discusses several further techniques, including a new plot of main effects and interactions that is attuned to the corresponding mean squares. Chapter 12 presents an unfamiliar but very helpful plot that aids in judging the significance of pairwise differences.

Chapter 10 introduces the pigeonhole model as a way of subsuming the fixed- and random-effects models within one very general framework. It also illustrates two ways of formulating the idea of interaction.

In Chapter 11 we note that fixed effects and random effects are not adequate to describe all the circumstances that commonly use ANOVA. We introduce a special downsweeping approach that gives us a practical routine way of pooling effects of various orders, leading ultimately to finding and evaluating the sizes of the effects and interactions that the investigator wants to focus on and appreciate.

Familiar procedures based on the Studentized range yield inferences for a variety of effects. Chapter 12 presents these in a framework of qualitative and quantitative inferences, and it introduces a new procedure, the

Studentized birange, to study the double differences that describe the detail of two-factor interactions.

In preparing a monograph such as this, a number of research problems inevitably demand attention. We have pursued some of these and included the results in this volume.

In more than one sense the present volume extends the attitude and some of the techniques in the two volumes of "The Statistician's Guide to Exploratory Data Analysis": *Understanding Robust and Exploratory Data Analysis* (1983) and *Exploring Data Tables, Trends, and Shapes* (1985). Both were edited by David C. Hoaglin, Frederick Mosteller, and John W. Tukey and published by John Wiley & Sons. This volume refers to them by the acronyms UREDA and EDTTS. In the present cooperative project, as in the earlier ones, each chapter has gone through several revisions on the basis of critiques from participants, including all three editors. Thus, even though the chapters are written and signed by different people, we feel that the book has maintained stylistic integrity.

DAVID C. HOAGLIN
FREDERICK MOSTELLER
JOHN W. TUKEY

Cambridge, Massachusetts
Princeton, New Jersey
November 22, 1990

Acknowledgments

The chapters in this book have benefited from the advice and criticism of a number of colleagues, many of whom participated in the working group on exploratory data analysis that produced the two volumes of "The Statistican's Guide to Exploratory Data Analysis." For their help in this form we are grateful to Thomas Blackwell, Constance Brown, Richard D. DeVeaux, John D. Emerson, Colin Goodall, Katherine Taylor Halvorsen, Peter J. Kempthorne, Lincoln E. Moses, R. Bradford Murphy, Anita Parunak, James L. Rosenberger, Christopher Schmid, Nae Kyung Sung, and Cleo Youtz.

These activities received partial funding from the National Science Foundation through grants SES 8401422 and SES 8908641 to Harvard University. John W. Tukey's activities at Princeton University also received partial support from the U.S. Army Research Office (contract DAAL03-86-K-0073). John D. Emerson's participation was made possible in part by support from the Middlebury College Faculty Leave Program. Judith D. Singer received support from a Spencer Fellowship from the National Academy of Education.

Cleo Youtz helped in countless ways with the tasks of assembling, checking, and coordinating the manuscript.

Marjorie Olson handled many administrative aspects of the project.

Alexia A. Antczak and Richard S. Brown generously provided expert advice in connection with the data on hardness of dental gold.

Thomas J. Broadman kindly clarified the position of curves in Figures 9-4 and 9-5.

Philip Converse suggested the data on political parties, prepared and analyzed by Donald Stokes, that we use as an illustration in Section 9G; and Donald Stokes supplied additional materials.

Philip T. Lavin provided a helpful discussion of the definition of cross-sectional area of solid tumors.

John W. McDonald III gave permission to use his unpublished data on acidity of homemade sour cream and reviewed the discussion in Section 6C.

Bernard Rosner kindly made available additional data from studies of blood pressure for the presentation in Section 9B.

Marjorie Olson and Cleo Youtz prepared the manuscript with great care and efficiency.

Ralph B. D'Agostino, Jr. gave us timely help with computation for an example in Chapter 10.

D. C. H.
F. M.
J. W. T.

Contents

CHAPTER 1

Concepts and Examples in Analysis of Variance

John W. Tukey
Princeton University

Frederick Mosteller and David C. Hoaglin
Harvard University

Through examples, this chapter introduces some of the ideas of analysis of variance treated in this book. The examples run from simple through complex, illustrating the ideas of levels (or versions) of factors associated with responses and the usual kinds of interactions—both additive and multiplicative. The responses are quantitative, and we study systematic ways of breaking the responses chosen into effects and interactions, associating these with levels of factors or their combinations.

Except for the discussion of interaction, the reader should regard the chapter more as illustrating our approach to analysis of variance (ANOVA) than as a guide to the steps in the execution. This chapter offers a somewhat journalistic overview of material that later chapters will treat in much more detail.

The first example describes a point system used by the U.S. Army to determine the order of discharging soldiers in 1945. By introducing several factors and assigning points for levels of the factors, each soldier received a score. The structure actually had several factors, and the response (total points) was additive in that points assigned to levels of factors were cumulated. Because no interaction effects were present, this is the simplest kind of example.

In Section 1A, we introduce and explain the ideas behind interaction, and we illustrate them for additive and for multiplicative situations.

In Section 1B, we give a rather complicated example concerning the hardness of gold fillings used in dentistry. Using some tabular (graphical) methods, we illustrate some kinds of findings about the factors that analysis-of-variance methods may deliver.

In Section 1C, we illustrate the application of confidence intervals to the analysis of some presidential election data. Again, the reader's main purpose should be to see the kinds of results that an analysis may produce.

In Section 1D, we explain the benefits of analyzing the data using several factors simultaneously. In doing so, we indicate some circumstances where we can learn a lot from analyzing the response to a few factors even though many factors may simultaneously influence the size of the response.

Example: a point system for U.S. soldiers

In 1945 many U.S. soldiers had to be brought back from Europe and discharged. The administratively convenient way to determine an order of returning home was a "point system." Each soldier was awarded points for such characteristics as length of service, dependency, and number of campaigns. Those soldiers with the largest total of points were the first to be sent back and released. Careful opinion polling had determined how soldiers thought it best to award points for each characteristic (Stouffer 1949, p. 529).

If the aim had been to use the points to describe a quantitative *response*, rather than just to provide an order, this process would have been an example of ANOVA as we see it. The points awarded for individual characteristics would have been values of *main effects*—and, had there been points for certain combinations, these would have been values of an *interaction*. The *factors* would have been the characteristics for which points were assigned, with differing points for differing levels of the factors. For example, soldiers got one point per month for length of service (the levels are numbers of months), and five points per campaign star or combat decoration (the levels are numbers of campaigns and combat decorations).

One difficulty with the point system used in the demobilization was the area of combat and medals. The difficulty of defining "actual combat" and the differential rate at which medals were awarded by different branches of service suggested the need for interaction effects for certain combinations of characteristics. This refinement was not used by the Army, and its lack turned out to be the largest source of complaints about the procedure (Stouffer 1949, pp. 529–530). Thus the method used had no interaction terms, though they might have been useful.

This book views ANOVA as a process that describes a quantitative response by a point scheme—often as a *sum* of main effects and interactions. This description is valuable for itself and is also the basis for making verbal statements about the "points" assigned to different levels of the basic characteristics (factors) or their combinations. These statements are initially about appearances—what the data seem to say descriptively, without statistical inference. Later we often find expressions of (statistical) confidence, sometimes only about direction of difference but also, where feasible, about the amount of difference. (Throughout this book we use "level" or "version" for the categories of factors interchangeably. Level may be preferable when the category has a numerical value, and version when the label is qualitative such as male and female.)

Although the best practitioners using analysis of variance have never regarded ANOVA as a bundle of significance tests, the beginning student might unfortunately get such a perception, especially when our computer packages deliver tests so speedily. Part of our dissatisfaction with the excessive emphasis on significance testing for main effects and interactions comes from the belief that all of us share when we have our wits about us, namely, that the true value of any quantity in question is not zero, though it may be small. Admittedly, we cannot usually get the true value, but we do not expect different varieties of wheat to produce the same average yields, nor different surgical operations to produce the same success rates. Indeed, people whose business is mass-production find it very difficult to produce nearly identical objects.

In addition, even when we find significance, such a statement as "the AB interaction is significant at the 0.1% level" is at best a start on a further search for implications.

Conclusions need to be more sharply focused on the versions (levels) of the factors. For example, we want to say things like "version 1 of factor A produces larger scores than version 3 of factor A." Going further, as in "the difference in values for version 4 and version 2 of factor B lies between -7 and -3 with 95% confidence," can be very helpful too.

For readers with a background in the rudiments of traditional analysis of variance—perhaps for one-way and two-way layouts—our views about ANOVA may include some surprises. Statistical software packages often reinforce an attitude that places ANOVA tables—with their degrees of freedom, sums of squares, F's, and P values—at the center of attention. Instead, we give a central role to the breakdown of a table of measurements into components belonging to each factor, components belonging to each interaction among factors, and components that contain whatever variability is left over. We break up each measurement in this way, and we study the component values both separately and together with those for other measurements. An ANOVA table can provide valuable support for these activities, but it is never at center stage. For those who are accustomed to examining ANOVA tables, the methods of examining variability presented in this book should give insights about the underlying sources of the various numbers in an ANOVA table. Our contact with the component parts of variation is direct and explicit.

Before turning to illustrative examples of what we expect ANOVA to accomplish, we take time out to clarify the idea of interaction as used in ANOVA, because the expression has a variety of meanings in different fields and is sometimes misunderstood.

1A. INTERACTION IN ANOVA

Suppose that we are studying the effects of two factors on some response. In the simplest case each factor comes in only two levels—$A1$ and $A2$ for factor A, $B3$ and $B5$ for factor B—and we naturally begin at the pairing $A1B3$, where the typical response happens to be 130. If we move to $A1B5$, the typical response may well be different, say 195, and if we move to $A2B3$ instead, we shall suppose it to be 260. The picture to date looks like this

	$B3$	$B5$
$A1$	130	195
$A2$	260	??

What is it most natural to expect at the fourth corner, $A2B5$?

One plausible view says changing from $A1$ to $A2$ increased the typical response by $+130$ $(= 260 - 130)$, and changing from $B3$ to $B5$ increased the typical response by $+65$ $(= 195 - 130)$. Since

$$130 + 65 = 195,$$

it might possibly be that making both changes, which together change from $A1B3$ to $A2B5$, would change the typical response by 195, thus ending up at a typical response of

$$130 + 195 = 325.$$

The world does not have to behave so simply—in fact it is rare that it is exactly this simple—but simplest behavior is often a good reference point. So if we study empirically the fourth combination ($A2B5$) as well as the other three, we shall often find a deviation from such simplicity, as when we find 345 instead of 325.

When we do find a deviation that we cannot ascribe to chance, we have a word for it: we say that the two factors *interact*, that there is a nonzero *interaction*.

A Simple Breakdown

Let us follow up the additive analysis of this example with greater care and more detail. We have four values

	$B3$	$B5$
$A1$	130	195
$A2$	260	345

The arithmetic mean of the two typical responses at $A1$ is 162.5, and that of those at $A2$ is 302.5. Take each pair apart thus:

$$\boxed{130 \quad 195} = \boxed{162.5 \quad 162.5} + \boxed{-32.5 \quad 32.5}$$

$$\boxed{260 \quad 345} = \boxed{302.5 \quad 302.5} + \boxed{-42.5 \quad 42.5}$$

We can think of this breakdown as taking the whole 2×2 table apart thus:

$$\begin{array}{c|cc|} & B3 & B5 \\ \hline A1 & 130 & 195 \\ A2 & 260 & 345 \\ \hline \end{array} = \begin{array}{|cc|} B3 & B5 \\ \hline 162.5 & 162.5 \\ 302.5 & 302.5 \\ \hline \end{array} + \begin{array}{|cc|} B3 & B5 \\ \hline -32.5 & 32.5 \\ -42.5 & 42.5 \\ \hline \end{array}$$

We carry out a parallel step vertically on the four columns on the right-hand side, as in

$$\begin{array}{cccc} \boxed{\begin{array}{c}162.5\\302.5\end{array}} & \boxed{\begin{array}{c}162.5\\302.5\end{array}} & \boxed{\begin{array}{c}-32.5\\-42.5\end{array}} & \boxed{\begin{array}{c}32.5\\42.5\end{array}} \\ \| & \| & \| & \| \\ \boxed{\begin{array}{c}232.5\\232.5\end{array}} & \boxed{\begin{array}{c}232.5\\232.5\end{array}} & \boxed{\begin{array}{c}-37.5\\-37.5\end{array}} & \boxed{\begin{array}{c}37.5\\37.5\end{array}} \\ + & + & + & + \\ \boxed{\begin{array}{c}-70\\70\end{array}} & \boxed{\begin{array}{c}-70\\70\end{array}} & \boxed{\begin{array}{c}5\\-5\end{array}} & \boxed{\begin{array}{c}-5\\5\end{array}} \end{array}$$

We can now use these breakdowns to rewrite

	$B3$	$B5$
$A1$	130	195
$A2$	260	345

as shown in Table 1-1, where the original table is taken apart, entry by entry, into four tables of structure similar to the original table; we shall call such component tables *overlays*. For the moment our attention should be focused

Table 1-1. Breakdown of Original 2 × 2 Table into Four Overlays

$$
\begin{array}{cccccccc}
 & \begin{array}{cc} B3 & B5 \end{array} & & & \begin{array}{cc} B3 & B5 \end{array} & & \begin{array}{cc} B3 & B5 \end{array} \\
\begin{array}{c} A1 \\ A2 \end{array} & \begin{array}{|cc|} \hline 130 & 195 \\ 260 & 345 \\ \hline \end{array} & = & & \begin{array}{|cc|} \hline 162.5 & 162.5 \\ 302.5 & 302.5 \\ \hline \end{array} & + & \begin{array}{|cc|} \hline -32.5 & 32.5 \\ -42.5 & 42.5 \\ \hline \end{array} \\
 & & & & \| & & \| \\
 & & & \begin{array}{c} A1 \\ A2 \end{array} & \begin{array}{|cc|} \hline 232.5 & 232.5 \\ 232.5 & 232.5 \\ \hline \end{array} & + & \begin{array}{|cc|} \hline -37.5 & 37.5 \\ -37.5 & 37.5 \\ \hline \end{array} \\
 & & & & + & & + \\
 & & & \begin{array}{c} A1 \\ A2 \end{array} & \begin{array}{|cc|} \hline -70 & -70 \\ 70 & 70 \\ \hline \end{array} & + & \begin{array}{|cc|} \hline 5 & -5 \\ -5 & 5 \\ \hline \end{array}
\end{array}
$$

Note: The original table is the sum of the four overlay tables in the last two rows; for example, 345 = 232.5 + 37.5 + 70 + 5. The natural interpretation of each of the four tables is as follows:

(common)	(*B*3 vs. *B*5)
(*A*1 vs. *A*2)	(interaction!)

Because the entries in the upper left-hand box are all identical, we label the value "common." That value is the grand mean of the original four observations.

on the lower right-hand component table

$$
\begin{array}{|cc|} \hline 5 & -5 \\ -5 & 5 \\ \hline \end{array}
$$

which we have said has the interpretation of an interaction. Why do we say this?

Let us put the other three overlays of Table 1-1 back together and write

$$
\begin{array}{cccccc}
 & \text{data} & & \text{fit} & & \text{interaction} \\
 & \begin{array}{cc} B3 & B5 \end{array} & & \begin{array}{cc} B3 & B5 \end{array} & & \begin{array}{cc} B3 & B5 \end{array} \\
\begin{array}{c} A1 \\ A2 \end{array} & \begin{array}{|cc|} \hline 130 & 195 \\ 260 & 345 \\ \hline \end{array} & = & \begin{array}{|cc|} \hline 125 & 200 \\ 265 & 340 \\ \hline \end{array} & + & \begin{array}{|cc|} \hline 5 & -5 \\ -5 & 5 \\ \hline \end{array}
\end{array}
$$

The first box (labeled "fit") to the right of the " = " sign has a nice property. If we change from $A1$ to $A2$, at either version of B, the entry rises by exactly 140 (because $265 - 125 = 140$ and $340 - 200 = 140$). The difference in response does *not* depend on which B-version we work at. Similarly, for each version of A, the change from $B3$ to $B5$, in the same table, produces the difference in response of 75 ($200 - 125 = 75$; $340 - 265 = 75$). All this is exactly the behavior we observe when a table is free of interaction.

Thus, whatever interaction is in the parent table

130	195
260	345

cannot have gone into the "fit" table

125	200
265	340

so it must all be in

	$B3$	$B5$
$A1$	5	−5
$A2$	−5	5

Note too that this last overlay (interaction) has the property that, if you average the two entries for $A1$—or for $A2$, or for $B3$, or for $B5$—you get zero. In this sense, this final overlay is pure interaction.

We shall soon learn how to do similar things with larger tables, but what matters now is that we can supplement the idea of interaction—of a failure of the effects of changes to add up—by a simple process to pull out a numerical array that is the essence of the interaction in the particular instance. The table is a table of interactions, and the value in a cell is the interaction effect for the cell. We often speak about interaction without emphasizing whether we are pointing to one cell or to a whole table.

Multiplicative Interaction

Let us go back to our original three values

	$B3$	$B5$
$A1$	130	195
$A2$	260	??

and think about a different kind of simplicity—a simple behavior in the multiplicative scale. Changing from $A1$ to $A2$ at $B3$ doubled the value. It would be simple if this happened at $B5$ also. This would mean that the open corner would have to be

$$2(195) = 390.$$

Thus there is a sense in which

	$B3$	$B5$
$A1$	130	195
$A2$	260	390

is a very simple table too, because the change from $A1B3$ to $A2B5$ changes the response by a multiple of

$$\frac{260}{130} \cdot \frac{195}{130} = 2(1.5) = 3.$$

We could work with such a table directly, but addition seems simpler than multiplication to most of us. Accordingly, in the face of this second kind of simplicity, we usually take logarithms and work with

	$B3$	$B5$			
$A1$	log 130	log 195	=	2.11	2.29
$A2$	log 260	log 390		2.41	2.59

We see the logarithms showing the first kind of simplicity: changing from $A1$ to $A2$ gives an additive effect of $+0.30$, and changing from $B3$ to $B5$ gives 0.18.

Once we realize that the fitting of estimates to cells can be carried out in more than one way, we see that the concept of interaction, just like the concept of residuals, is not unique. The values depend on the model being fitted and even on the method used to carry out the fitting. Thus interaction is not a property of a set of measurements alone, but also of the devices we use to describe the measurements. The general concept is that two factors interact when their combined contribution to the response goes beyond their separate contributions.

Although no hard and fast rules hold, practitioners generally prefer additive models. They prefer models whose fits seem to assign more of the effects to the single factors than to the interaction, and in complicated problems with several factors they prefer interactions involving few factors rather than many. Perhaps we could sum up by saying that when we have a choice, we prefer small interactions.

When we have a mathematical theory or a structure that tells us how the data might have been formed, whether a scientific hypothesis to be examined or a well-established law, we are likely to use the theory to construct a model. Then the residuals and interactions would give us information about the lack of fit, and maybe even suggestions about adjustments needed. For example, if we were looking into the weight (or mass) of materials stored in cylinders of various sizes, the fact that $\pi r^2 h$, where r is the radius of a cross section and h is the height, gives the volume of the cylinder might influence us to use a multiplicative model with factors r^2 and h rather than an additive model with factors r and h. To aid in the calculations we might well take logarithms. (Then $\log \pi r^2 h = \log \pi + 2 \log r + \log h$, and we would be dealing with an additive model based on logarithms of radii and of heights. Thus the calculations associated with the original multiplicative model can be converted to additive calculations.)

Sometimes our choices are driven by the search for a formula that would be instructive. We discuss purposes of ANOVA in Chapter 2; here we merely want to clarify the variety of possibilities in interactions.

1B. A GRAPHICAL ANALYSIS OF A COMPLEX EXPERIMENT ON THE HARDNESS OF DENTAL GOLD

A later chapter gives a fuller description of an example where five dentists produced gold fillings using three methods of treating the gold and two alloys, each sintered at three temperatures (the full data are somewhat more extensive). Table 1-2 shows the $5 \times 3 \times 2 \times 3 = 90$ values of measured hardness. The goal is to have high values of hardness.

This table contains all the data to be used, but scanning the table conveys only a little information. We can understand the grosser effects in the data by making an appropriate semigraphical picture, as in Table 1-3.

In this table vertical position is divided into blocks of values, about 10 units of hardness long, horizontal position is divided into five dentists, each subdivided into the two alloys, and the character used to represent the appearance of a data value is the number of the method (1, 2, or 3) employed. Because we want hard fillings, the two most prominent features of this three-factor picture, involving $2 \times 5 = 10$ parallel stem-and-version plots, are:

Method 3 behaves differently from the other two methods, often straggling low and doing very poorly for Dentist 5 and for Dentist 4.

Two individual values stand out (their gold tested much harder than did the other 88). Method 2 for Dentist 1 with Alloy 2 (at Temperature 1) and Method 1 for Dentist 5 with Alloy 2 (at Temperature 2) are worth noting.

Table 1-2. Dental-Gold Example (Raw Data): Two Alloys (A1, A2), Three Temperatures (T1 = 1500, T2 = 1600, T3 = 1700), Three Methods (M1, M2, M3), and Five Dentists (D1, D2, D3, D4, D5) with Observed Hardness for Each Combination of the Four Factors (A, T, M, and D)

		A1			A2		
D	M	T1	T2	T3	T1	T2	T3
1	1	813	792	792	907	792	835
	2	782	698	665	1115	835	870
	3	752	620	835	847	560	585
2	1	715	803	813	858	907	882
	2	772	782	743	933	792	824
	3	835	715	673	698	734	681
3	1	743	627	752	858	762	724
	2	813	743	613	824	847	782
	3	743	681	743	715	824	681
4	1	792	743	762	894	792	649
	2	690	882	772	813	870	858
	3	493	707	289	715	813	312
5	1	707	698	715	772	1048	870
	2	803	665	752	824	933	835
	3	421	483	405	536	405	312

Source: Morton B. Brown (1975). "Exploring interaction effects in the ANOVA," *Applied Statistics*, 24, 288–298 (data from Table 1, p. 290). Data reproduced by permission of the Royal Statistical Society.

Such a picture can show us much about the behavior of a few factors. We have managed to get a good look at dentists and alloys—in their 10 (= 5 × 2) combinations—and a useful look at the three methods. But trying to get temperatures into the same picture seems almost certain to lead to (graphical) catastrophe. There are 24 kinds of such pictures for the dental-gold example, of which this one is probably the most revealing. (There are four factors. Consequently there are four ways to choose the factor for broad columns, three for the factor that splits columns, and two for the entry in the table, and 4 × 3 × 2 = 24.) We can start with any one of the four factors as the outer classification (here dentists) and subclassify by any other (here alloys). Then we can use any one of two remaining factors (here method) as the identifier in our stem-by-identifier plots.

What our computers will do easily for us determines whether we begin with the 24 possible pictures (going on to the analysis of variance) or do an analysis of variance first, as a guide to which of the 24 pictures are likely to be promising. We ought to regard doing one or both as important ways to become acquainted with the data.

Table 1-3. Stem-and-Version Plot for the Example of Hardness of Dental Gold for Factors Dentist, Alloy, and Method[a]

Hardness axis: 1100, 1000, 900, 800, 700, 600, 500, 400, 300

Dentist	Alloy	Method digits plotted (top = highest hardness, to bottom)
D1	A1	3, 1, 11, 2, 3, 2, 2, 3
D1	A2	2, 1, 2, 3, 12, 1, 3, 3
D2	A1	3, 1, 1, 2, 2, 2, 13, 3
D2	A2	2, 1, 1, 1, 2, 2, 3, 3, 3
D3	A1	2, 1, 1233, 3, 1, 2
D3	A2	1, 2, 23, 2, 1, 1, 3, 3
D4	A1	2, 1, 2, 1, 1, 3, 2, 3, 3
D4	A2	1, 2, 2, 23, 1, 3, 1, 3
D5	A1	2, 2, 1, 1, 1, 2, 3, 3, 3
D5	A2	1, 2, 1, 2, 2, 1, 3, 3, 3

[a] The number plotted is that for Method. D, dentist; A, alloy; column digit, Method.

Table 1-4. Results (Point Scheme) for the Dental-Gold Example with Method 3 Set Aside

Common (grand mean hardness of 60 measurements)	799
Effect for Alloy 1:	−51
Effect for Alloy 2:	+51
Among the 60 residuals	
Half are between	−41 and 44
Three-quarters are between	−68 and 57
Seven-eighths are between	−88 and 83
All are between	−201 and 265

Two large observations (both for Alloy 2), if correct, might lead to adjustment and reanalysis as a next step. They are 1115 at D1, A2, M2, T1 and 1048 at D5, A2, M1, T2.

In Chapter 11 we apply methods described later to these data, omitting all results for Method 3 because its values are consistently low, whereas high values are desired. There we find that only the alloy effects need to be distinguished and that the other main effects and the two- and three-factor interactions do not need to be distinguished, though the four-factor interaction does. Here we simplify slightly: after omitting Method 3, we take out the common term and the alloy effects, compute residuals, and summarize the results in Table 1-4.

Although by themselves the 60 individual values did little to tell a coherent story, Table 1-4 gives a simple picture, telling nearly all that the 60 values can tell us about Methods 1 and 2, Alloys 1 and 2, Temperatures 1, 2, and 3, and Dentists 1, 2, 3, 4, and 5. (The identities of these different versions are, of course, very important in practice and are discussed in Section 6B.)

Except for some very unusual values, once Method 3 is set aside, the only thing worthy of separate attention appears to be the difference between the two alloys, which is substantial.

To the sequence just illustrated in Table 1-4, where we first "take it apart as far as we can," then "put back together those parcels that do not deserve to be kept separate," and, third, "express in words what the remaining parcels seem to tell us," we often want to add a fourth step, "express what appearances we have confidence in, either as to direction or as to amount."

In the dental-gold example the only surviving difference worth noting is "Alloy 2 MINUS Alloy 1" with an observed value of +102 and, as it will turn out, a standard error of 19.4 on 58 degrees of freedom. With the 5% point of $|t|$ at about 2.00, this means we have 95% confidence that "Alloy 2 MINUS Alloy 1" is between $+63.2 = 102 - 2.00(19.4)$ and $+140.8 = 102 + 2.00(19.4)$. Now we have summarized the results of the four basic steps of ANOVA for one part of the dental-gold data.

1C. AN ELECTION EXAMPLE

We now look at state-to-state behavior in the four U.S. presidential elections having Franklin Delano Roosevelt as a candidate. Although elections represent a planned form of data gathering, their design does not involve the sort of control that one expects in investigations in science, medicine, and agriculture. Even so, we may learn much by applying ANOVA techniques to such data. The present example illustrates a basic attitude. We are prepared to explore almost any set of data, taking it apart into a sum of pieces determined by its design. When patterns emerge, we can consider how to distinguish them from chance fluctuations, but that can be difficult when we lack a carefully implemented sampling plan.

Table 1-5 shows Roosevelt's share of the vote (for him and his Republican opponent, omitting votes for other candidates) in the four elections in 39 of the 48 states. To take into account the tendency of nearby states to show similar voting patterns throughout the country, we have (without attention to voting patterns) arranged these 39 states into 13 groupings of 3 states each, keeping each grouping within the structure of regions and divisions used by the U.S. Census Bureau. The remaining nine states either are isolated geographically or cannot go into a grouping without breaking census division boundaries.

The main structure in the data table comes from the groupings and the elections, and each grouping has three states within it. Thus we consider the voting behavior of each state relative to the typical pattern in its grouping. We also take a long-run attitude toward variation among states. We treat each triple of states as if they were a sample from a large population of the kind of states representing the grouping.

The $156 = 4 \times 39$ data values can be taken apart into a sum of six pieces:

(a) An overall constant (common value or grand mean),
(b) An effect for each election,
(c) An effect for each grouping,
(d) An effect for each state within its grouping,
(e) An interaction effect for each combination of an election with a grouping, and
(f) A set of residuals.

Table 1-6 shows all six of these pieces and indicates how they may be computed. The sum of the six pieces gives back the original cell value. For example, for New Hampshire, 1936 we add the effect from each of (a), (b), (c), (d), (e), and (f): $610 + 43 + (-142) + 46 + (-54) + 6 = 509$.

Confidence statements for effects in these pieces might well take into account the problem of multiplicity. That is, within each set of effects we must compare more than two effects in order to say which effects differ from

Table 1-5. Results, for Three States in Each of 13 Groupings, of the Presidential Elections in Which Franklin D. Roosevelt Was Elected[a]

Grouping	State	Election 1932	1936	1940	1944
1	Maine	436	428	488	475
	New Hampshire	493	509	532	521
	Vermont	416	434	451	429
2	Connecticut	494	578	536	527
	Massachusetts	521	551	534	529
	Rhode Island	560	568	568	587
3	New Jersey	510	601	518	507
	New York	567	602	518	525
	Pennsylvania	471	582	535	514
4	Illinois	568	592	512	517
	Indiana	560	575	493	471
	Ohio	515	608	522	498
5	Minnesota	623	666	519	528
	North Dakota	713	692	445	458
	South Dakota	649	560	426	417
6	Kansas	548	539	427	394
	Missouri	645	614	524	515
	Nebraska	641	584	428	414
7	Delaware	488	549	548	546
	Maryland	631	627	588	519
	Virginia	695	705	683	625
8	Georgia	922	874	851	817
	North Carolina	705	734	740	667
	South Carolina	981	986	956	952
9	Alabama	857	871	856	817
	Kentucky	595	594	576	546
	Tennessee	672	691	675	606
10	Arkansas	872	821	789	701
	Mississippi	964	972	958	936
	Oklahoma	733	672	576	557
11	Idaho	606	655	545	517
	Montana	620	715	594	547
	Wyoming	579	618	530	488
12	Arizona	687	722	638	590
	New Mexico	637	632	567	535
	Utah	579	699	624	605
13	California	610	679	581	568
	Oregon	611	685	541	525
	Washington	629	690	589	574

[a]Each entry is 1000 times the ratio of the vote for Roosevelt to the sum of the vote for Roosevelt and the vote for his Republican opponent.

Source: Richard M. Scammon (1965). *America at the Polls: A Handbook of American Presidential Election Statistics 1920–1964*. Pittsburgh: University of Pittsburgh Press, pp. 7–14.

Table 1-6. The Six Pieces that Result from Analyzing the Roosevelt Election Data

(a) The overall constant (common value or grand mean)

610

(b) The election effects (mean for year MINUS grand mean)

Election:	1932	1936	1940	1944
Effect:	21	43	− 20	− 44

(c) The effects for the groupings (mean of grouping MINUS grand mean)

Ordered by Grouping		Ordered by Value	
Grouping	Effect	Effect	Grouping
1	−142	239	8
2	−64	186	10
3	−72	87	9
4	−74	16	12
5	−52	−3	13
6	−87	−9	7
7	−9	−25	11
8	239	−52	5
9	87	−64	2
10	186	−72	3
11	−25	−74	4
12	16	−87	6
13	−3	−142	1

(d) The effects for the states within grouping. Within each grouping the states appear in alphabetical order (see Table 1-5) (state mean MINUS grouping mean).

Grouping	States		
1	−11*	46	−35
2	−12	−12	25
3	−4	16	−12
4	11	−11	0
5	26	19	−45
6	−46	52	−6
7	−68	−9	77
8	17	−137	120
9	154	−119	−35
10	0	162	−161
11	−4	34	−31
12	33	−34	0
13	3	−16	14

*Example: Mean for Maine (436 + 428 + 488 + 475)/4 = 456.75; Maine mean MINUS grouping 1 mean = 456.75 − 467.67 = −11.

Table 1-6. *(Continued)*

(e) The interaction effects for groupings by elections (grouping-year mean MINUS grouping mean MINUS year mean PLUS grand mean).

Grouping	Election 1932	1936	1940	1944
1	−40*	−54	43	51
2	−42	−24	20	46
3	−43	14	7	22
4	−9	12	−6	3
5	83	38	−74	−46
6	67	13	−43	−38
7	−17	−17	26	7
8	−1	−28	21	7
9	−9	−21	26	4
10	39	−18	−1	−21
11	−4	35	−8	−23
12	−13	15	4	−6
13	−11	34	−16	−7

*Example: Grouping 1 mean for 1932 = 448.33 = (436 + 493 + 416)/3; Grouping 1 mean = 467.67 = (436 + 493 + ··· +521 + 429)/12; 1932 mean = 630.85 = (436 + 493 + ··· +611 + 629)/39; grand mean = 609.76;
448.33 − 467.67 − 630.85 + 609.76 = 40.43.

(f) The residuals: cell entry MINUS grand mean MINUS election effect MINUS grouping effect MINUS effect for the state within grouping MINUS interaction.

Grouping	State	Election 1932	1936	1940	1944
1	Maine	−1	−18	9	11
	New Hampshire	−1	6	−4	0
	Vermont	3	12	−4	−11
2	Connecticut	−19	25	2	−8
	Massachusetts	8	−2	0	−6
	Rhode Island	10	−22	−3	15
3	New Jersey	−2	10	−2	−5
	New York	36	−8	−21	−6
	Pennsylvania	−33	−1	23	11
4	Illinois	9	−11	−8	10
	Indiana	24	−6	−5	−13
	Ohio	−32	16	13	3
5	Minnesota	−65	1	30	34
	North Dakota	32	34	−37	−29
	South Dakota	32	−34	8	−6
6	Kansas	−18	6	13	−1
	Missouri	−18	−17	13	22
	Nebraska	36	11	−26	−21

Table 1-6. *(Continued)*

Grouping	State	Election 1932	1936	1940	1944
7	Delaware	−49	−10	9	50
	Maryland	35	9	−9	−35
	Virginia	14	1	0	−15
8	Georgia	35	−8	−15	−12
	North Carolina	−27	7	28	−8
	South Carolina	−8	1	−13	20
9	Alabama	−5	−2	0	7
	Kentucky	6	−6	−8	8
	Tennessee	−1	8	8	−15
10	Arkansas	16	0	15	−30
	Mississippi	−54	−11	22	43
	Oklahoma	38	12	−37	−13
11	Idaho	8	−4	−8	3
	Montana	−16	18	3	−5
	Wyoming	8	−14	4	1
12	Arizona	20	5	−5	−20
	New Mexico	36	−19	−9	−8
	Utah	−56	14	14	28
13	California	−9	−8	8	10
	Oregon	11	17	−13	−14
	Washington	−1	−8	5	5

which others. Without going into the details, we handle the multiplicity so that, when all the effects in the set are zero, the probability is at most 5% that we would declare one or more comparisons among those effects significant. This 5% error rate applies to each set of effects separately. We do not explain here how to get the appropriate allowances. We merely supply them.

In panel (b) of Table 1-6, an allowance of ± 40 for each difference yields 95% simultaneous confidence. Thus we are confident that the Roosevelt percentage is higher in 1932 or 1936 than in either 1940 or 1944. For example, the allowance of 40 around 1932's 21 gives -19 to 61. This interval does not include the values -20 for 1940 or -44 for 1944, and so they are declared to differ; whereas 1936 has an effect of 43 that falls in the interval, and it does not confidently differ. Also we can be confident that (1936) MINUS (1944), observed at $43 - (-44) = 87$, would (on the average) be between $47 = 87 - 40$ and $127 = 87 + 40$.

To give 95% confidence for the effects of the groupings in panel (c), we use an allowance of ± 238. Looking at the ordered effects, we can be confident that grouping 8 was more favorable to Roosevelt than groupings 13,

Table 1-7. Stem-and-leaf Display for the Effects of States Within Groupings[a]

H	120, 154, 162
7	7
6	
5	2
4	6
3	34
2	56
1	14679
0	0003
−0	9644
−1	622211
−2	
−3	5541
−4	65
−5	
−6	8
L	−119, −137, −161

[a]A difference of 78 or more has simultaneous confidence of at least 95% for the $39 \times 38/2 = 741$ pairs of differences. If the true means are all alike, there is only 1 chance in 20 of observing at least one difference of more than 78.

7, 11, 5, 2, 3, 4, 6, and 1, and that grouping 10 was more favorable than 2, 3, 4, 6, and 1. If we set aside groupings 8 (North and South Carolina and Georgia) and 10 (Arkansas, Mississippi, and Oklahoma), we are not confident of the order of those remaining, nor of those groupings themselves.

For convenience we have arranged the effects for states, panel (d), in a two-way table. The column position of a particular state, however, matters only within a row, where it corresponds to the state's position in alphabetical order within the grouping. For a more informative look at these effects, we use a stem-and-leaf display (Table 1-7). One analysis (overconservative at 95% simultaneous confidence for differences of states within groupings) gives an allowance of ± 78. Thus, for example, any of the six states with observed effects (within grouping) of $+46$ or higher was confidently more favorable to Roosevelt than any of the nine states with effects of -34 or lower.

When we analyze the EG (Elections by Groupings) interaction effects (Table 1-6(e)), we are led to distinguish a set of two groupings (both New England) that trended toward Roosevelt (#1, #2) from a set of two groupings (both upper Midwest) that trended away from Roosevelt (#5, #6). When we ask about confidence, we are led to be quite confident about this distinction.

In summary, the data indicate that Roosevelt won by smaller (but still comfortable) margins in 1940 and 1944 than in 1932 and 1936; that two groupings of states, both in the South, generally gave him a larger percentage of their vote; that, relative to their groupings, some states were more favorable, and other were less favorable; and that, relative to the pattern of the election effects, the New England states trended over time toward Roosevelt, while the West North Central states trended away from Roosevelt.

1D. WHY MAIN EFFECTS AND INTERACTIONS?

At one time the dogma of experimentation—at least in some fields—was "change one variable at a time!" It is unlikely that a dogma will be widespread unless it works in some circumstances; it is also unlikely that a dogma will disappear unless it sometimes fails. What about "one variable at a time"?

If the simplest form of point scheme (as illustrated in the demobilization example at the beginning of the chapter) were to apply, where points for the versions of all the factors needed only to be added up—main effects but no interactions—we would get correct answers by changing one factor at a time. This merit gives that dogma some credence.

If there are large interactions, changing one factor at a time will give confused, ineffective answers. This is one major reason for the disappearance of "one at a time"; another is that, even without interactions, one-at-a-time wastes effort by giving less precision in assessing the main effects than a factorial pattern provides.

Except for the point masses of freshman physics—or the hard round spheres of the earliest models for gas molecules—few phenomena involve only a few variables. Sometimes only a few variables matter much in important recognizable circumstances. More often, increasing insight increases the number of variables that matter to a meaningful degree. How can we get at simple truths when several (or even a dozen) factors matter and some interactions are important?

If we can isolate a few factors—perhaps two, three, or four—that may interact among themselves, but whose interactions with the factors left out are small, the corresponding data gathering, with its two-way, three-way, or four-way table of results, can—and should—give us very helpful information. Usually our data gathering will be for a single combination of versions for the left-out factors. If it is not, then additional variability or noise will be distributed from cell to cell. When measurement error or sampling fluctuations play a substantial role, as they often do, repeated measurements—replication—will help distinguish measurement error from interaction. The more factors an investigation uses, the less likely it is that replication will be carried out. With as many as four factors, replication is very unlikely unless measurements are remarkably inexpensive to gather.

We may not know much about the *value* of the response in future circumstances, where both included factors and left-out factors have levels different from those in our data collection, because we do not know either sizes or directions of the effects of changing the left-out factors. But we shall know a lot about *differences* in response *associated with changes in (combination of) levels of* the included factors.

By assumption—and in the real world if we know enough (about the subject-matter field) to choose the included factors wisely—interactions between included and left-out factors are almost negligible. So *changes* in response associated with specified changes in combinations of levels of included factors will be almost the same for future combinations of levels of the left-out factors as they were for the single combination that occurred in our data collection. In many circumstances much preliminary work will have discovered a few variables known to be important, and these variables are probably the ones entering the investigation. If other variables have relatively small effects all told, they are likely also to have small interactions with the included ones.

It is the hope of, and frequent success in, having small interactions between included factors and left-out factors that supports the present great emphasis on comparative study, both in experiments and more passive data gatherings. If we work with the right collection of included factors, we can learn much about differences.

Because important factors often have substantial interactions with one another, "one at a time" is likely to fail, but in many instances "few at a time" can succeed—and in more "several at a time" will do so. If we need to work with actual levels, rather than being able to confine ourselves to differences, we are likely to need to include even more factors in our data gathering.

When up to four or five factors suffice, the efficiency advantages of data in a factorial pattern make four-way or five-way data as effective for separate main effects as one-way or two-way data (with the same total number of observations) would have been. Our difficulties with several-at-a-time data gathering then will often be confined to (1) doing the analysis, for which this book offers methods that can be computerized, and (2) the substantial sizes to which more-way tables easily grow (where fractional factorial patterns, *not* to be discussed here, offer some relief).

To keep simple what we learn about the world, we need to reduce the importance of interaction as much as we can—both among included factors and, especially, across the gulf between included and left-out variables.

REFERENCE

Stouffer, S. A. (1949). "The point system for redeployment and discharge," Chapter 11 in *The American Soldier: Combat and Its Aftermath*, by Samuel A. Stouffer, Arthur A. Lumsdaine, Marian Harper Lumsdaine, Robin M. Williams, Jr., M. Brewster Smith, Irving L. Janis, Shirley A. Star, and Leonard S. Cottrell, Jr., *Studies in Social Psychology in World War II*, Vol. II. Princeton, NJ: Princeton University Press.

EXERCISES

1. In the discharge plan for the U.S. soldiers at the close of World War II, suppose only two factors were used. Factor A gave 5 points for each combat campaign up to three, and a soldier may have had 0, 1, 2, or 3 campaigns. Factor B gave 12 points for each year of service (rounded off to whole years) with 0, 1, or 2 years being possible. Scores were additive. Form the two-way table showing the number of discharge points for each possible campaign-year service, and identify the factors and their levels.

2. Break the table constructed in Exercise 1 into its four overlays and label each.

3. Given the table

	$B1$	$B2$
$A1$	35	45
$A2$	55	65

do factors A and B have an additive interaction? How can you tell?

4. Use one-digit numbers to form a two-way table of your own that has nonzero effects for both factors A and B, but no additive interaction.

5. Break the table in Exercise 3 into its four overlays and identify each.

6. What is meant when two factors interact?

7. Break the table

	$B1$	$B2$
$A1$	35	55
$A2$	55	75

into its four overlays and label them.

8. Given the four overlays:

common: $\begin{array}{|cc|} 60 & 60 \\ 60 & 60 \end{array}$; $A1$ vs. $A2$: $\begin{array}{|cc|} -5 & -5 \\ 5 & 5 \end{array}$;

$B1$ vs. $B2$: $\begin{array}{|cc|} -10 & 10 \\ -10 & 10 \end{array}$; interaction: $\begin{array}{|cc|} 2 & -2 \\ -2 & 2 \end{array}$,

construct the original table.

9. **(a)** In the 2×2 table

	$B1$	$B2$
$A1$	10	20
$A2$	30	u

replace u by a number that produces no additive interaction.

(b) Replace u by a number that produces no multiplicative interaction.

10. Create the four additive overlays for the table

	$B1$	$B2$
$A1$	10	20
$A2$	30	80

11. (Continuation) Use logarithms to get the multiplicative overlays for the table of Exercise 10. That is, taking logarithms, use the additive approach to get the logarithmic overlays.

12. (Continuation) Add the logarithms for the three overlays *excluding the interaction overlay* and take antilogarithms to get the multiplicative fit.

13. (Continuation) Subtract the fit from the original measurements in the table of Exercise 10. These are the multiplicative interaction effects. Compare them with the additive interaction effects of Exercise 10. (Strictly speaking, they are not entirely comparable because the interactions of Exercise 10 are to be viewed according to their absolute size, whereas those of Exercise 13 are to be thought of in terms of percentage error.)

14. Find a 2×2 table with at least two different positive numbers that has zero interaction additively and also multiplicatively.

15. Using the dental-gold data in Table 1-2, make a stem-and-version plot similar to Table 1-3, using temperatures instead of dentists as the outer classification. In other words, plot the hardness values for the three temperatures, two alloys, and three methods.

16. (Continuation) From your plot for Exercise 15, what can you say about Method 3? Method 1? Method 2?

17. Table 1-6 gives the six pieces that result from analyzing the Roosevelt election data in Table 1-5. Use data from these six pieces to reconstruct the following data values given in Table 1-5:

(a) Grouping 4, Illinois, 1932 election.
(b) Grouping 6, Nebraska, 1936 election.
(c) Grouping 9, Kentucky, 1940 election.
(d) Grouping 11, Wyoming, 1944 election.

18. If the cells of a 2×2 table are u, v, w, x

	$B1$	$B2$
$A1$	u	v
$A2$	w	x

write out the three overlays for the common, the factor A effects, and the factor B effects. Now add these up to get the fits for the four cells. (That for the upper left cell is $\frac{1}{4}(3u + v + w - x)$.)

19. (Continuation) Subtract the fitted values from the original entries in the table of Exercise 18. This gives the additive interaction terms. (For example, the upper left interaction term is $\frac{1}{4}(u - v - w + x)$.)

20. (Continuation) Check that the interaction in Exercise 19 has the form

	$B1$	$B2$
$A1$	t	$-t$
$A2$	$-t$	t

CHAPTER 2

Purposes of Analyzing Data That Come in a Form Inviting Us to Apply Tools from the Analysis of Variance

Frederick Mosteller
Harvard University

John W. Tukey
Princeton University

In preparing to study tools and structures used in analysis of variance, we first review some purposes of studying such bodies of data. Although these techniques and concepts have been valuable for nearly all kinds of applications of statistics:

- The same body of data may call for different tools and analyses when the analysts have differing goals.
- The same approach may be profitable for one investigator and uninformative for another.
- The same tool may play different roles for analyses with differing purposes.

Readers with a special purpose in mind may suddenly find an analysis in an example puzzling or superfluous because it responds to goals other than their own. Thus, just as in an art museum, where we get more by asking "Why might the artist have created this work?" rather than "Do I like it?", we will profit from appreciating the many goals of data analysis.

In Section 2A, we describe the kinds of purposes that an analyst or a viewer of a set of data may have, including understanding the data so that we know more about how a process behaves. We may want to know about the relative importance of the factors that influence the outcome of the process. Ideally, some factors do not matter, and we are glad to dismiss them. We may want to know about the measurement process being used, its systematic and random variation. Or for practice or policy we may want to know how factors relate to costs and benefits. Simple optimization may not be the only goal even then; we may wish to ensure that a floor is set under the outcome, and thus we want forgivingness.

After these concepts have been introduced, we turn in Section 2B to a numerical illustration of understanding in a circumstance where we are well aware of the scientific facts and can therefore see how the analysis of variance behaves when an approximate fitting method is applied.

We discuss in Section 2C the more classical purposes, including the introduction of confidence limits and kinds of forecasting.

Often we want to distinguish the comparison of performances or outcomes observed under two different conditions from the changes in outcome wrought when we change the conditions from one to another. Thus we distinguish between differences in performance *associated* with levels of a factor and differences in performance *caused* by the changes in levels of a factor. We discuss these distinctions in Section 2D, where we also discuss the distinction between experiments and observational studies and how they relate to our studies of causation. When an investigator imposes the level of a factor, we think of the investigation as an experiment, but investigators who observe or collect the information related to factors as they occur in nature are said to be carrying out observational studies.

2A. PURPOSES CAN BE BOTH DIVERSE AND UNFAMILIAR

Understanding

Understanding must lead our list of purposes. We create structures and offer tools to improve our understanding of the phenomenon—or bundle of phenomena—under study, of the science related to it, and of its engineering and practice. By focusing on a phenomenon, we may gain a qualitative or even a partially quantitative understanding of how, in some average sense, each individual factor influences the outcome and often how much—and in what ways—the factors interact. For example, in considering a consumer's satisfaction with food, how much does improvement in taste contribute, how much does increase in price contribute, and how much can be attributed to both together? (As we make explicit below, we can often affect the answers about interaction by how we express the response.)

Understanding contributes not only to *what* we think but also to *how* we think. Recognizing different possibilities of approach is important. For example, shall we think in terms of the raw data, or would explanations be simpler for the logarithm of the data? When recognition is forced on us by some understanding, we have gained theory. And if understanding teaches us which approach(es) to prefer, we have gained even more.

We need to say much more about understanding, but we shall leave it aside for a while, turning to more diverse purposes.

Relative Importance

Among these efforts at understanding, we may wish to assess the relative importance of factors in a system. When factors interact, the idea of such relative importance can turn out to be very complicated, as we shall be forced to discuss in later chapters, and sometimes it makes no sense at all. In comparatively simple situations, we can speak intuitively. For example, Gould et al. (1988) measured on a seven-point scale (1 = none, 7 = very great) risks of six technologies as perceived by samples of people from populations in Connecticut and Arizona. Table 2-1 gives the response, average perception of risk, associated with two factors: technology and state.

The table shows that the responses from Connecticut and Arizona are very similar, Arizona's being generally a little lower (perceived safer)—by about .3 for most technologies, but by .9 for handguns. Thus, in this table, the factor "technology" is relatively more important in determining the perceived risk score than the factor "state." Of course, by constraining or expanding the factors, one may be able to change the relative importance. Had the technology factor consisted only of nuclear power and automobile travel, for instance, we would have found the factors "technology" and "state" about equal in importance.

This fact, that relative importance depends on how wide a variety of versions (levels) should be considered for each factor, is a general one. It is rare that a comparison of relative importance is similar for all possible situations, though it can sometimes be similar for all reasonable ones. For example, in a menu if a side order of cole slaw adds 50 cents to the price of any meal, then all comparisons of the importance of other variables (e.g., type of entree, beverages) will be similar. But if cole slaw is already included in some meals and not others, then the factors will not be so similar in their comparisons. In either thinking or communicating about relative importance, then, it is almost always important to attend to the scope of variation appropriate to—and observed for—each of the factors.

The Importance of Being Unimportant

In studying the importance of variables, some special situations are worth noting. Sometimes the response is essentially unchanged for every version of a factor, so that the factor seems to be capable of being omitted from estimation and other analytic procedures. As a general proposition, such outcomes have great value for science because being able to eliminate

Table 2-1. Average Perceived Risk as Assessed on a Seven-Point Scale, for Six Technologies in Probability Samples Drawn from Two States

Technology	State	
	Connecticut	Arizona
Nuclear weapons	6.7	6.4
Handguns	6.5	5.6
Industrial chemicals	5.7	5.4
Nuclear power	5.1	5.0
Automobile travel	5.0	4.7
Air travel	4.0	3.6

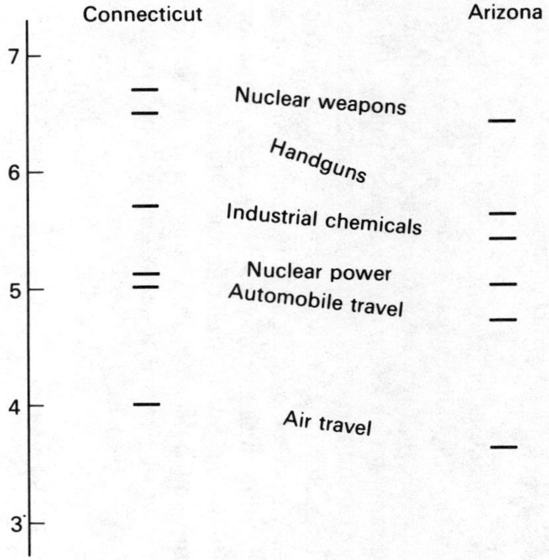

Source: Leroy C. Gould, Gerald T. Gardner, Donald R. DeLuca, Adrian R. Tiemann, Leonard W. Doob, and Jan A. J. Stolwijk (1988). *Perceptions of Technological Risks and Benefits*. New York: Russell Sage Foundation (data from Table 5.2, p. 76).

variables from consideration simplifies both empirical research and theory. Admittedly for the single investigator when the basis for the prime hypothesis proposed in the research turns out to be the omittable factor, the result must often be a disappointing temporary setback (unless the investigator really wanted to show unimportance), but for the rest of us it is a pleasure.

L. J. Savage used to joke that the world is full of functions, most of which are constants, a few being straight lines, and rare quadratic ones. He'd say: "All physics is built on this principle." He well knew that he was exaggerating or maybe minimizing, but in stressing the value of the search for simplicity and in streamlining a problem by setting aside the sufficiently unimportant variables, he gave a positive tone to findings that would disappoint many researchers.

Measurement Performance

We may wish to compare laboratories that carry out chemical analyses of the composition of body fluids for their performance both for level and for variability. Separating systematic errors from variable errors, along with careful calibration, helps our understanding of the collective system of standards used by industries and by disciplines. The information forms the first step in a process of improvement that leads to quality control and assurance. Knowing that one's laboratory has a systematic error on a test, or knowing that the measurement error is larger than other laboratories are getting on the same test, gives the investigator directions to start looking for the causes of these problems.

Cost–Benefit Situations

The process of making decisions often uses data analysis, usually to promote optimization or to aid in a cost–benefit analysis in one of its many forms. These purposes distinguish themselves from the ones we have already described because they point more definitely toward a next action. A variety of choices might be made, and the policymaker or the farmer wants to choose among them. Let us take a simple example.

A study of weight gains in young pigs deals with diets based on three factors: type of protein, amount of protein, and a chemical additive. In the work we have described thus far we would be concerned with total amounts of weight gained and its variation as the levels of the three factors changed. And we would be able to estimate weight gains for pigs like these on such diets. Essentially this part of the work is a study of nutrition. Let us suppose also that some of these diets produce more weight gain than diets have in the past.

If one merely wanted the best diet, the cell with the highest weight gain would be a good choice, possibly improvable by fitting a function. But the farmer will want to know more. For example, how will these diets improve profits? To the information available from the analysis, we have to add further data on diet costs and on prices of pork when the farmer sells the pigs. Indeed, understanding the uncertainties of costs and prices can be of central importance.

Some users of data suitable for analysis of variance may essentially be concerned only with part of the whole table and not with all its alternatives. Whether they work with an analysis of part of the data or with an analysis of all of it, they may need to add calculations related to the response and information about costs, risks, and benefits not included in the table.

Exclusions and Forgivingness

Policymakers may not wish to use some alternatives. For example, they may know that fierce police action, however effective, or high taxes, however attractive, will lead to consequences that they will not like or that they have scruples about. As a result, some alternatives in a table may not be relevant. And again many aspects of decisions will not be apparent in the data available.

In making decisions, we may want forgiving solutions rather than the very best ones. George Box has spoken of a homey example of cake mixes. He explains that a large baking company knew a lot about how to make the best cake from the point of view of the preferences of the customers. At the same time the company also knew that in mass-producing something complicated like a cake mix with many variables (for a start, amounts of flour, milk, sugar, and shortening; baking temperature; length of baking; and waiting time till sale), even with the finest quality control something may go a little wrong, and this deviance may lead even the "best" cake mix recipe to produce a low-scoring cake. Therefore the company asked whether there was among the recipes a high-scoring one that was also forgiving; that is, slight changes in the several variables (factors) produced cakes that were still high-scoring.

The field of agriculture recognizes a similar issue of forgiving solutions, especially in making plans for improved yields in less developed countries. The idea is to choose a program, not necessarily the highest yielding variety with related fertilizers and management, to make sure that a bad year won't make the farmer destitute. More realistically the program should have a high probability of achieving a certain minimum yield in spite of droughts, floods, or extreme temperatures.

2B. A QUANTITATIVE MICROCOSM

This section illustrates the idea of developing understanding of a data set and the importance of contributions of factors and residuals. It may help many of us to turn next to a very extreme example, one where the factors are measured and where the numerical value of the result can be expressed, at least approximately, in a simple way. The way associates numerical values with the levels of the factors. After we look at that example, we can relax its restrictions, step by step, and gain much more generally applicable insight and understanding. Accordingly, we turn to the relation of speed, duration (time) of motion, and distance traveled.

Some Raw Fits

If speed means average speed, we would expect

$$\text{distance} = \text{speed} \times \text{time};$$

and even if speed were to mean maximum speed, we might expect some similar relation. Trying to describe (raw) distance in terms of main effects and interactions seems unpromising when written down in words. We turn therefore to approximate formulas.

Table 2-2 offers three instances where distance is exactly speed × time, and it shows the data split into main effects and interactions. In Panel (a), where speed and time vary by moderate factors, interaction is moderate, and the main effects tell most of the story. Subsequent future understanding based on

$$\text{fitted distance} = (50)(\text{time}) + (2.5)(\text{speed}) - 125$$

would be relatively useful. (For refined questions we would probably need to bring in the interaction.)

Some motivation for fitting such a formula can come from the expansion of st around some central values s_0 and t_0 of speed s and time t. In Table 2-2, s_0 and t_0 were taken to be the average speed 50 and the average time 2.5. The expansion gives

$$\begin{aligned} d = st &= [s_0 + (s - s_0)][t_0 + (t - t_0)] \\ &= s_0 t_0 + s_0(t - t_0) + t_0(s - s_0) + (s - s_0)(t - t_0). \end{aligned} \quad (1)$$

Neglecting the final term and multiplying out the rest and gathering terms give the approximation

$$d \approx s_0 t + t_0 s - s_0 t_0. \quad (2)$$

In discussing Table 2-2 we speak of main effects. Our approximate formula combines the main effects of speed and time. Our fitted main effects given in equation (1) are

$$\text{main effect for speed:} \quad t_0(s - s_0),$$

$$\text{main effect for time:} \quad s_0(t - t_0).$$

Table 2-2. Three Instances of an Exact Relationship "Distance = Speed × Time" and the Corresponding Additive Fits (Common Term and Main Effects Only) and Residuals (Here All Interaction)[a]

Panel (a) A Moderate Case

Time, t	s, Average Speed 45	50	55		$s_0t + t_0s - s_0t_0$ = $50t + 2.5s - 125$ = fitted distance				Residual Distance		
	Distance										
2	90	100	110		87.5	100	112.5		2.5	0	−2.5
2.5	112.5	125	137.5	=	112.5	125	137.5	+	0	0	0
3	135	150	165		137.5	150	162.5		−2.5	0	2.5

Panel (b) A Squeezed-Together Case

	Average Speed 49	50	51								
2.4	117.6	120	122.4		117.5	120	122.5		0.1	0	−0.1
2.5	122.5	125	127.5	=	122.5	125	127.5	+	0	0	0
2.6	127.4	130	132.6		127.5	130	132.5		−0.1	0	0.1

Panel (c) An Expanded Case

	Average Speed 20	50	80								
1	20	50	80		−25	50	125		45	0	−45
2.5	50	125	200	=	50	125	200	+	0	0	0
4	80	200	320		125	200	275		−45	0	45

[a]All panels give distance, fitted distance, and residual distance; all fitted values are 50(time) + 2.5(speed) − 125.

Thus in Panel (a) of Table 2-2 the deviation of the entries in the first fitted column from those in the second is always −12.5 or $t_0(s - s_0) = 2.5(45 - 50) = -12.5$. Differences between second and third column entries are also −12.5. Similarly, the difference between the entries in the first and second fitted rows is −25 or $s_0(t - t_0) = 50(2 - 2.5) = -25$. Similarly for the second and third rows. Thus the main effects for speed are −12.5, 0, and 12.5, and the main effects for time are −25, 0, and 25, in the orders given by Panel (a).

In Panel (b), where speed and time are still more nearly constant than in Panel (a), the main effects tell essentially all the story. But in Panel (c) of Table 2-2, where both speed and time vary by much larger ratios, getting along with main effects and without the residual is hardly conceivable.

One aspect of what are often called natural laws is that they extrapolate well: Newton's extrapolation from the falling apple to the earth going around the sun is a classical example. If, in Table 2-2, we compare a tighter instance Panel (b) with a broader one Panel (c), we see that

$$(50)(\text{time}) + (2.5)(\text{speed}) - 125$$

will not extrapolate well because the residuals are huge (one fit even sending us in the opposite direction), although we know that

$$\text{time} \times \text{speed}$$

will extrapolate well.

Re-expression

Because the logarithm of a product is the sum of the logs,

$$\text{distance} = \text{time} \times \text{speed}$$

goes over to

$$\log \text{distance} = (\log \text{time}) + (\log \text{speed}).$$

This last relation tells us four distinct things:

1. Log distance can be written exactly as a sum of common term and main effects.
2. The time main effect can be taken as log time MINUS the mean of the log times.
3. The speed main effect can be taken as log speed MINUS the mean of the log speeds.
4. The common term, the mean of the log times PLUS the mean of the log speeds, is the grand mean of

$$\log \text{time} + \log \text{speed} = \log \text{distance}.$$

An exercise asks for an application of these findings.

If all we do is pull out main effects—as we can for any factorial pattern of data—it will not matter how we express time or speed: the same values will appear for the main effects (those in items 2 and 3 above). Being free of interaction does not depend on how we express the factors. When we can have it, it depends on how we express the response.

Is Engineering Different?

Sometimes people object to using re-expression by saying that although scientists could live with logarithms and other choices of re-expression, engineers and other practical workers (including agriculturalists, doctors, and applied economists) need their answers in the good old raw terms. Today if they need responses expressed on the original scale instead of log responses, they or we need only push one more key on the hand calculator, or write a part of a line of code for the PC or workstation. (Undoing any useful choice of re-expression is not likely to require more than one line of code.) Today engineering is not different in this way.

Much more, engineering is different because it is more likely to involve more factors—not all studied in any one data set—and this can matter. Our concern about interaction with unstudied factors needs to be greater in engineering. Hence our concern with re-expression—especially in connection with interactions that we are not able to look at—has to be greater, and not less.

Part or Whole?

A factorial experiment often covers a wide variety of alternatives. This need not mean that we have to think about all of them—or make an analysis that involves all of them. Fitting a straight line to data can give us better values, even for x's where we have data, but only if the data actually being fitted resemble a straight line sufficiently.

If we have data scattered from $x = 0$ to $x = 100$, and the relation of y to x is closely linear only from $x = 50$ to $x = 90$, *and* if we want to estimate y for $x = 60$ or $x = 75$, we would do better to fit the line to the more limited range of x values (50 to 90). (If, on the other hand, the straight-line behavior went all the way from 0 to 100, we would do better to fit the line to all the data.) Similarly, if there is a piece of the data where interactions are negligible and if what we need to know falls within that piece, we are likely to get a more useful analysis by working with only that piece. We focused on the straight-line fit to avoid nonlinearities; we can sometimes focus on the analysis of variance to avoid interaction nonadditivities. If chosen with good judgment, such focusing can help a lot.

Many experiments are done to seek good performance. (As we noted above, cost, selling prices, forgivingness, and long-term undesirability of alternatives may all have to be considered in saying what is "best.") It is not infrequent that the part of the experiment that provides excellent, very good —or even just good—performance is relatively free of interaction. Once we stray into regions of poor performance, however, sizable interactions do appear. In such circumstances, it is likely (1) that restricting the analysis to regions of high performance may produce a closer, more useful fit and (2) that overall analyses can guide the choice of an expression for the response that will perform even better in the analysis of the data from the more restricted region. We want to use the analysis of variance wisely—as an aid, not as an automatic constraint.

2C. MORE CLASSIC PURPOSES

We discussed or mentioned previously some less familiar purposes for the sorts of analysis offered by "the analysis of variance." We emphasized understanding, as we shall continue to emphasize it. Earlier we explained why we often have to emphasize differences—because unstudied factors may drive levels up or down in ways that are at best poorly foreseeable.

When we ask "How sure are we?" or "What aspects are we sure about?", we have to go to confidence procedures, both for direction of difference and amount of difference. In many circumstances we can use the confidence procedures developed for data analysis to get confidence results for a cost–benefit analysis. Once we have tied down the two ends of a confidence interval, we need not be abashed in the face of cost–benefit machinery, even when that machinery is complicated. As long as the evaluated outcome is a monotone function of the response we have studied, putting through the cost–benefit values based on the extremes of the confidence interval for the difference in responses will give us reasonable approximations to confidence intervals for the cost–benefit comparison. We can usually tell whether we should be confident about the direction of the difference in cost–benefit evaluations.

Understanding of interactions depends on our ability to express their behavior in words. So we need to learn how to do this. *Usefulness* of quantitative knowledge about interactions may depend on having clear tables, whether two-way or more-way, and, where possible, good interpolation schemes.

Fitted Values Versus Better Forecasts

Among our purposes, estimation, fitting, or forecasting may be high on our agenda. If so, analysis-of-variance methods aid us in figuring out how to get good fitted values of the response variables. Simultaneously, we produce estimates of the uncertainty of the resulting values. It can well happen that we can get a better estimate for the result at levels of a pair of factors by fitting a surface to all the data rather than by taking our fitted value of the response to be the average value of the measurements for the chosen levels of the factors. In linear regression of y on x, we are used to the idea that, if we fit a line to a set of observations (x_i, y_i), where the b's are regression coefficients, the fitted regression value

$$\hat{y}_x = b_0 + b_1 x$$

can be a better estimate for the average value of y for the population at that x than the single y observation taken at that value of x, or even the average of all the y's taken at that value of x. The same idea extends in myriad, sometimes complex, ways to the estimates in the analysis of variance. In such situations, we say that we are borrowing strength from the other observations to improve the estimates.

Lateral Forecasting

Most of the problems of forecasting forward over time arise in forecasting sidewise over conditions. Thus, for example, farmers involved in agricultural experiments are not a random sample of farmers, and differences in practices can ruin otherwise clear predictions.

2D. CAUSATION, EXPERIMENTATION, AND OBSERVATION

In discussing causation, we need to point to two kinds of factors (or variables). In *experiments*, the levels of some of the factors, often called *treatments*, are ordinarily assigned by the investigator. In *observational studies*, the levels of the factors are usually observed (imposed by nature) and are not under the direct control of the investigator. To illustrate the important distinction between assignment and observation, we describe two studies designed to look very similar but to differ in this one crucial aspect. A factor (or variable) is more than a collection of levels. It includes also the way the factor is assigned. Thus in some investigations some factors are controlled and lead to experiments; when none are controlled, merely observed, we have observational studies.

Example of an experiment. To teach pupils to spell, two methods are proposed, a method self-chosen by the pupil (excluding computers), such as spelling the words aloud letter by letter, and a computer method. From a collection of pupils, half are randomly assigned to the self-chosen group and half to the computer method group for a semester. Their scores on many spelling lessons are recorded and compared. If the investigators find a substantial difference in performance between the two methods, they will be fairly confident that at least part of the difference in performance is due to the treatment assigned. If the computer was associated with higher scores, it will likely be credited with *causing* the improved performance unless something obvious has gone awry.

We contrast this scenario with one where the investigator does not assign the spelling methods but finds them in the population:

Example of an observational study. To find out whether a self-chosen method of studying spelling or a computer method produces better scores, some investigators search out two groups. One group they find consists of people who teach themselves spelling by using their own method (excluding computers). The other group uses a computer method of studying. Suppose that the pupils using the computer produce better spelling scores than those using their own teaching methods; what should the investigators conclude?

They should be cautious. Why did some people use the computer method and others not? One possibility is that richer people are more likely to have computers available. Richer people are likely on average to be better educated, and better educated families may be better at supporting the home learning process. Thus the investigators may not have discovered that the computer method caused better spelling, but rather that its use selected people with a better training environment.

From the point of view of the data observed in these two investigations, the spelling experiment and the spelling observational study, the records of the data could be indistinguishable, yet the interpretations are very different. We need, then, to distinguish between variables or factors whose levels occur as found in nature and those assigned by investigators.

In many studies nature supplies the factors and their levels: the family income, the age of the patient, the disease needing treatment—more generally, for individuals, the demographic variables. In experiments, we may assign to various groups different exercises or diets or, more generally, treatments intended to cause some beneficial outcome. Similarly, in a manufacturing process for cakes we may want to control and therefore assign different temperatures, lengths of baking time, relative amounts of flour, sugar, and shortening to try to discover the recipe yielding the tastiest cake. The levels of these factors are imposed by the investigators.

Sometimes we cannot carry out experiments, and then we have to make what we can of observational studies in lieu of experiments. For example, we cannot randomly assign amounts of cigarette smoking to study health outcomes.

Sometimes nature provides what are called natural experiments by exposing different groups to different treatments. But from our point of view, these situations are observational because the "experimenter" does not assign the treatment to the subjects. For example, nature provided water in different parts of the country that had different concentrations of fluoride. Dentists trying to explain why the amount of tooth decay varied from place to place observed that more fluoride led to fewer caries in the teeth of the residents. Much later some experiments were done to verify the causal implications of these observational studies.

Experiments alone are not sure to be able to establish causality. They must be well executed. The populations treated have to be comparable. Investigators have to be wary of possible extra actions that may destroy the implication of causality. For example, a very well-designed experiment in training soldiers to send and receive Morse code compared the additional effect of sleep learning (with earphones) with control groups given no extra training. The sergeants in charge of the control groups did not want their groups bested by the sleep learners, and so they added extra hours of wide awake practice, to the chagrin of the investigators and the ruination of the experiment.

We often use the same methods of analysis on data involving either kind of factor, "assigned" or "found," and so the distinction needs to be kept in mind, especially when we interpret the analyses done.

When we change the level of a factor for individuals in comparable populations, and this changes the output of the populations on an output variable, then the change ine in output variable is often said to be caused by the change in the level of the factor. Thus turning the handle on a working water faucet may be said to cause the increase in the rate of water flowing through the faucet (even though it takes a whole municipal water system to make this happen). Similarly, vaccination may reduce the rate of flu in a group of people, though not guaranteeing absolute immunity.

We are not trying here to discuss the usual tests for causation:

Consistency—whether you can make the change happen at will;

Responsiveness—changing the level of the factor in some individuals changes the output for those individuals (taken as a group);

Mechanism—a mechanism that someone might sometime understand relates the change in the factor, step by step, to the output.

In many circumstances, we wish to identify and measure the effects or results of different levels of factors. When we do this in experiments, we expect to interpret the effects as *caused* by the changes in levels of the variables. In some investigations such as observational studies, although we have difficulty drawing causal inferences, we can still make estimates that associate changes in output with the changes in the levels of the factors. Thus we often can make predictions about amount of change in output from group to group, even though we may not wish to claim that the levels of the factor cause the change. For example, when we take a random sample of adult men and group them according to their weights in 10-pound intervals, we find that heavier men are taller, on average. We do not infer from that relation that changing a man's weight changes his height. We merely recognize that, in the population as it stands, heavier men are taller on the average than lighter men.

To summarize, the distinctions between the outcomes that arise from experiments and from observational studies need not be evident from the data table or the analysis of the data's variability. Either kind of data requires a careful analysis to evaluate behavior and size of contributions associated with various sources of variation. The distinctions are especially relevant when we seek to attribute causality.

2E. SUMMARY

To sum up, this book emphasizes as its main purposes for analyzing tables of the sort used in analysis of variance the understanding of phenomena, their science, their engineering, and the practice of the topic of the investigation. We do this by studying the size and importance of factors and their interactions, and by fitting functions of the factors to the observations, and by breaking measurements into components. Thus we study both the response levels achieved and the contributions of the factors, as well as other sources of variation such as measurement error. These estimates can feed into forecasts.

The areas of decision and further action, such as policy choices, optimization in business decisions, and cost–benefit analyses, also use the results of such data analysis, though they ordinarily require further input not usually available in the original data used for analysis of variance.

REFERENCE

Gould, L. C., Gardner, G. T., DeLuca, D. R., Tiemann, A. R., Doob, L. W., and Stolwijk, J. A. J. (1988). *Perceptions of Technological Risks and Benefits*. New York: Russell Sage Foundation.

EXERCISES

1. When might two different investigators use different methods to analyze the same set of data? Give an example.

2. When do investigators find it beneficial to be able to set factors aside? Why is it beneficial?

3. Varying purposes. Consider the following distribution of scores, where $P(x)$ is the probability that the next number that appears is x.

$P(x)$	.4	.3	.2	.1
x	0	1	2	3

(a) What value of x would you choose to maximize the probability of choosing exactly the correct number?

(b) What value of x would you choose if you want to be correct on the average?

(c) What value of x would you choose to minimize the largest possible deviation from your choice?

4. In the dental-gold example (Table 1-2) are there any factors that an investigator might want to dismiss? Why?

5. Describe an experiment and an observational study for which the resulting data might be very similar but where the argument that a key factor was causing the difference in outcome was strong for the experiment but weak for the observational study.

6. List three purposes of investigations and explain what sort of analysis each might require.

7. From your experiences in mathematics, physics, science, and everyday life, illustrate L. J. Savage's joke by providing

(a) a useful function that is a constant;

(b) a useful relation that is linear;

(c) a useful relation that is quadratic.

8. Explain the idea of forgivingness as it applies in the following example. In baking cakes, suppose the data show that taste tests give scores shown as entries in the table for various baking temperatures and amounts of a key ingredient. High scores are preferable. What would be good choices for the recipe when dealing with amateur cooks and unreliable ovens?

Amount of a Key Ingredient: Parts of a Cup	Temperature				
	350	375	400	425	450
0.2	4.2	5.0	6.8	8.4	8.5
0.3	5.5	6.0	6.1	8.3	8.3
0.4	6.0	8.7	7.1	8.2	8.4
0.5	6.5	6.8	6.0	6.4	7.4
0.6	6.8	6.0	5.5	5.0	4.5

9. The example explored in depth in Table 2-2 could be used to discuss the adequacy of an approximate formula for the area of a thin, long rectangle whose width in feet is given by t and whose length is given by s. The approximate formula is $-t_0 s_0 + t_0 s + s_0 t$, where $t_0 = 2.5$ and $s_0 = 50$.

(a) What is the correct formula for the area of the rectangle?

(b) What is the error in the area when $t = 2$ and $s = 55$?

(c) Explain what the residuals in the panels show about the approximation.

(d) If t is between 1 and 2 and s is between 45 and 50, what is the largest error in the formula?

10. (Continuation) In the area example of Exercise 9, we could use logarithms. If we take the logarithm of the area, what will be its relation to the width and the length of the rectangle?

11. In Panel (a) of Table 2-2 the times are 2, 2.5, and 3 and the average speeds are 45, 50, and 55. Use logarithms to find the common term for the logarithmic table. Find the main effects for time. Find the main effects for average speed.

CHAPTER 3

Preliminary Examination of Data

Frederick Mosteller and David C. Hoaglin
Harvard University

Whatever purpose motivates the analysis, we often gain from examining the data in a variety of ways before systematic analysis begins. The present chapter briefly explains some useful strategies for editing the data (a process sometimes called "data cleaning") and appreciating the general features of the data in relation to the structure that will guide the analysis. We then review three versatile techniques for exploring the data beyond the initial phase: plotting, boxplot, and stem-and-leaf display.

Some of the matters spoken of here were handled so automatically 50 years ago that they would scarcely have been worth mentioning; but now that we stuff our numbers into computers, and move them from one place to another without seeing them, some active effort may be necessary to bring them into view.

3A. EDITING

Without being formal about our definition of editing, we can still profitably list under this heading a number of routine activities that are worthy of effort before launching extensive analyses.

Counting the House

Are all the data present that you expected to have? Some common troubles come from losing the final page of a set of data, or not having a few final columns because of programming or space problems. Sometimes data may be missing in less routine ways. For example, an analysis of a set of data in a two-way table revealed some surprising discontinuities. It turned out that, to save space, the source had eliminated the middle third of the columns and then closed up the gap, and nothing in the labeling of the columns gave the trick away. Thus, when you are not acquainted in advance with a data set, you may not find it easy to check that all the data are present. An easier but important situation arises when row numbers go along ..., 26, 27, 28, 29, 31, 32, ...; then it is worth asking whether row 30 is legitimately absent.

Anomalies

Are any of these data impossible? In data that must represent only part of a mile, do some values amount to more than 5300 feet? Was the year longer than 366 days? If columns or rows or entries in cells or their sums are bounded in some way, try to check that the conditions were met.

Lots of wrong numbers aren't out of bounds. Our telephone often tells us that "the number you have dialed, 999-999-9999, is not in service." Of course, we never dial this number, but apparently our phone tells the system that we do. In data, when you see a set of numbers repeating or oscillating or patterning, it is often worth asking whether something has come unstuck. Two rows alike in real data must raise the suspicion of a copying mistake.

If independent cross-checks are possible, these may be valuable. One small book of data included a set of summary tables. Cross-checking these summaries against the data located six errors, mostly omissions.

When numbers are in bounds but extreme, it's worth checking whether something has gone awry—a moved decimal, a wrong multiplier. When numbers have been constructed by hand through some calculations, the very early numbers to be computed (often those in the upper left corner) may have a mistake because the person constructing the entry hadn't gotten a routine down properly. A machine is more likely to get them all wrong systematically. If you have a way to check a few numbers, seize it.

We have systematic ways of looking for outliers, and such methods can be used to locate possibly erroneous entries. Section 3D mentions one of them.

Missing cells may need attention, and a missing entry can be the source of a great deal of trouble when a space is not properly left for it. Then all later data may be in wrong cells, even though sometimes they look very smooth.

Data that go in the wrong direction—the faster the projectile, the longer it took it took to get to its target—happen a lot, and one has to look out for them. As one example, various scales used in educational, sociological, and psychological tests, as well as in marketing, ask a respondent to check a number from 1 to 10. Subjects sometimes get mixed up as to whether 1 is their top choice or lowest rating, and the whole questionnaire needs recoding because x should have been $10 - x + 1$. And similarly, when marking a point on a line to indicate a position between low and high, the respondent may mix up the ends. Sometimes internal consistency can help sort this out, sometimes the respondent can be called back or the data set aside, and sometimes we are stuck.

Discontinuities

When rows or columns of a table move smoothly but then suddenly jump, make sure you understand why that happens. That's how one of us found that the middle third of the table was missing. The rise between two successive columns was too big for the other differences.

3B. APPRECIATING THE DATA

It is helpful to get a feel for a set of data. Make friends.

1. *Size.* About how big are these numbers? All around 1000? All around 3.14? All near 0.002? Are they all over the place, with a smallest and largest number?
2. *Variability.* Are numbers near one another in the table close in value? How much do they jump around?
3. *Slant.* How do the data tend? Do they get bigger (or smaller) monotonically from left to right? Does the whole table move in the same way? Are the numbers higher around the edges and lower in the middle (or the reverse)?
4. *Shape.* Is it practically constant, linear, or nonlinear? What does it take to try to fit the data with a function?
5. *Random Variation.* Do the data show an element of random behavior?

3C. PLOTS OF THE DATA

For appreciating the data in a table or listing, graphical displays often prove very helpful. The simplest of these put the response variable on the vertical axis and some feature of the structure on the horizontal axis and then plot one point for each observation. When the data consist of several groups and follow no further structure, we often assign numbers to the groups $(1, 2, \dots)$ and use these numbers as the horizontal coordinate. By spreading out the groups side by side, this display (sometimes known as a *dotplot*) brings all the data into view.

EXAMPLE: FUEL CONSUMPTION

In its general report on the cars of the 1990 model year, *Consumer Reports* included a measure of fuel consumption for selected cars: the number of gallons that the car would use in going 15,000 miles. (The figure is an estimate, extrapolated from data collected in a standard test drive.) The report groups the cars into seven categories: small cars, sporty cars, compact cars, medium cars, large cars, small vans, and sport/utility vehicles. Figure

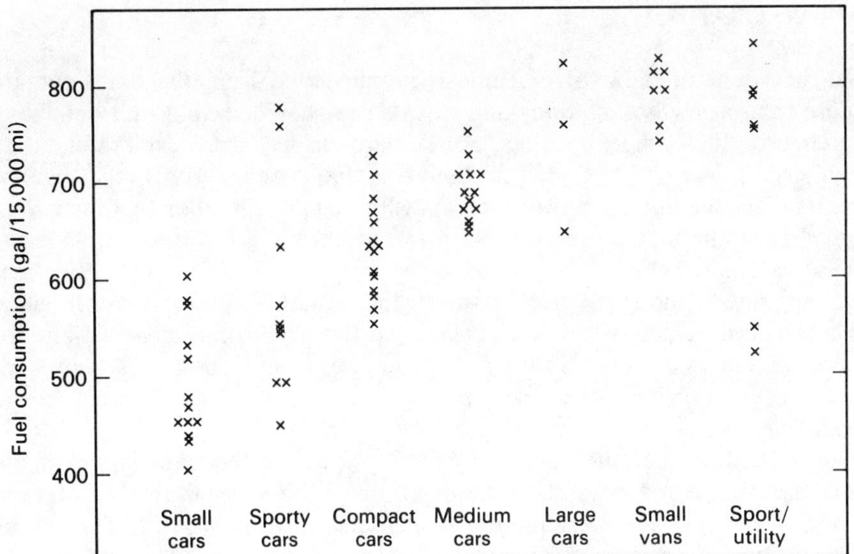

Figure 3-1. Side-by-side plot of estimated fuel consumption (gallons per 15,000 miles) for selected 1990 cars.

3-1 plots the data for the groups side by side. For each group we readily get a rough idea of the typical level of fuel consumption and the variation among cars. In addition, the definite trend toward greater fuel consumption as one moves from left to right reflects the general ordering of the groups from smaller, lighter cars to larger, heavier ones. Among the sport/utility vehicles, two stand out as having lower fuel consumption. They turn out to be substantially smaller than the other vehicles in the group. Similarly, two of the sporty cars have somewhat higher fuel consumption. These have larger engines. Before any analysis, this display has provided a useful look at the data.

When the data have more structure, it is often possible to adapt the basic plotting approach to fit the particular situation. For example, if the data come in a two-way table, so that we may denote the observation in row i and column j by y_{ij}, many analysts customarily plot y_{ij} versus i and include line segments to link points with the same value of j. Sometimes one sees the behavior of the data more clearly by making both this plot and the plot of y_{ij} versus j.

As a further example, with three (or more) factors one could choose two factors whose combination would define the groups and then use plotting symbols that show the level of a third factor (instead of the $\times$ that serves as the plotting symbol in Figure 3-1). This type of display has the same ingredients and purpose as the stem-and-version plot illustrated in Table 1-3.

3D. BOXPLOTS

When groups of data (or of summaries calculated from the data) contain more than a few observations and we wish to show less than full detail, we often use side-by-side boxplots. A standard display from exploratory data analysis (Tukey 1977, Chap. 2; Emerson and Strenio 1983), the *boxplot* summarizes the middle half of the data with a rectangular box that includes a crossbar at the median. Figure 3-2 gives an example of a boxplot, with key features labeled. From each end of the box a dashed line (or solid line) extends out to the farthest observation that is not beyond the cutoff value. These cutoff values (which we explain shortly) aim to flag observations for investigation as possible outliers. Thus any observation that lies below the lower cutoff or above the upper cutoff appears as a separate point in the boxplot.

The definition of the cutoffs uses the *fourths* of the data (approximate quartiles that mark the ends of the box). From the lower fourth F_L and the upper fourth F_U we determine the *fourth-spread*, $d_F = F_U - F_L$. The cutoffs lie 1.5 times d_F beyond the fourths:

$$F_L - 1.5d_F \quad \text{and} \quad F_U + 1.5d_F.$$

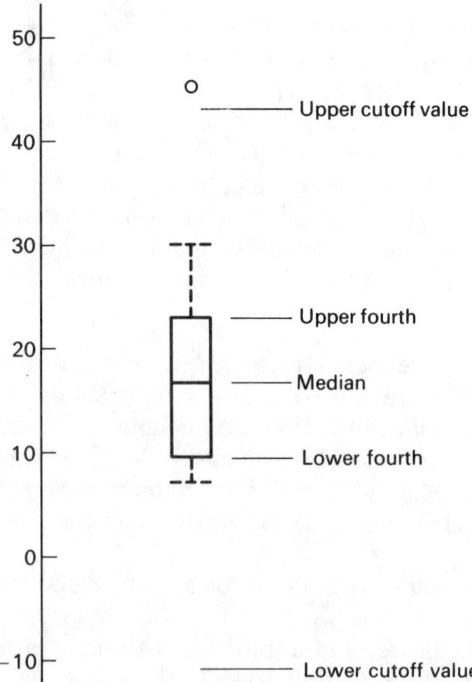

Figure 3-2. An illustrative example of a boxplot, showing the key features.

At a glance a boxplot show the location, spread, and skewness of the data, along with observations that may be outliers. The bar across the box at the median summarizes location; the length of the box shows spread, using the fourth-spread (approximately the interquartile range); and the relative positions of the median, the upper fourth, and the lower fourth provide an indication of skewness.

Some popular computer software produces boxplots that depart from the above standard in unexpected ways (Frigge et al. 1989). To avoid getting a mistaken impression of the data, it may be necessary to check the documentation that accompanies the software.

Example: fuel consumption

For the data on fuel consumption of 1990 cars, plotted in detail in Figure 3-1, Figure 3-3 shows the side-by-side boxplots. In this display it is easier to focus on the typical relation of fuel consumption to the categories of cars. The small numbers of sporty cars and sport/utility vehicles allow the apparently deviant observations to affect the upper fourth and the lower fourth, respectively, and so those observations do not lie beyond the corresponding cutoff values. Thus this example also illustrates the benefit of looking at both a dotplot and a boxplot before choosing one of them to display a set of data. Also, with so few cars we would often choose (in displays that combine dotplots and boxplots) to plot the points individually, instead of concealing them beneath a boxplot.

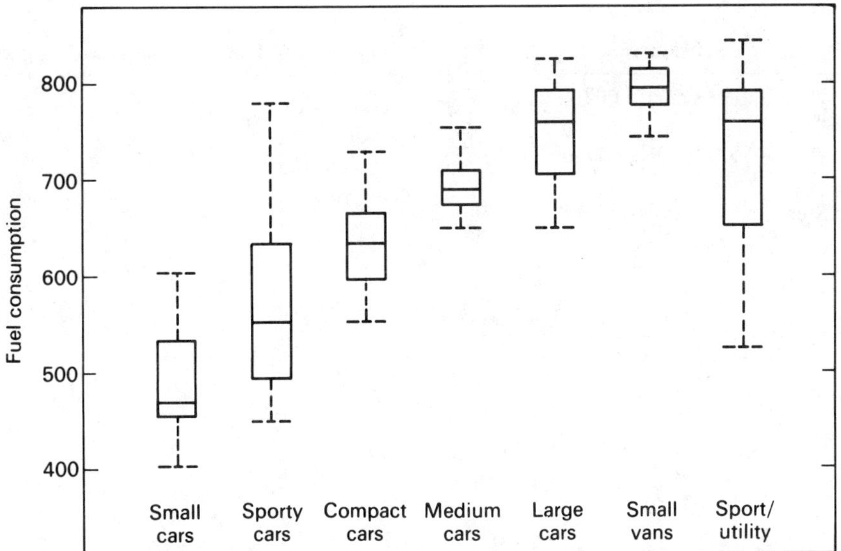

Figure 3-3. Side-by-side boxplots for the data on fuel consumption of selected 1990 cars.

3E. STEM-AND-LEAF DISPLAYS

For a first look at a batch or sample of numbers, we often use a *stem-and-leaf display* (Tukey 1977, Chap. 1; Emerson and Hoaglin 1983). In this technique from exploratory data analysis, the basic idea is to let the most significant digits of the data values themselves do most of the work of sorting the data into numerical order and displaying them. To illustrate the process, Table 3-1 lists the fuel consumption data for medium cars and shows one possible stem-and-leaf display.

The strings of digits that appear to the left of the vertical line $(65, 66, \ldots, 75)$ are *stems*. Most of the stems arise as sets of initial digits among the data values, and the others (here 70, 72, and 74) are included to complete the coverage of the range of the data. Each data value is represented by a single-digit *leaf* to the right of the vertical line. Thus, for example, 755 separates into 75 and 5, and the display show the leaf (5) on the line corresponding to the stem (75). This stem-and-leaf display has a total of 13 leaves because the medium cars yielded 13 observations on fuel consumption.

By using the display in Table 3-1 to help appreciate these data, we readily see that the observations range from 650 to 755 and that the data are somewhat more concentrated in the lower half of the range. A value around 690 seems typical. We also note that each leaf is either 0 or 5—perhaps the result of a deliberate decision to round the estimates in this way and thus avoid making them appear more accurate than they actually are.

Table 3-1. Fuel Consumption Data for Medium Cars and One Stem-and-Leaf Display

Data (gallons per 15,000 miles):

755, 710, 710, 680, 690, 690, 675, 655, 650, 730, 675, 710, 660

Stem-and-leaf display (unit = 1 gallon):

Stem	Leaf
65	05
66	0
67	55
68	0
69	00
70	
71	000
72	
73	0
74	
75	5

Source: Consumer Reports, April 1990 (data from pp. 247–251).

Two Basic Variations

The stem-and-leaf display in Table 3-1 uses one line for each stem. That is, each line could receive any of the 10 leaf digits 0 through 9, according to which data values lie in that interval. For some sets of data this format looks too crowded: too many leaves pile up on a single line. One variation in format relieves such crowding by using two lines for each stem. Then the first three stems in Table 3-1 would expand into

```
65* |
65· |
66* |
66· |
67* |
67· |
```

with leaves 0 through 4 on the * line and leaves 5 through 9 on the · line.

A second variation in format uses five lines per stem, for example:

```
6* |
 t |
 f |
 s |
6· |
```

Now leaves 0 and 1 go on the * line, 2 and 3 on the t line, 4 and 5 on the f line, 6 and 7 on the s line, and 8 and 9 on the · line. This format may occasionally serve to relieve extreme overcrowding, but more often it handles undercrowding (leaves scattered out over too many stems), in conjunction with a shift of the split between stem and leaf one digit position further to the left. In this format the display in Table 3-1 could become

```
6* |
 t |
 f | 55
 s | 677
6· | 899
7* | 111
 t | 3
 f | 5
 s |
7· |
```

with the leaves now coming from the tens digit of the data values (i.e., stem and leaf together give a multiple of 10 gallons). Alternatively, we note that each line in the above display corresponds to an interval of width 20 gallons, whereas a line in Table 3-1 covers an interval of 10 gallons.

Data Beyond the Cutoffs

When a batch contains data values below the lower cutoff or above the upper cutoff (as defined in Section 3D), we usually list these on special lines just beyond the ends of the stem-and-leaf display. In place of a numerical stem these two lines have "LO" (or "L") and "HI" (or "H"), respectively. The data values on them appear without being split into stem and leaf. Table 1-7 in Chapter 1 shows an example of such a stem-and-leaf display.

Side-by-Side Stem-and-Leaf Displays

In a way that resembles the plot in Figure 3-1, we can use a form of stem-and-leaf display to show several groups of data side by side. Table 3-2

Table 3-2. Side-by-Side Stem-and-Leaf Display of the Fuel Consumption Data

	Small Cars	Sporty Cars	Compact Cars	Medium Cars	Large Cars	Small Vans	Sport/ Utility
4*	0						
t	3						
f	4555	5					
s	7						
4·	8	99					
5*							
t	23						2
f		455	5				5
s	7	7	7				
5·	8		89				
6*	0		01				
t		3	333				
f			4	55	5		
s			67	677			
6·			8	899			
7*			1	111			
t			3	3			
f				5		4	5
s		6			6	6	6
7·		8				99	99
8*						11	
t					2	3	
f							4

Source: Consumer Reports, April 1990 (data from pp. 234–257).

uses the fuel consumption data to illustrate this form. Once we allow for the fact that the large values are at the bottom, we see in this visual display essentially the same features that we saw in Figure 3-1.

Stem-and-Identifier Displays

The side-by-side arrangement of stem-and-leaf displays opens the way for us to base the groups on the factors in a set of data that has two or more factors. For example, if Factor A has two versions (A_1 and A_2) and Factor B has three versions (B_1, B_2, and B_3) and the data have several observations at each combination of a version of A with a version of B, the side-by-side display could use the six combinations as groups. The order might run A_1B_1, A_1B_2, A_1B_3, A_2B_1, A_2B_2, A_2B_3 with a wider horizontal space between A_1B_3 and A_2B_1.

When the data have three or more factors, we can introduce a further modification. We choose two factors to define the groups (as above) and then use an identifier such as the version number of a third factor in a place of the leaf digit. We call the result a *stem-and-identifier display* (or, if appropriate, a stem-and-version display). Table 1-3 in Chapter 1 illustrates a stem-and-version display for a set of data with four factors.

REFERENCES

Emerson, J. D. and Hoaglin, D. C. (1983). "Stem-and-leaf displays." In D. C. Hoaglin, F. Mosteller, and J. W. Tukey (Eds.), *Understanding Robust and Exploratory Data Analysis*. New York: Wiley, pp. 7–32.

Emerson, J. D. and Strenio, J. (1983). "Boxplots and batch comparison." In D. C. Hoaglin, F. Mosteller, and J. W. Tukey (Eds.), *Understanding Robust and Exploratory Data Analysis*. New York: Wiley, pp. 58–96.

Frigge, M., Hoaglin, D. C., and Iglewicz, B. (1989), "Some implementations of the boxplot," *The American Statistician*, 43, 50–54.

Tukey, J. W. (1977). *Exploratory Data Analysis*. Reading, MA: Addison-Wesley.

CHAPTER 4

Types of Factors and Their Structural Layouts

Judith D. Singer
Harvard University

The data we examine with analysis of variance involve a *response* variable and structural components called *factors*, which identify the observations under study as collected from specific subgroups or under particular circumstances. An epidemiologist studying the correlates of cardiac function, for example, might collect data for each respondent on factors such as GENDER, AGE, WEIGHT, DIET, SMOKING STATUS, and STRESS. A psychologist examining the effects of positive reinforcement on the speed of maze solving by rats might collect data on factors such as MAZE DIFFICULTY, MAZE LENGTH, and TYPE OF REWARD. (In this chapter only, we capitalize factor names to emphasize that we are speaking of factors.)

Each factor defines specific subgroups or particular circumstances under which the observations were collected. These are called the factor's *levels* or *versions*. Of these two terms, "level" is the more common in the literature of analysis of variance and design of experiments. Some writers use "version" to cover a broad range of circumstances, and they reserve "level" for use with measured factors. In this book we do not attempt to maintain such a distinction. Instead, we ordinarily use "level" and "version" interchangeably.

The levels of some factors are well established; for example, the factor GENDER customarily has two levels, male and female. The levels of most other factors are determined by the researcher. The epidemiologist studying cardiac function can record SMOKING STATUS as a dichotomy (smoker or nonsmoker) or as the number of cigarettes smoked per day.

Two characteristics of factors convey important information that helps shape the data analysis. A factor's *type* describes the amount of quantitative information in its levels. The more quantitative information contained in a factor's levels, the easier it is to develop models of the factor's relationship to the response. For example, if an epidemiologist records SMOKING

STATUS as the number of cigarettes smoked per day, the factor gives more information than if the epidemiologist records only whether the respondent smokes. This information could show not only whether cardiac function is worse among smokers than among nonsmokers, but also *how much* worse cardiac function seems to be on average for every additional cigarette smoked per day.

The second characteristic that helps shape the data analysis is the *structural layout* of the factors. One popular layout, known as completely crossed, presents data at each of the possible combinations of factor levels. With such a layout, a researcher can determine whether the relationship between each factor and the response is the same regardless of the levels of the other factors under study, or whether the relationship differs when specific levels of the factor are observed in combination with specific levels of the other factors—for example, whether the relationship between SMOKING STATUS and cardiac function is the same regardless of GENDER, AGE, WEIGHT, or DIET. Analysis of variance applies to many other data structures; some of these structures facilitate exploration of relationships between the response and the factors, and others inhibit this exploration.

In this chapter, we show how two characteristics of factors—the factor type and the structural layout—help guide the analysis of variance. We begin by describing the three different types of factors that arise when conducting an analysis of variance and continue with examples of several common structural layouts.

4A. TYPES OF FACTORS

Factors differ in the amount of quantitative information that underlies their levels. We find it helpful to distinguish three types of factors—unordered, ordered, and measured—each of which occurs in the following example.

Example: The Great Books

Until the late nineteenth century, the curricula of English courses in U.S. secondary schools focused primarily on the principles of grammar and secondarily on comprehension and appreciation of literature. This began to change when, in 1869, Harvard College introduced knowledge of selected authors and works as a requirement of admission. Other colleges soon followed suit, and junior and senior high schools responded by including the study of the classics in their English literature courses.

Believing that the study of literature should be coordinated with each student's understanding of the material, Burch (1928) examined how students' comprehension of 35 classics taught in the California public schools varied by the students' grade and ability. For each book, she selected a sample passage and developed a multiple-choice test that assessed a student's comprehension of the passage. She administered the test to 452 students in

Table 4-1. Reading Comprehension (Mean Percentage of Items Answered Correctly) by Book, Grade, and Ability Group

		Book					
Grade	Ability Group	*Black Beauty*	*Gulliver's Travels*	*The Iliad*	*Lorna Doone*	*The Aeneid*	Milton
7	I	59	40	19	26	16	5
	II	78	60	31	25	40	13
	III	83	69	42	39	46	10
	IV	90	71	55	56	47	27
8	I	70	46	31	23	30	23
	II	84	72	41	34	39	17
	III	91	79	54	57	53	15
	IV	99	85	64	68	67	31
9	I	62	53	30	24	26	18
	II	81	66	46	40	39	11
	III	83	78	56	50	48	24
	IV	94	89	73	71	71	41
10	I	72	47	39	21	29	12
	II	78	59	47	40	44	10
	III	88	75	66	62	55	27
	IV	96	88	85	83	67	47

Source: Mary Crowell Burch (1928). "Determination of a content of the course in literature of a suitable difficulty for junior and senior high school students," *Genetic Psychology Monographs*, 4, 165–293 (data from Table 9, p. 231).

seventh through twelfth grades in one school district in California. The students in each grade were ranked by ability into four groups, labeled I, II, III and IV, with I indicating the lowest ability and IV indicating the highest. Table 4-1 and Figure 4-1 present a portion of her data: the mean percentage of items answered correctly for six books by ability level for students in grades 7, 8, 9, and 10.

All three factor types are represented in this data set. BOOK is an *unordered* factor because its observed levels merely identify books; each level contains no information about that level's (book's) relationship to the other levels (the other books). The levels of an unordered factor do not contain quantitative information. Because the levels of an unordered factor simply name things—groups, treatments, or categories—S. S. Stevens, in his classic 1946 paper on scales of measurement, said such factors have *nominal* scales.

GRADE is a *measured* factor because each of its levels (7, 8, 9, and 10) conveys precise quantitative information—the student's year in school. Each observed level of a measured factor represents a specific position along a continuum defined by the factor. Other examples of measured factors include AGE (in years) and LENGTH OF EXPOSURE TO A TREATMENT (in

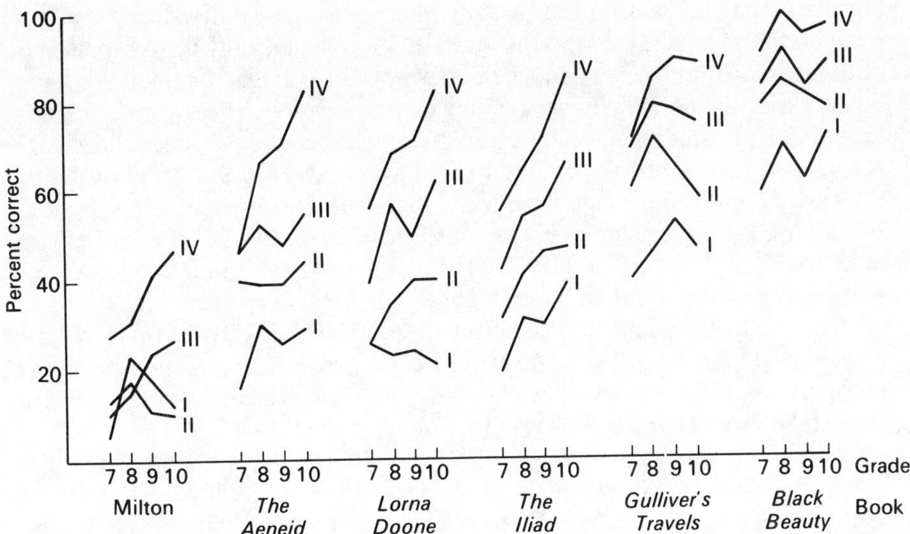

Figure 4-1. Reading comprehension (mean percentage correct) by book, grade, and ability group.

hours, days, or number of sessions). In most school populations, as in this data set, AGE and GRADE are nearly equivalent measures.

The amount of information in the third factor, ABILITY GROUP, falls between these two extremes. Its levels convey some quantitative information: students in ability group I have the lowest abilities, and students in ability groups II, III, and IV have progressively higher ability. But unlike a measured factor, each level of ABILITY GROUP does not indicate a precise measurement, but only the position of that level relative to the other levels. Although we know that the students in ability group II are of higher ability than those in ability group I, we cannot say precisely how much higher their abilities are, nor can we say whether the distance between ability groups III and IV is the same as the distance between ability groups I and II. When a factor's levels are rankings (often along a continuum), we say the factor is *ordered*, or that it has an *ordinal* scale. Other examples of ordered factors include LEVEL OF EDUCATION (primary school only/some high school/high school graduate/some college/college graduate), SOCIAL CLASS (lower/middle/upper), POLITICAL PHILOSOPHY (conservative/moderate/liberal), and FAMILIAL RELATIONSHIP (unrelated/siblings/twins).

How does a factor's type help guide the analysis of variance? During preliminary analyses, the particular factor type actually has little importance. Regardless of the amount of quantitative information contained in the factor, our first analytic question is usually the same: Does the factor have a recognizable effect? Does the response differ systematically across the levels of the factor? In the Great Books data, we would ask, for example: Are some books easier than others? Does reading comprehension differ by the student's grade? By the student's ability group? Comparing the six panels of Figure 4-1, we see that the books do differ in difficulty: *Black Beauty* is consistently the easiest, and the passage from Milton's poems is consistently the hardest. Examining the six individual books more closely, we see the expected effects of GRADE and ABILITY GROUP: students in higher grades and students of higher abilities tend to have higher scores than younger students and students of lower abilities. The effect of ABILITY GROUP is particularly pronounced; for each book and within each grade, scores rise with ability group. A careful analysis may reveal that ABILITY GROUP is actually more important than GRADE.

If an initial analysis reveals systematic variation in the response across the levels of a factor, as it would for each factor in the Great Books data, we then ask a second question: *How* does the response differ across levels? Which levels have responses that differ from the responses at other levels? The factor type guides our investigation of this second question.

When a factor is unordered, our examination of systematic differences in responses can be guided, at most, by background information about the levels. No quantitative information is available to direct the search. In the Great Books data, we might ask whether students find it more difficult to understand epic poems, such as *The Iliad* and *The Aeneid*, than to understand prose texts, such as *Gulliver's Travels* and *Lorna Doone*.

When a factor is ordered, we can ask whether the response generally follows the same rank order as the levels. For example, because the ordered factor ABILITY GROUP is associated with reading comprehension, we would expect not only that reading comprehension scores would differ among the four groups, but also that the order of the scores would approximately follow the order of ability group. With a few exceptions, this is precisely the trend displayed in Figure 4-1. Although we may not expect a linear pattern of responses across levels of ordered factors—indeed, we often expect nonlinear relations—we usually anticipate that the order of responses will be associated with the order of the levels, and that the response will increase or decrease monotonically.

Measured factors enable us to extend the search for a systematic association between the order of the levels and the order of responses at these levels, incorporating the actual numeric value assigned to the level in the analysis. With measured factors, we can determine not only whether the responses follow the same *pattern* as the levels, but also whether the pattern of responses is associated with the *distance* between levels. We might ask, for example, *how much* increase in the response is associated with a specified increase in the factor. So in the Great Books data, we might ask how much higher mean reading comprehension scores are for every additional year of education. Different mathematical functions can be used to summarize the relationship between reading comprehension and GRADE, and we can look for both linear and nonlinear relationships. (Thus a measured factor also allows us to analyze the data by regression methods.)

Some measured factors offer a further advantage: when the relationship between the response and the factor follows a smooth function, we can use this function to predict what the response would be at values of the factor that fall between the actual levels observed. This behavior enables us to predict the response under circumstances for which we have not even collected data, as the following experiment illustrates.

Example: shelf space and sales volume

Supermarket executives and marketing representatives generally assume that food sales, especially for impulse items, are sensitive to the amount of shelf space given to a product. Cox (1964) examined this conjecture by conducting an experiment in six randomly selected supermarkets in Texas. During a 6-week period that did not coincide with any holiday season, he systematically varied the number of shelves in each store allocated to each of four products—baking soda, hominy, Tang, and powdered coffee creamer. The number of shelves used differed among products; the results for hominy, for which he varied shelf space from 4 to 14 in increments of 2, are tabulated in Table 4-2 and graphed in Figure 4-2.

The stores showed the largest differences in sales volume; this possibility explains why Cox collected data in six locations. Over and above these differences, sales volume did vary a modest amount by the number of shelves allocated to the product: the more shelves allocated, the higher the sales volume. Because NUMBER OF SHELVES is a measured factor, however, not only can we *describe* the relationship for the six levels observed, but we can also *quantify* this relationship using a mathematical function. This

Table 4-2. Sales Volume for Hominy by Supermarket and Number of Shelves

	Number of Shelves						
Supermarket	4	6	8	10	12	14	Mean
A	133	126	130	154	188	131	143.7
B	71	121	111	150	140	127	120.0
C	109	123	84	96	134	127	112.2
D	93	49	67	112	161	94	96.0
E	27	58	71	62	49	59	54.3
F	36	51	38	37	52	58	45.3
Mean	78.2	88.0	83.5	101.8	120.7	99.3	95.3

Source: Keith Cox (1964). "The responsiveness of food sales to shelf space changes in supermarkets," *Journal of Marketing Research*, 1, 63–67 (data from Data Appendix, p. 66). Reprinted from *Journal of Marketing Research*, published by the American Marketing Association.

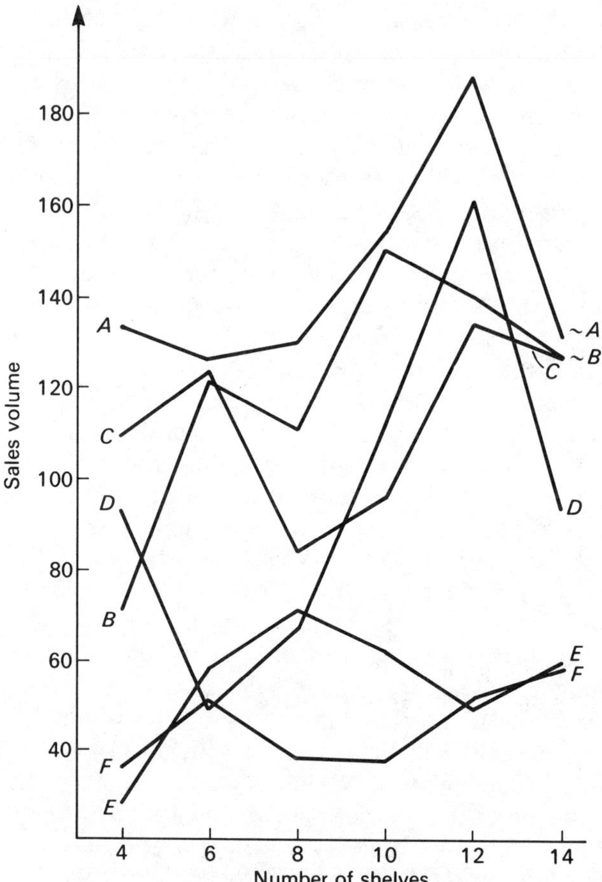

Figure 4-2. Sales volume of hominy by supermarket and number of shelves.

mathematical summary gives predicted sales volume for the intermediate values of shelf space not observed—5, 7, 9, 11, and 13. When the effect of a measured factor can be summarized by a smooth mathematical function, the payoff of using measured factors is clear: they carry information about levels of a factor not even present in the data.

The types of the factors determine some of the directions an analysis of variance may take. But to appreciate the contribution of the data layout, we must also consider the structural relationships between the factors.

4B. RELATIONSHIPS BETWEEN FACTORS

When data layouts include two or more factors, the factors can have different structural relationships to each other. We describe these in terms of which levels of one factor occur in combination with which levels of the other factor. These structural relationships determine which effects we can calculate in the analysis of variance. We distinguish two types of structural relationship—crossing and nesting.

Crossing

The most common structural relationship between two factors is called crossing. Factor A is *crossed* with Factor B if every level of Factor A occurs in combination with every level of Factor B. In the Great Books data, BOOK and GRADE are crossed: Burch gave a passage from every book to students of every grade. In the supermarket shelf space study, SUPERMARKET and NUMBER OF SHELVES are crossed: every supermarket used all six levels of shelf space.

Crossed factors form a very desirable and popular building block for research design. One reason is that when two factors are crossed, we can assess the *main effects* of each individual factor—the overall effects of each factor, averaging over the other. In the Great Books data, for instance, we can determine whether average reading comprehension scores vary by book (the main effect of BOOK) and whether average reading comprehension scores vary by grade (the main effect of GRADE).

Crossed factors offer a further advantage: when two factors are crossed, the data actually permit us to ask how the response varies as we change one factor and hold the other factor fixed. In addition, when we have more than one observation within every combination of crossed levels, we can examine the *statistical interaction* between the factors—whether the effects of one factor differ according to the level of the other factor—and separate these interactions from chance fluctuations in the data. Because the Great Books data contain four observations for each combination of the crossed factors BOOK and GRADE (the mean reading comprehension scores for the students at each level of ABILITY GROUP), we can examine the interaction between these two factors, telling us whether the effect of GRADE differs by BOOK or, alternatively, whether the effect of BOOK differs by GRADE. The possibility of examining interaction effects is the major payoff from having crossed factors in a data layout (as discussed in Chapter 1).

What precisely is a statistical interaction? The term itself is general, intentionally broad enough to encompass the myriad ways in which the effects of one factor may differ by levels of another factor. In any data layout, many different patterns of interactions are possible. In the Great Books data, for example, one possibility would be that because some books are too easy for all grades, reading comprehension is stable and high across grades, whereas for other books, reading comprehension increases monotonically with grade. If this were true, we would say that the effect of BOOK differs by GRADE. As another possibility, the effect of GRADE may be more or less pronounced (characterized by a steeper function) for certain books. If this were true, we would say that the effect of GRADE differs by BOOK. The general characterization of both types of effects is that BOOK and GRADE interact.

The examination of interaction effects—which can reveal complex relationships between the response and the factors—is a major goal of analysis of variance. When two factors are crossed, we can investigate many possible patterns of interaction. Had BOOK and GRADE not been crossed (i.e., if Burch had not given passages from every book to students of every grade), it would have been difficult to determine whether these two factors interacted.

Nesting

Practical constraints may make it impossible to cross factors in a multi-factor study; instead, one factor must nest within another. Factor B *nests* within Factor A if each level of Factor B occurs only within a single level of Factor A. Data layouts composed of nested factors typically arise when it is simply not possible to study the levels of the nested factor within more than one particular level of the other factor. Consequently, the data do not allow us to say how the response varies as we change Factor A and hold Factor B fixed. Educational researchers often encounter nested factors, for example, because for each subject, an individual student usually is enrolled in only one class, and an individual class is found within only one school. As a practical matter, students are nested within classes, and classes are nested within schools. The following example illustrates another nested layout, in which the nesting necessarily follows from practical constraints.

EXAMPLE: REGIONAL VARIATIONS IN CRIME RATES

Are all urban areas in the United States equally safe, or do crime rates vary across regions and states? Table 4-3 and Figure 4-3 present the 1983 rates of serious crime per 100,000 residents by region, for selected states and urban areas. Within each of the four U.S. census regions—Northeast, North

Table 4-3. Annual Rates of Serious Crime[a] Known to Police per 100,000 Residents in 1983, by Region, State, and SMSA

Region	State	SMSA	Crime Rate
Northeast	Pennsylvania	Allentown	2981
		Pittsburgh	3042
		Erie	3286
		Philadelphia	4538
	Massachusetts	Worcester	3738
		New Bedford	4627
		Springfield	4837
		Boston	5448

Table 4-3. (*Continued*)

Region	State	SMSA	Crime Rate
	New York	Syracuse	3909
		Buffalo	4642
		Rochester	5194
		New York City	8091
North Central	Ohio	Cincinnati	5075
		Dayton	5487
		Columbus	5548
		Toledo	6547
	Illinois	Chicago	5898
		Rockford	6139
		Champaign-Urbana	6989
		Springfield	8109
	Michigan	Kalamazoo	5506
		Battle Creek	7168
		Detroit	7201
		Flint	7640
South	North Carolina	Raleigh-Durham	5608
		Fayetteville	6005
		Charlotte	6049
		Wilmington	6349
	Texas	Amarillo	6166
		Houston	6820
		Dallas	7610
		Lubbock	8333
	Florida	Tampa	6433
		Daytona Beach	7103
		Miami	8467
		West Palm Beach	8538
West	Washington	Richland	5008
		Spokane	5282
		Bellingham	5581
		Seattle	7007
	Colorado	Colorado Springs	6282
		Pueblo	6581
		Greeley	6787
		Denver	7424
	California	San Francisco	6433
		Los Angeles	7030
		Sacramento	7424
		Bakersfield	7557

[a]Serious crimes known to police include seven offenses: murder and nonnegligent manslaughter, forcible rape, robbery, aggravated assault, burglary, larceny-theft, and motor vehicle theft.

Source: U.S. Bureau of the Census, *State and Metropolitan Area Data Book, 1986* (Table B, Entry 32).

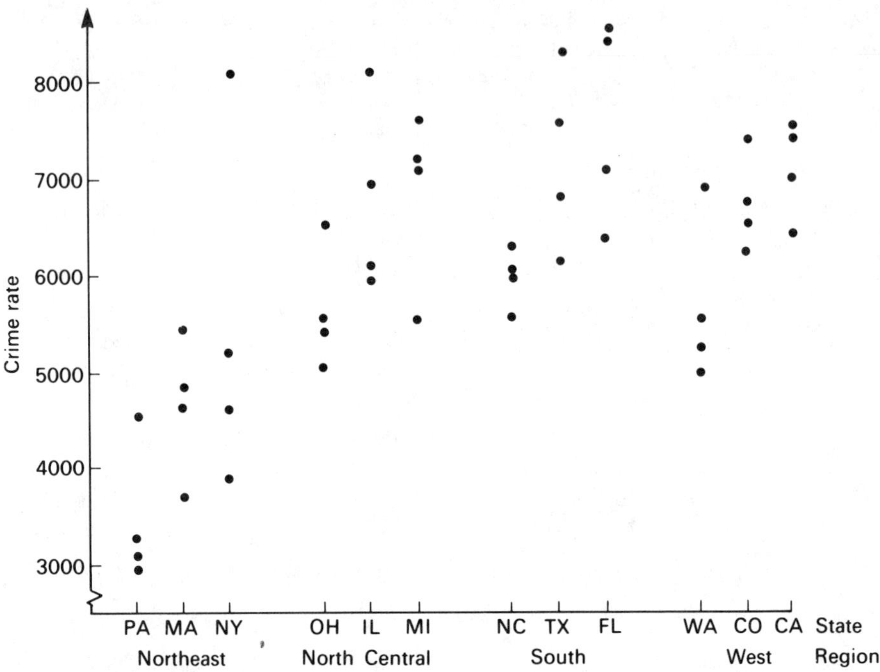

Figure 4-3. Annual rates of serious crime per 100,000 residents, in 1983, by region and state, for selected SMSAs.

Central, South, and West—we have selected three states, and within each state, we have selected four Standard Metropolitan Statistical Areas (SMSAs) for study. SMSA is nested within STATE, and STATE is nested within REGION.

The analysis and interpretation of the outermost factor, which is not nested within any other factor—REGION in the crime rate example—are similar to those of crossed factors. Examining Figure 4-3, for example, we would say that crime rates tend to differ across regions, with the Northeast being the safest. The North Central, the South, and the West are more crime-ridden, or at least more crime is reported in these regions. These comparisons among regions (and hence the main effects of REGION), however, depend somewhat on differences among the sets of states within the regions. We cannot say how the crime rate would behave if we held a set of states fixed and varied the region, as we could if it were possible to cross REGION and STATE.

The analysis and interpretation of nested factors, in contrast, differ from those of crossed factors. When one factor is nested within a second factor, its effect can be interpreted only in the context of a level of that second factor. For example, any level of the factor STATE has meaning only within a level of REGION, and any level of the factor SMSA has meaning only within its REGION and STATE. After all, an SMSA is located in a specific state and region; we cannot predict what the crime rate for an SMSA would be if it were located in a different state or region.

We cannot determine whether a nested factor interacts with the factor within which it nests. The crime rate data illustrate this. To examine the interaction between STATE and REGION, we would need to observe what the crime rates of the selected states would be if they were located in other regions of the country. Because this is impossible, we cannot examine this interaction. We can examine only the nested effect of STATE within REGION.

4C. ONE-WAY AND TWO-WAY LAYOUTS

Having described types of factors and structural relationships among factors, we now consider some common data layouts. This section presents simple arrangements, and the following section progresses to more complex layouts. Our goal is to give an overview of the types of data layouts that arise throughout this volume. The variety of research designs extends well beyond the few that we have space to mention. For additional details, we refer the reader to Box et al. (1978), Cochran and Cox (1957), Hicks (1987), John (1971), Kempthorne (1952), Kirk (1982), or Winer (1971).

One-Way Layout

The simplest data structure is the one-way layout, which has one or more observations at every level of a single factor. We call each level of a one-way layout a *group* or a *cell*. If we focus on a single region, the crime rate data of Table 4-3 form a one-way layout. The single factor is STATE, and the layout has three cells, each containing four observations (the SMSAs).

One-way layouts derive their benefits from a structural feature called within-cell *replication*. When a layout involves replication, more than one observation is gathered for each cell of the data structure. The number of data values in each cell is called the number of replications. Each cell in the one-way layout of the crime data for a single region has four replications. Because the number of replications is equal in all cells, this layout is *balanced*. If the number of replications varies among cells, the layout is *unbalanced*.

How does replication affect the analysis? Because we have more than one observation within each cell, we can distinguish between two types of effects: those attributable to the particular level (or levels) of the factor defining the cell (STATE) and those attributable to the units (SMSAs) that happen to occur within the cell. To understand this distinction, suppose we had gathered data on only one SMSA within each state. All our information on a state's crime rate would come from that one SMSA; we could not disentangle the contribution of the particular SMSA from the contribution of the particular state. If we have data for two or more SMSAs within each state, we can separate these two contributions—one for the state (still somewhat perturbed because we have only a few SMSAs in each state) and one for the SMSAs within state. The more replications within cells, the better the separation. Of course, the figures for the SMSAs can tell us only about serious crime in urban areas. Thus the response measure is the annual rate of serious crime in urban areas. Also, a state has only a limited number of SMSAs, thus limiting the number of replications within that cell.

On the surface, replication may appear little more than a special type of nested factor. After all, replications behave just like nested factors: we cannot disentangle completely the effect of the replication from the effect of the level it nests within. For example, we cannot completely separate the effect of SMSA from the effect of STATE. But replications are conceptually distinct from other nested factors. When a layout involves replication, we usually do not care to examine the effect of "replications." Instead, we are content to assume that the variation among replications reflects errors of measurement and sampling. In the results of an analysis such variation shows up in the residuals, one residual for each replication (here, the difference between the replication and the mean for the cell). We often study measurement error by combining the residuals from all the cells into a single batch, where the identity of the original replication is either absent or secondary. With other nested factors, in contrast, we are specifically interested in examining the factors' effects—after all, that is why we chose to study those factors.

Two-Way Crossed Layout Without Replication

Another simple data structure is the two-way crossed layout without replication. In layouts involving more than one factor, we use the term *cell* to refer to particular combinations of factor levels. Each cell in this layout is defined by a combination of crossed levels. Because we collect data on only one observation within each cell, we have no replications. The shelf space experiment is such a data structure: NUMBER OF SHELVES and SUPERMARKET are crossed to produce 36 cells, and the sales volume was collected only once for each cell.

The presence of crossing in a two-way layout ensures that we can examine the main effects of both factors. In the shelf space study, for example, we can examine the main effect of SHELF SPACE (averaging over specific supermarkets) and the main effect of SUPERMARKET (averaging over specific amounts of shelf space). We may also begin investigating whether the two factors interact detectably—whether the effect of one factor differs detectably according to values of the other factor. For example, we might explore whether the effect of SHELF SPACE differs across supermarkets. If this were the case, some of the lines in Figure 4-2 connecting the sales volumes in each supermarket would be steeper than the others. Because we have no replications within each cell, however, we cannot be certain whether differences in sales volume for particular combinations of supermarkets and shelf space (the different slopes) would be attributable to an interaction between these two factors or whether they would be attributable to some other source of variation. A two-way layout without replication only sets the stage for examining interactions. If a researcher suspects an interaction between two factors, other layouts for data collection are preferable, unless the size of the differences between replications is known to be small or its variance is known, at least approximately.

Two-Way Crossed Layout with Replication

Two-way crossed layouts with replication combine the best features of the previous two data structures. Together, crossing and replication allow a researcher to examine fully the main effects of both factors and their interaction.

Example: hostility after unprovoked verbal abuse

Atkinson and Polivy (1976) studied people's reactions to unprovoked verbal abuse. Their main hypothesis was that people would be especially hostile after experiencing such attacks. However, they also hypothesized that this effect could be mitigated by an opportunity to retaliate.

The researchers conducted an experiment with a sample of 40 male college students. Each student was asked to sit alone in a cubicle and answer a brief questionnaire. After they had waited far longer than it took to fill out the form, a graduate research assistant returned to collect their responses. For half the subjects, the research assistant apologized for the long delay; for the other half, the research assistant insulted the subject, by telling him, among other things, that he couldn't even fill out the form properly. Each group of 20 was further divided into two groups of 10: half got to retaliate against the research assistant by giving her a bad grade; the other half did not get an opportunity to retaliate. The 40 subjects were randomly assigned to the four treatments. The data presented in Table 4-4 are the scores on a standardized measure of hostility obtained after treatment.

Subjects who were insulted after the long delay tended to have higher hostility scores, whether or not they were given the chance to retaliate, than subjects who received an apology. The effect of the opportunity to retaliate, however, interacted with this treatment. Among those who were abused, retaliation produced lower hostility ratings; among those who received apologies, retaliation produced higher hostility ratings. The structural features of crossing and replication allow us to explore these three effects.

Table 4-4. Hostility Scores by Verbal Abuse and Retaliation

Retaliation	Insult	Apology
No	37	32
	43	33
	51	38
	53	47
	56	49
	71	50
	80	53
	82	67
	88	68
	90	99
Mean	65.1	53.6
Yes	30	41
	51	42
	52	50
	55	54
	55	55
	57	58
	69	59
	76	60
	77	67
	89	80
Mean	61.1	56.6

Source: Carolyn Atkinson and Janet Polivy (1976). "Effects of delay, attack, and retaliation on state depression and hostility," *Journal of Abnormal Psychology*, 85, 570–576. Raw data are given in: Bonnie H. Erickson and T. A. Nosanchuk (1977). *Understanding Data*. Toronto, Canada: McGraw-Hill Ryerson (data from Table IIR.2, p. 189). Reproduced with permission of Carolyn J. Atkinson.

4D. THREE-WAY AND MORE-WAY LAYOUTS

Many phenomena are associated with more than two factors. Multi-factor layouts involve data collected under conditions defined by several factors simultaneously; they are usually more informative and sometimes more economical than a series of one-way or two-way designs.

Multi-way designs can combine crossed and nested factors, with or without replication. In the Great Books study, for example, ABILITY GROUP is crossed with GRADE and with BOOK. In the crime rate data, SMSA is nested within STATE, which is nested within REGION. In this section, we present examples of two multi-way layouts, complete factorial designs and Latin squares, to illustrate the variety of possibilities and possible economies when studying several factors.

Complete Factorial Designs

Because of the advantages of crossing factors, one popular multi-way design is the fully crossed layout, with or without replication. When every factor in a multi-way layout is crossed with every other factor and every other combination of factors, the data structure is called a *complete factorial design*. Such structures allow us to examine not only the main effects of all factors, but also all the two-way and more-way interactions among them. Although within-cell replication may be absent, its use further improves analyses because we can distinguish among all of the main effects, interactions, and within-cell variability.

Example: hostility after unprovoked verbal abuse (continued)
The Atkinson and Polivy (1976) study of people's reactions to unprovoked verbal abuse actually included a third factor—GENDER. In addition to the 40 men whose responses are displayed in Table 4-4, 40 women also participated in the study. Table 4-5 presents results for the full three-factor factorial design; the response this time is the difference in hostility scores obtained before the experiment and obtained after the experiment.

This three-factor design allowed Atkinson and Polivy to examine seven effects: the three main effects of VERBAL ATTACK, RETALIATION, and GENDER; the three two-factor interactions, one for each pair of factors; and the one three-factor interaction, for all three factors taken together. This last effect, which can be examined only when three factors are fully crossed, allows us to determine whether the palliative effect of retaliation differed by gender (it did not). Such economy of effort—the ability to examine many more effects than there are factors—makes factorial designs an attractive layout. Unfortunately, the number of cells in such a layout grows rapidly with the number of levels in the factors.

Latin Squares

On occasion, a researcher can cross all *pairs* of factors in a multi-way layout but cannot fully cross all factors together. For example, when Cox was developing his shelf space study, he was concerned about the effects of a third factor—WEEK—on sales volume. Even though he had chosen a nonholiday season for his study, previous research suggested that sales of the four target products might differ across weeks of the project.

The problem was how to control for this potential source of variation while not increasing the period of data collection. The situation was especially complex because each supermarket could assign only a given number of shelves to each product during each week of the study. In structural terms, WEEK had to be nested within each combination of a SUPERMARKET and a NUMBER OF SHELVES. But to separate the main effect of WEEK from the effects of the other two factors, WEEK had to be crossed individu-

Table 4-5. Changes in Hostility by Verbal Attack, Retaliation, and Gender

	Male		Female	
Retaliation	Attack	Apology	Attack	Apology
No	−4	−16	−6	−21
	−3	−3	0	−17
	1	−3	3	−1
	4	−1	5	0
	6	0	7	1
	9	3	10	4
	11	4	11	6
	14	6	23	9
	20	6	25	10
	34	45	25	11
Mean	9.2	4.1	10.3	0.2
Yes	−23	−17	−4	−6
	−13	−4	1	−5
	−6	−1	5	−3
	−1	−1	5	−2
	2	0	6	0
	6	0	7	1
	10	0	8	3
	13	6	8	6
	15	6	32	11
	24	31	57	11
Mean	2.7	2.0	12.5	1.6

Source: Carolyn Atkinson and Janet Polivy (1976). "Effects of delay, attack, and retaliation on state depression and hostility," *Journal of Abnormal Psychology*, 85, 570–576. Raw data are given in: Bonnie H. Erickson and T. A. Nosanchuk (1977). *Understanding Data*. Toronto, Canada: McGraw-Hill Ryerson (data from table RIV.1, p. 323). Reproduced with permission of Carolyn J. Atkinson.

ally with SUPERMARKET and with NUMBER OF SHELVES. Cox needed a design in which all three factors were pairwise crossed, but not fully crossed. One such design is called a Latin square.

Mathematically, a Latin square is a square matrix of letters (or numbers, or objects) such that each letter appears once, and only once, in each row and column. Table 4-6 presents sample Latin squares of dimensions 3 × 3, 4 × 4, 5 × 5, and 6 × 6. When applied to experimental design, the rows of the matrix represent one factor, and the columns represent a second factor. In the shelf space study, for example, the rows would represent supermarkets, and the columns would represent the number of shelves allocated to the product. Letters would represent weeks and therefore tell us the week in which each supermarket used that particular number of shelves. Cox actually used the second 6 × 6 Latin square presented in Table 4-6, with letters

Table 4-6. Sample Latin Square Plans

3 × 3	A B C B C A C A B	X Y Z Y Z X Z X Y
4 × 4	A B C D B C D A C D A B D A B C	X Y Z W W X Y Z Y Z W X Z W X Y
5 × 5	A B C D E B C D E A C D E A B D E A B C E A B C D	X W Z Y V W Z Y V X Z Y V X W Y V X W Z V X W Z Y
6 × 6	A B C D E F B C D E F A C D E F A B D E F A B C E F A B C D F A B C D E	Y V W Z X U W Y X V U Z U X Z W V Y X W U Y Z V Z U V X Y W V Z Y U W X

Table 4-7. Sales Volume of Hominy by Supermarket, Number of Shelves, and Week of Testing

	Number of Shelves					
Supermarket	4	6	8	10	12	14
A	133 (*Y*)	126 (*V*)	130 (*W*)	154 (*Z*)	188 (*X*)	131 (*U*)
B	71 (*W*)	121 (*Y*)	111 (*X*)	150 (*V*)	140 (*U*)	127 (*Z*)
C	109 (*U*)	123 (*X*)	84 (*Z*)	96 (*W*)	134 (*V*)	127 (*Y*)
D	93 (*X*)	49 (*W*)	67 (*U*)	112 (*Y*)	161 (*Z*)	94 (*V*)
E	27 (*Z*)	58 (*U*)	71 (*V*)	62 (*X*)	49 (*Y*)	59 (*W*)
F	36 (*V*)	51 (*Z*)	38 (*Y*)	37 (*U*)	52 (*W*)	58 (*X*)

Note: Letters in parentheses indicate the week in which the shelf allocation was used (U = week 1; V = week 2; W = week 3; X = week 4; Y = week 5; Z = week 6).

Source: Keith Cox (1964). "The responsiveness of food sales to shelf space changes in supermarkets," *Journal of Marketing Research*, 1, 63–67 (data from Data Appendix, p. 66). Reprinted from *Journal of Marketing Research*, published by the American Marketing Association.

corresponding in sequential order to weeks of the study. Table 4-7 presents Cox's data again, this time with an indication of the week that each supermarket used the particular number of shelves. Ignoring any single dimension of a Latin square—rows, columns, or letters—yields a fully crossed two-way layout without replication. To produce the two-way layout of Table 4-2 with just SUPERMARKET and NUMBER OF SHELVES, for example, we ignored the third factor represented by the letters of the Latin square in Table 4-7.

Researchers using a Latin square design can examine three factors using no more observations than were needed to examine only two. As evidence, the data given in Table 4-7, which shows three factors, are identical to those given in Table 4-2. In the proper context, the use of a Latin square can add important information at very little cost. However, as a form of two-way layout without replication, a Latin square has the same disadvantage we identified earlier: it is very difficult (often impossible) to detect interactions between the factors. To obtain information on interactions, other layouts, such as factorial designs, are preferable.

4E. SUMMARY

We have described the types of factors that typically arise in analysis-of-variance problems, and we have presented some of the more common structural relationships among them. We have distinguished between three types of factors—unordered, ordered, and measured—and we have discussed how the presence of quantitative information in the levels of a factor suggests different analytic directions. We have also explored the difference between crossed and nested factors, and we have shown how these structural relationships between factors affect the types of effects we can examine. Finally, we have reviewed the structural elements of some common data layouts, including the merits of replication. The next chapter begins to demonstrate how to analyze the data from these designs using a technique called value splitting.

REFERENCES

Atkinson, C. and Polivy, J. (1976). "Effects of delay, attack, and retaliation on state depression and hostility," *Journal of Abnormal Psychology*, 85, 570–576.

Box, G. E. P., Hunter, W. G., and Hunter, J. S. (1978). *Statistics for Experimenters*. New York: Wiley.

Burch, M. C. (1928). "Determination of a content of the course in literature of a suitable difficulty for junior and senior high school students," *Genetic Psychology Monographs*, 4, 165–293.

Cochran, W. G. and Cox, G. M. (1957). *Experimental Designs*, 2nd ed. New York: Wiley.

Cox, D. R. (1957). *Planning of Experiments*. New York: Wiley.

Cox, K. (1964). "The responsiveness of food sales to shelf space changes in supermarkets," *Journal of Marketing Research*, 1, 63–67.

Hicks, C. (1987). *Fundamental Concepts in the Design of Experiments*, 3rd ed. New York: Holt, Rinehart and Winston.

John, P. W. M. (1971). *Statistical Design and Analysis of Experiments*. New York: Macmillan.

Kempthorne, O. (1952). *The Design and Analysis of Experiments*. New York: Wiley. [Reprinted in 1972 by Robert Krieger, Huntington, NY.]

Kirk, R. (1982). *Experimental Design*, 2nd ed. Belmont, CA: Brooks-Cole.

Stevens, S. S. (1946). "On the theory of scales of measurement," *Science*, 103, 677–680.

Winer, B. J. (1971). *Statistical Principles in Experimental Design*, 2nd ed. New York: McGraw-Hill.

EXERCISES

1. In Table 2-1, what are the factors? Levels? What do the observations 6.7, . . . , 3.6 measure?

2. In Table 2-1, state whether each factor is unordered, ordered, or measured.

3. In Table 2-1, state whether the levels of each factor contain quantitative information.

4. In Panel (a) of Table 2-2:

(a) Identify the response, the factors, and their levels.

(b) State whether each factor is ordered, unordered, or measured.

(c) Do the levels contain quantitative information?

5. *The World Almanac and Book of Facts*, 1990, p. 339, gives the percentage of voter population that voted for President of the United States in the election years 1932 through 1988 (52.4% for 1932, etc.). Identify the response, the factors, and their levels.

6. In 1972 the U.S. Department of Health, Education and Welfare published data on infant mortality rates for the period 1964–1966, by regions of the United States (Northeast, North Central, South, and West) and by father's education in years (≤ 8, 9–11, 12, 13–15, and ≥ 16). In an analysis of variance, what would you consider to be the factors? What are the levels of each factor? What are the responses to be analyzed? (UREDA, p. 168.)

7. What are some types of structural relationships between factors in data?

8. Illustrate the response to Exercise 7 with concrete examples.

9. In ANOVA, when is one factor said to be crossed with another factor?

10. Is Table 2-1 an example of crossing? How can you tell?

11. In Panel (a) of Table 2-2, are the factors crossed?

12. How do crossed factors aid in the analysis of statistical interaction?

13. In Table 1-5, what factors are crossed? What factors are nested?

14. How do the analysis and interpretation of nested factors differ from those of crossed factors?

15. Give an example of a table having nested factors. Emphasize the feature leading to the nesting.

16. Give an example of a one-way layout having two observations at every level of a single factor. What is the factor? How many cells are in your table? How many observations are in each cell?

17. In a one-way layout, how do replications in a cell affect the analysis?

18. What benefits does replication provide?

19. Give an example of a two-way crossed layout without replication.

20. What are the advantages of a two-way layout with replication?

21. What is meant by a complete factorial design? Give a small example.

22. What are the advantages of a complete factorial design?

23. How many factors does a Latin square design have?

24. What is special about the numbers of levels in the factors of a Latin square?

25. Explain the distinction between a Latin square design and a complete factorial design.

26. If in Table 4-7 there were still six numbers of shelves and six different weeks, but only four supermarkets, could you create a Latin square design? Why or why not? Could you have a complete factorial design? If so, how many cells would it have?

27. What are some advantages of a Latin square design?

CHAPTER 5

Value Splitting: Taking the Data Apart

Christopher H. Schmid
Harvard University and BBN Software Products

Exploratory analysis of variance differs from the classical inferential approach by focusing on the components of the data decomposition. Those components reflect the structure of the data layout (including the relations among the factors). Within this framework the analysis splits each data value into pieces that reveal how the factors, singly or in combination, contribute to the variation in the data. Looking at the individual pieces provides information unavailable from statistical tests and may also suggest more insightful ways to approach the data. This chapter and the next describe the mechanics of the decomposition and some simple methods for studying the results. The present chapter discusses data with no more than two factors; Chapter 6 introduces designs with more factors.

Section 5A introduces overlays, the basic building blocks of the decomposition, with the simple example of pulling out a mean. For the balanced one-way layout, Section 5B describes a process of sweeping through the data to obtain the overlays. Section 5C shifts attention from the overlays of fitted values to the residual overlay and to the role of the residuals in determining how well the model fits the data. Section 5D introduces sums of squares and mean squares to summarize the sizes of overlays.

A two-way layout requires an additional overlay for the contribution of the second factor. Section 5E discusses the sweeping procedure when each cell in the two-way layout contains only one observation. Multiple observations in each cell allow us to estimate the joint effect of the factors. Section 5F discusses such interaction effects, and Section 5G proceeds with the analysis. As discussed in Chapter 4, a different kind of design arises when one factor nests within a second factor. Section 5H analyzes these nested designs. Finally, Section 5I briefly discusses descriptions for unbalanced data and descriptions that do not involve main effects and interactions.

5A. FORMING OVERLAYS

In a table of data, each cell, representing a combination of levels of factors, contains one or more observations. We may think of each cell as associated with a different population, so that the cell mean estimates the mean of that population. The more observations, or replications, the better the estimate of the mean. The replications also yield an estimate of the within-cell variability. Because the cells of the table are linked by a design that relates the factors, each cell mean reflects contributions from those factors. Analysis-of-variance techniques summarize the variation in the data by breaking each observation into a sum of components, each describing a contribution to the response from one or more factors.

Chapter 4 introduced some common designs, classifying them by the number, type, and relationship of the factors involved. In analyzing the data, we try to separate variation that can be explained by contributions of the factors from variation that cannot. The former comprises the *structural* component and the latter a *residual* component. Ultimately, we want a simple description, preferably additive, in terms of contributions from a small number of the factors. This description by components neatly summarizes the way the factors relate in a manner that the description by individual cells cannot, particularly as the number of factors increases.

EXAMPLE: BASEBALL

A small example illustrates this decomposition. Suppose that a baseball fan attends 5, 7, 4, 6, and 8 games per year over a five-year span. To summarize her attendance, the fan reports that she has attended an average of six games each year for the past five years. Description by the mean condenses a table of five numbers into a one-number summary, but it gives up information about yearly attendance. To recover this information, the fan could say that she has attended an average of six games a year but attended one game less than the average in the first year, one game more in the second, two less in the third, and so forth. Here the mean is the structural component, the deviations form the residual component, and the sum of the two components gives the data.

Value Splitting and Overlays

We call this descriptive technique *value splitting* because it splits each value in the data table into components. Collecting all components of the same kind, we can place them in a table in the same position occupied by the corresponding response in the table of data from which they were derived. Each such table contains exactly as many entries as the data table. Because we can recover the data by overlaying the tables of components and adding the entries at the same location in the stack, we call the new tables *overlays*.

Table 5-1. Value Splitting for the Five Data Values of the Baseball Example into Two Overlays: A Systematic Component and a Residual Component

	5	7	4	6	8	data value
			=			
a stack	6	6	6	6	6	common value
			+			
	−1	1	−2	0	2	residual

We refer to the resulting collection of overlays as a value splitting for the data.

Example: baseball
The overlays for the value splitting in the baseball attendance example appear in Table 5-1. Each overlay has five entries, one for each observation in the data, and adding the entries at a specific location in the two overlays gives the data value at that location. For example, in the first position, corresponding to the first year, the sum of 6 and −1 is 5. The deviations from the mean form the residual overlay. In general, the residual overlay contains whatever variation remains after removing all components of the structure associated with the factors.

Other Descriptions with a Constant Structural Component

We could use a different decomposition. For example, instead of describing the structural component by the mean, we could employ the median. In the baseball example, the result would not change; the median is also 6. On the other hand, suppose the fan wanted to compare her attendance in each previous year to her attendance in the most recent year. Then the structural component would consist of the number of games attended in year 5, as the overlays in Table 5-2 show.

All these decompositions share the feature that the structural component consists of a single value. The associated overlay is the simplest one possible, having a single constant value for every observation in the table. We call this kind of overlay a *common overlay* and its single value the *common value*.

We describe this value splitting symbolically by the equation

$$y_i = m + e_i, \qquad i = 1, 2, \dots, n. \tag{1}$$

In the baseball example, y_i represents the number of games attended in year

Table 5-2. Value Splitting for Five Yearly Baseball Attendance Figures Using the Fifth Year as a Base Year[a]

5	7	4	6	8	data value
		=			
8	8	8	8	8	common value
		+			
−3	−1	−4	−2	0	residual

[a]The year-5 attendance forms the common value overlay, and deviations from this base year form the residual overlay.

i, m stands for the mean number of games attended per year (the value common to each observation), and the e_i are the residual deviations from this mean. The plus sign indicates that the terms combine by addition and form an *additive* description.

The two terms m and e_i on the right-hand side of equation (1) contribute a component to every observation. All the components for a single term make up an overlay, in which an observation's component occupies the same position that the observation has in the table of data. In the common overlay all entries equal the common value, m; and in the residual overlay, each entry, e_i, represents the deviation of y_i from the common value. Reassembling the pieces for the first year from Table 5-1 yields

$$\begin{aligned} y_1 &= m + e_1 \\ &= 6 + (-1) \\ &= 5. \end{aligned}$$

Although value splitting can accommodate a description by many different structural components, classical ANOVA uses means because they provide a least-squares fit to the data. This method has provided both mathematical simplicity (necessary in the age of hand computation) and a convenient theoretical framework with well-developed tools for analysis.

More complex designs have two or more factors related by crossing or nesting. Although the common value summarizes the data in a single number, it does not explain any variation associated with the factors in the design. For this we need to introduce new components, new overlays, and new terms into our descriptions.

5B. OVERLAYS FOR ONE-WAY DATA

When the design has a single factor with several levels or versions and a number of replications at each level, the data form a *one-way layout*. Each level determines a cell (or group), and each cell ordinarily has multiple entries. Within each cell we summarize the typical value of the data by calculating the mean. The residuals then tell us about variation within that cell. To study variation among the cell means, we often average them to get a grand mean and then calculate the deviation of each cell mean from the grand mean of the data. We view this variation among cells relative to the variation within cells. Large variation within cells makes differences among cell means seem less impressive, less obvious.

Balanced and Unbalanced Designs

If the number of replications is the same in each cell, the design is *balanced*; if the number varies among the cells, the design is *unbalanced*. Because a balanced design often offers simpler calculations and more straightforward interpretations, researchers usually aim for balance. Sometimes, however, this proves impossible. Analysts might intend to have balance but lose data by accidental failure, death, or withdrawal from the study. They may find that at one or more levels data are expensive to collect, and so they decide to take fewer observations there than at other levels. Most of this chapter treats balanced designs, although the techniques apply to some special kinds of imbalance.

Value Splitting for the One-Way Design

The additive description of data from a one-way layout takes the cell means as the structural component. The remainder, the deviation of each data value from the mean in its cell, comprises the residual. We can split the structural component further into a sum of two pieces: one for the value common to every entry (here the mean of all the values in the table) and another for the difference between the cell mean and the common value. We call these latter deviations the *effects* of the factor. (Later, when the data involve more than one factor, we refer more specifically to effects related to one factor as *one-factor effects* or *main effects*). Negative effects come from means less than the common value; positive effects come from means greater than the common value. The number of effects equals the number of levels, and because we use means, the effects add to zero. Together, the two parts of the structural component, common value and effects, represent the *fit*, and we call the sum of the individual pieces (omitting the residual) the *fitted value*.

In a one-way analysis, each data value is a sum of three components: the common value, the factor effect, and the residual. In other words, the data are represented as fit + residual. A description for this additive decomposition of the one-way replicated design has

$$y_{ij} = m + a_i + e_{ij}. \tag{2}$$

Here m is the common value, the single factor has I levels ($i = 1, 2, \ldots, I$) with effects a_i adding to zero ($\Sigma a_i = 0$), and e_{ij} is the residual for replication j at level i ($j = 1, 2, \ldots, J_i$; balanced data have $J_i = J$ for each i). Arranging the components from one of the terms in the same format as the data table produces one of the three overlays. All entries in the common overlay are the same: m in equation (2). Entries in the factor overlay are the same within each level of the factor but differ among levels; the value for level i is a_i. Numerically, some a_i may be equal, but each represents a different cell. Residual entries differ between and within cells; their algebraic representation, e_{ij}, uses two subscripts: i for the cell and j for the replication. Once again, two e_{ij} may have the same numerical value, but each represents a different residual.

To calculate the overlays, we introduce *sweeping*. From the table of data, first compute cell means, called $y_{i\cdot}$, and subtract them from each observation in that cell. The remainders are the residuals, data MINUS cell mean. Then compute the mean of the cell means. This is the common value, and the cell mean MINUS the common value is the effect for that cell.

Before presenting an example of a one-factor analysis of variance, we caution the reader that sometimes the components of the fit are defined differently. For example, at least one statistical package defines the mean of group I as the common value and then defines the group effects as deviations from this common value. In this framework, the effects do not sum to zero; the effect for group I is zero instead. When the data include more factors, the common value is defined as the cell mean for one combination of factors, and effects are calculated as deviations from this common value. Mainly, this alternative choice facilitates computation in a particular framework. The resulting fit is the same.

Example: new-car loan interest rates

A 1985 article in *Consumer Reports* noted that "banks are in the business of lending money, [but] consumers are not in the business of borrowing money." It concluded that, because consumers don't shop for credit the way they might shop for consumer goods, "banks have an inherent advantage over consumers in need of a loan." To demonstrate that buyers can save a great deal on interest rates by shopping, the magazine presented interest rates for four types of loans from a survey of the largest banks in each of 10 U.S. cities. Table 5-3 gives the annual percentage rate of interest charged for new-car loans in six of the cities. Within each city, the table orders the interest rates from highest to lowest. The cities form the columns of our data table and represent six levels of one factor called "city." Because all banks

Table 5-3. Annual Percentage Rate of Interest Charged on New-Car Loans at Nine of the Largest Banks in Six American Cities[a]

Atlanta	Chicago	Houston	Memphis	New York	Philadelphia
13.75	14.25	14.00	15.00	14.50	13.50
13.75	13.00	14.00	14.00	14.00	12.25
13.50	12.75	13.51	13.75	14.00	12.25
13.50	12.50	13.50	13.59	13.90	12.00
13.00	12.50	13.50	13.25	13.75	12.00
13.00	12.40	13.25	12.97	13.25	12.00
13.00	12.30	13.00	12.50	13.00	12.00
12.75	11.90	12.50	12.25	12.50	11.90
12.50	11.90	12.50	11.89	12.45	11.90

[a] *Consumer Reports* surveyed the ten largest banks in each city. In Atlanta, Chicago, Houston, Memphis, and New York, only nine of the ten banks reported their rates for new-car loans. In Philadelphia all ten banks reported rates. To make this a balanced design, we omitted one bank at random from the Philadelphia data. The omitted bank reported a new-car loan interest rate of 12.25%.

Source: Consumer Reports, October 1985, pp. 582–586.

are different, the rows represent replications within each column and have no factor associated with them.

Table 5-4a illustrates the first step of the sweeping process for the interest rates on new-car loans. We compute column means and subtract them from each entry in their respective columns. Thus the data for Chicago have mean 12.61. Subtracting this mean from the largest rate, in the first row, yields the remainder 14.25 − 12.61 = 1.64. We note two features of Table 5-4a. The residuals in each column add to zero, and adding the residuals in each column to the column mean reproduces the original data. (Rounding error may cause discrepancies in the last decimal place.) These two characteristics hold throughout the calculations: the remainders after subtracting out a mean add to zero, and, as the term "value splitting" implies, we can always get back to the original data by adding up the pieces. Checking these two characteristics is a good practice. The remainders from this first step are the residuals because they are the deviation of each replication from the cell mean. One step suffices to split the data into fit + residual. The next step uses only the fitted values, the cell means, and separates the factor effects from the common value.

Subtracting that common value, here the grand mean, from each city mean leaves the city effects. Because these effects are deviations from the common value, they must add to zero (except for rounding). Table 5-4b shows the results of these calculations.

We can now create the overlays for the three components in Table 5-5. The common overlay contains the common value in each entry. The common value tells us about the typical interest rate over all cities. To obtain the

Table 5-4. The Two Steps of the Sweeping Process for the New-Car Loans Data

(a) Sweeping out city means. The nine entries in each column are the remainders after subtracting the column means from the data values in Table 5-3.

	Atlanta	Chicago	Houston	Memphis	New York	Philadelphia
	0.55	1.64	0.69	1.75	1.01	1.30
	0.55	0.39	0.69	0.75	0.51	0.05
	0.30	0.14	0.20	0.50	0.51	0.05
	0.30	−0.11	0.19	0.34	0.41	−0.20
	−0.20	−0.11	0.19	0.00	0.26	−0.20
	−0.20	−0.21	−0.06	−0.28	−0.24	−0.20
	−0.20	−0.31	−0.31	−0.75	−0.49	−0.20
	−0.45	−0.71	−0.81	−1.00	−0.99	−0.30
	−0.70	−0.71	−0.81	−1.36	−1.04	−0.30
Column mean	13.19	12.61	13.31	13.24	13.48	12.20

(b) Sweeping the common value from the city means. The remainder is the city effect.

	Atlanta	Chicago	Houston	Memphis	New York	Philadelphia	Common Value
City effect	0.19	−0.40	0.30	0.24	0.48	−0.81	13.01

overlay for cities, we replace each entry in the original data table by its city effect. Finally, the residuals come from the first step of the sweeping process, which removed the city means.

Table 5-5 highlights the *principal entries* in each overlay. These represent the first occurrence of all the (potentially) distinct values in an overlay. The other entries are just repetitions of the principal entries. Once we have determined them, we can fill in all the other entries in the overlay. Thus the common overlay has one principal entry, the common value; the city overlay has six principal entries, one for each of the six cities; and the residual overlay has 54 principal entries, one for every bank represented in the table.

We can simplify the overlays by writing only the principal entries. Doing this gives us *reduced overlays*, depicted in Table 5-6. By comparison, the full overlays for the common value and the city effects in Table 5-5 contain much redundancy. In practice, we would show only the reduced overlays. As we discuss value splitting in this chapter, however, we emphasize visually the representation of each data value as a sum of components. Also, the sums of squares (Section 5D) involve all the entries in the full overlays. Subsequent chapters work only with reduced overlays.

The common value indicates that the mean annual interest on new-car loans in the six cities is 13.01%. Philadelphia has the lowest rates on average among these cities. Its rates are 0.81% less than the overall mean. New York

Table 5-5. Full Overlays for the New-Car Loans Example[a]

	Atlanta	Chicago	Houston	Memphis	New York	Philadelphia
			Original data (y_{ij})			
	13.75	14.25	14.00	15.00	14.50	13.50
	13.75	13.00	14.00	14.00	14.00	12.25
	13.50	12.75	13.51	13.75	14.00	12.25
	13.50	12.50	13.50	13.59	13.90	12.00
	13.00	12.50	13.50	13.25	13.75	12.00
	13.00	12.40	13.25	12.97	13.25	12.00
	13.00	12.30	13.00	12.50	13.00	12.00
	12.75	11.90	12.50	12.25	12.50	11.90
	12.50	11.90	12.50	11.89	12.45	11.90
			Common (m)			
	13.01	13.01	13.01	13.01	13.01	13.01
	13.01	13.01	13.01	13.01	13.01	13.01
	13.01	13.01	13.01	13.01	13.01	13.01
	13.01	13.01	13.01	13.01	13.01	13.01
=	13.01	13.01	13.01	13.01	13.01	13.01
	13.01	13.01	13.01	13.01	13.01	13.01
	13.01	13.01	13.01	13.01	13.01	13.01
	13.01	13.01	13.01	13.01	13.01	13.01
	13.01	13.01	13.01	13.01	13.01	13.01
			City (a_i)			
	0.19	**−0.40**	**0.30**	**0.24**	**0.48**	**−0.81**
	0.19	−0.40	0.30	0.24	0.48	−0.81
	0.19	−0.40	0.30	0.24	0.48	−0.81
	0.19	−0.40	0.30	0.24	0.48	−0.81
+	0.19	−0.40	0.30	0.24	0.48	−0.81
	0.19	−0.40	0.30	0.24	0.48	−0.81
	0.19	−0.40	0.30	0.24	0.48	−0.81
	0.19	−0.40	0.30	0.24	0.48	−0.81
	0.19	−0.40	0.30	0.24	0.48	−0.81
			Residual (e_{ij})			
	0.55	**1.64**	**0.69**	**1.75**	**1.01**	**1.30**
	0.55	**0.39**	**0.69**	**0.75**	**0.51**	**0.05**
	0.30	**0.14**	**0.20**	**0.50**	**0.51**	**0.05**
	0.30	**−0.11**	**0.19**	**0.34**	**0.41**	**−0.20**
+	**−0.20**	**−0.11**	**0.19**	**0.00**	**0.26**	**−0.20**
	−0.20	**−0.21**	**−0.06**	**−0.28**	**−0.24**	**−0.20**
	−0.20	**−0.31**	**−0.31**	**−0.75**	**−0.49**	**−0.20**
	−0.45	**−0.71**	**−0.81**	**−1.00**	**−0.99**	**−0.30**
	−0.70	**−0.71**	**−0.81**	**−1.36**	**−1.04**	**−0.30**

[a]Each overlay has an entry for each data value in the original table. At any one position, the sum over all the overlays equals the entry in that position in the data table. In each overlay the principal entries appear in bold type.

Table 5-6. Reduced Overlays for the New-Car Loan Interest Rates[a]

Common (m)

13.01

City (a_i)

0.19	−0.40	0.30	0.24	0.48	−0.81

Residual (e_{ij})

0.55	1.64	0.69	1.75	1.01	1.30
0.55	0.39	0.69	0.75	0.51	0.05
0.30	0.14	0.20	0.50	0.51	0.05
0.30	−0.11	0.19	0.34	0.41	−0.20
−0.20	−0.11	0.19	0.00	0.26	−0.20
−0.20	−0.21	−0.06	−0.28	−0.24	−0.20
−0.20	−0.31	−0.31	−0.75	−0.49	−0.20
−0.45	−0.71	−0.81	−1.00	−0.99	−0.30
−0.70	−0.71	−0.81	−1.36	−1.04	−0.30

[a]Each overlay shows only its principal entries.

has the highest rates, 0.48% more than the overall mean. Within a city, the spread between the highest and lowest interest rates varies between 1.25% in Atlanta and 3.11% in Memphis.

Figure 5-1 offers a pictorial view of the data. The side-by-side boxplots facilitate comparisons between and within cities. Though the boxplots use resistant measures (median and quartiles), we draw the same conclusions: New York has the highest "average" rates and Philadelphia the lowest. The boxplots also point out differences in variability of rates within cities. Among Memphis banks, rates vary more than those in any other city. The comparison with Philadelphia banks is striking. Except for one high value, rates in Philadelphia are almost identical from bank to bank. Although shopping for low interest rates may not be very rewarding in Philadelphia, it is important in Memphis, because rates among the largest banks differ by as much as 3.11%.

This variation is important to the consumer; even a 1% lower interest rate means both lower monthly payments and an overall lower cost to the buyer. On an $8000 loan at 13%, the monthly payments are $269.55, and the customer pays $1704 in interest over a 36-month term. If the interest rate is 12%, the payment is $265.71 per month with a total interest of $1566, a saving of $138 over the life of the loan.

We conclude, then, that the interest rates on new-car loans depend not only on the city but also on a customer's willingness to shop for the lowest

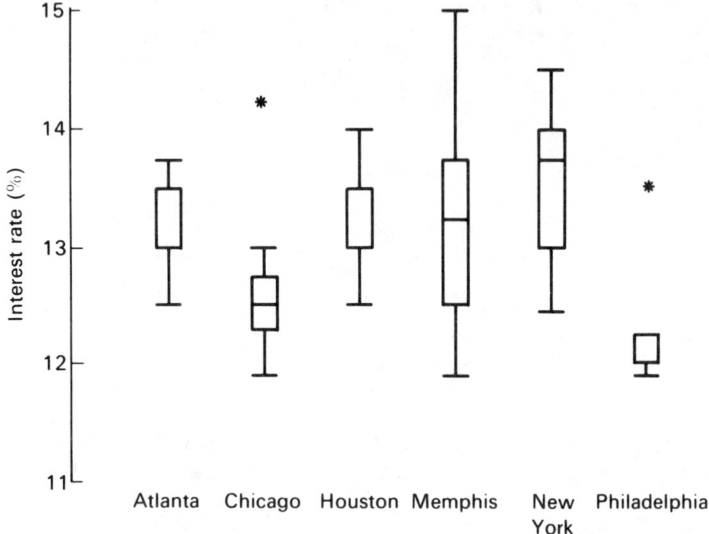

Figure 5-1. Side-by-side boxplots for the interest rates on new-car loans in six cities. In Atlanta, Houston, and Philadelphia the median and the lower fourth are equal. The asterisks indicate outliers.

rates available in a particular city. We have not tried to take the cost of shopping into account. Some people do not have the time or transportation to shop; on the other hand, shopping may have nonmonetary value for those who enjoy it.

5C. WHAT THE RESIDUALS TELL US

The preceding section focused on the overlays that make up the fitted values and ignored the residual overlay, except to define the residuals as data MINUS fit. We now explore the crucial role the residuals play in assessing the adequacy of the fit.

A good description of the data summarizes the systematic variation and leaves residuals that look structureless. That is, the residuals exhibit no patterns and have no exceptionally large values, or *outliers*. Any structure present in the residuals indicates an inadequate fit. Looking at the residuals laid out in an overlay helps to spot patterns and outliers and to associate them with their source in the data.

A useful description relates the systematic variation to one or more factors; if the residuals dwarf the effects for a factor, we may not be able to relate variation in the data to changes in the factor. Furthermore, changes in the factor may bring no important change in the response. Such comparisons of residuals and effects require a measure of the variation of overlays relative to each other.

The residuals are also an important diagnostic tool for checking common assumptions like symmetry, normality, additivity, and homoscedasticity (constant within-cell variance). We discuss some of these aspects here but leave others for later chapters.

Making Pictures of Residuals

Graphical displays of the residuals can tell a lot about the data. Many kinds of displays are available. Stem-and-leaf displays, scatterplots, dotplots, boxplots, and several others appear as diagnostic aids in this volume. For the present, we focus on dotplots and boxplots. A dotplot displays every residual. This detail becomes useful when the analysis seeks to pinpoint the location of some characteristic of the data. Boxplots summarize the data, focusing on location and spread without the distraction of the individual values. This parsimony gives up some information about the center of the distribution and about individual values. With large data sets, however, boxplots are easier than dotplots to construct and analyze. We now use these displays to examine the residuals from the new-car loans example.

EXAMPLE: NEW-CAR LOAN INTEREST RATES

No apparent patterns hold throughout the residual overlay from the new-car interest rates in Table 5-6, but some rather large values stand out. Among the Philadelphia banks, rates are very similar except for one bank, whose rate is 1.25% higher than any of the others. Chicago also has one rate that is much higher than the others. Memphis and New York provide other large residuals, both positive and negative. As more factors enter the data, the residual overlay and graphical representations of the residuals facilitate the identification of extreme values.

When the variation within each cell is fairly constant, we can pool the within-cell variance to compare residuals between cells. The sample variance, $s_i^2 = \Sigma_j(y_{ij} - y_{i\cdot})^2/(J - 1)$, offers the usual measure of the variation in cell i. The sample variances for the six cities are $s^2_{\text{ATL}} = .20$, $s^2_{\text{CHI}} = .50$, $s^2_{\text{HOU}} = .31$, $s^2_{\text{MEM}} = .94$, $s^2_{\text{NY}} = .52$, and $s^2_{\text{PHI}} = .25$. The sizable spread in the values (Memphis has more than four times as large a sample variance as Atlanta) may raise concern about the assumption of equal variances. We cannot, however, expect much stability from sample variances, as Chapter 7 explains. (A calculation whose details we do not give here suggests that six values like .19, .28, .36, .45, .54, and .71 would be typical of well-behaved random sampling fluctuations. Thus the variances .20, .25, .31, .50, .52, and .94 do not seem out of line.)

A dotplot of the residuals by city (Figure 5-2) does not suggest that interest rates among Memphis banks vary dramatically more than in the other cities. The values of the s_i^2 reflect the general variation within a city, as

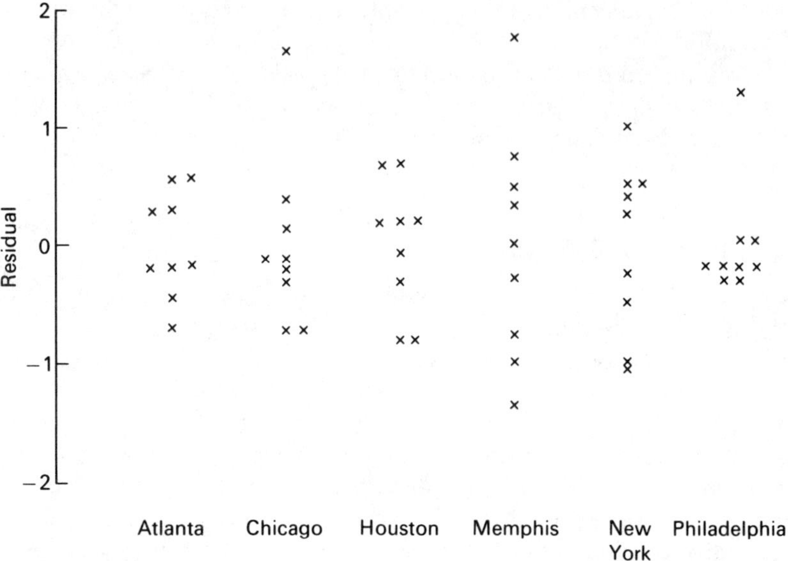

Figure 5-2. Dotplot of residuals by city in the new-car loans example.

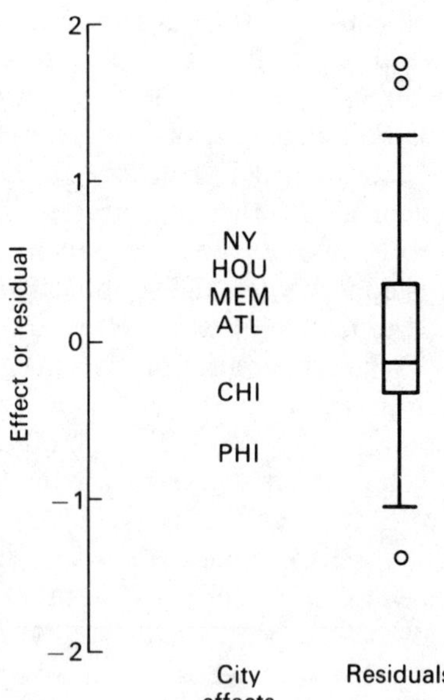

Figure 5-3. A side-by-side plot for auto interest rates. Placing a residual boxplot next to a dotplot of city effects allows comparison of the residual and city overlays.

well as the presence of isolated large residuals (e.g., in Chicago, as compared to Atlanta). Indeed, because we cannot determine from a large s_i^2 alone whether a single residual may be responsible for much of its size, we routinely look at a boxplot or dotplot.

It is often instructive to place the effects next to a boxplot of the pooled residuals from all the cells in a *side-by-side plot*, as in Figure 5-3. Two of the city effects (Philadelphia and Chicago) are negative, and the other four have roughly the same positive value. The residuals appear slightly skewed toward the positive side, and three of them are outliers. Because the factor effects are means, they have less variability than a single residual has. Recall that the variance of a single observation is n times the variance of a mean of n observations. If we want to compare the city effects with the residuals, we may want to scale the points on the side-by-side plot so that each has the same variance. Chapter 8 discusses methods of scaling side-by-side plots.

5D. COMPARING OVERLAYS: AN ANOVA TABLE

Just as s^2 measures variability in squared units, squares of the entries in an overlay lead to a useful measure of the overlay variation. For each overlay, we calculate a *sum of squares* by adding the squares of all the entries in the overlay.

The sum of squares (SS) partitions the squared variation, just as the overlays partition the data. More precisely, the sums of squares for the overlays add to the sum of squares for the data. For the one-way layout represented by equation (2),

RAW SUM OF SQUARES

$$\begin{aligned}\sum\sum y_{ij}^2 &= \sum\sum(m + a_i + e_{ij})^2 \\ &= \sum\sum m^2 + \sum\sum a_i^2 + \sum\sum e_{ij}^2 \\ &\quad + 2\sum\sum ma_i + 2\sum\sum me_{ij} + 2\sum\sum a_ie_{ij} \\ &= \sum\sum m^2 + \sum\sum a_i^2 + \sum\sum e_{ij}^2 \qquad (3a)\end{aligned}$$

because $\Sigma_i a_i = 0$ and $\Sigma_j e_{ij} = 0$ for each i. This shows that the data SS is the sum of the three overlay SS. Identity (3a) does not hold separately for each location in the overlays. We cannot add the squared entries in one location in all the overlays and get the square of the value in the data table.

Most discussions of partitioning the squared variation in the data start with the "corrected total sum of squares," here $\Sigma\Sigma(y_{ij} - m)^2$, which is algebraically the same as $\Sigma\Sigma y_{ij}^2 - \Sigma\Sigma m^2$. Because the value splitting includes an overlay for m, we usually start with $\Sigma\Sigma m^2$ on the right-hand side, as in equation (3a).

CORRECTED SUM OF SQUARES

$$\sum \sum y_{ij}^2 - \sum \sum m^2 = \sum \sum a_i^2 + \sum \sum e_{ij}^2 = \mathrm{SS}_{\mathrm{fit}} + \mathrm{SS}_{\mathrm{residual}}. \quad (3b)$$

The Multiple Correlation Coefficient

The sums of squares provide a measure of how well the fit accounts for the variation in the data. For this purpose the base is total variation about the overall mean—the corrected total sum of squares, $\Sigma\Sigma y_{ij}^2 - \Sigma\Sigma m^2$ in equation (3b). We measure the strength of the fit as the ratio of the sum of squares for fit to the corrected total sum of squares, $\mathrm{SS_T}$:

$$R^2 = \frac{\mathrm{SS}_{\mathrm{fit}}}{\mathrm{SS_T}} = 1 - \frac{\mathrm{SS}_{\mathrm{residual}}}{\mathrm{SS_T}}. \quad (4)$$

The second form simply exploits the identity $\mathrm{SS_T} = \mathrm{SS}_{\mathrm{fit}} + \mathrm{SS}_{\mathrm{residual}}$. In equation (3b), $\mathrm{SS}_{\mathrm{fit}} = \Sigma\Sigma a_i^2$. More generally, because the base is the corrected total sum of squares, $\mathrm{SS}_{\mathrm{fit}}$ is corrected also; that is, we do not include m in the fit for this purpose.

The measure in equation (4) occupies a prominent place in regression, where it is denoted by R^2 and known as the *multiple correlation coefficient*. This name derives from the measure's representation as the squared correlation between the data and the fit from the (multiple) regression. We use the same notation in this volume with the fit coming from the ANOVA.

Degrees of Freedom and Mean Squares

Comparisons of variation among overlays customarily take into account the number of entries that determine each overlay. We arrive at this number of *degrees of freedom* (d.f.) by starting with the number of principal entries (introduced in Section 5B) and subtracting the number of restrictions that apply. For example, in a one-way layout the $I \times J$ residuals are all principal entries, but the e_{ij} in cell i must satisfy $\Sigma_j e_{ij} = 0$—a total of I restrictions. Thus we count $IJ - I = I(J - 1)$ d.f.

By using this adjusted number of principal entries as a divisor for the corresponding sum of squares, we get a measure of variation that can treat the overlays differently. Thus it gives us a different picture of relative variation than we would get from dividing each sum of squares by the total number of entries in the full overlay, a number that is the same for all overlays and hence does not alter the impression we get from the sums of squares themselves. We now develop these ideas more fully, starting with some theoretical background for the residual sum of squares.

When the data in each cell are a sample from an associated population, we ordinarily use the sample mean to estimate the population mean and the sample variance to estimate the population variance. In cell i we have the mean $y_{i\cdot} = \Sigma_j y_{ij}/J$, the residuals $e_{ij} = y_{ij} - y_{i\cdot}$ $(j = 1, 2, \ldots, J)$, and the variance

$$s_i^2 = \sum_j (y_{ij} - y_{i\cdot})^2/(J-1) = \sum_j e_{ij}^2/(J-1). \tag{5}$$

Dividing by the degrees of freedom, $J - 1$, gives an unbiased estimate of the population variance for cell i.

In many one-way layouts it is reasonable to assume that all the cells have the same population variance, σ^2. Then, because we are working with a balanced layout, we average the s_i^2 to bring all the data to bear on estimating σ^2. The result is the *pooled within-cell variance*,

$$s^2 = \sum_i s_i^2/I = \sum_i \sum_j e_{ij}^2/I(J-1). \tag{6}$$

When all cell variances are identical, this s^2 is an unbiased estimate of σ^2 with $I(J - 1)$ d.f. (because each cell contributes $J - 1$ d.f.). In equation (6), $\Sigma_i \Sigma_j e_{ij}^2$ is the residual sum of squares that we formed in equations (3a) and (3b). Thus s^2 is known as the *residual mean square*. Other equivalent common notations include MS_{Resid}, MS_{Error}, and MS_e.

For a general overlay, a *mean square* is defined as the ratio of the sum of squares for that overlay to the associated degrees of freedom. The number of degrees of freedom for an overlay corresponds to the number of unconstrained values in the overlay. Thus in the residual overlay each cell has the one constraint that its values sum to zero; and so, once the first $J - 1$ residuals in the cell are known, the last residual is fixed to satisfy the constraint. This occurs in each of the I cells, so that I residuals are constrained and $I(J - 1)$ residuals are "free." Thus a mean square resembles a mean; its numerator is a sum of squared values, and its denominator is the number of unconstrained values.

The concept of degrees of freedom derives from the geometry of the sample space. The number of degrees of freedom represents the dimension of the space in which a sample point is free to move. On a coordinate plane, for example, the unconstrained point (y_1, y_2) can lie anywhere, but constraining the sum of its coordinates to equal zero forces $y_2 = -y_1$. Under this constraint, the sample point must lie on a line through the origin with slope -1. Restricted to a line, the point is free to move only in one dimension, and so it has only one degree of freedom. Similarly, placing a linear constraint on a point in three dimensions leaves it free to move only on a plane, so that the

Table 5-7. Symbolic Analysis-of-Variance Table for a One-Way Layout[a]

Source (Overlay)	Sum of Squares	Degrees of Freedom	Mean Square
Common	IJm^2	1	IJm^2
Effects	$J\Sigma a_i^2$	$I - 1$	$J\Sigma a_i^2/(I - 1)$
Residuals	$\Sigma\Sigma e_{ij}^2$	$I(J - 1)$	$\Sigma\Sigma e_{ij}^2/I(J - 1)$
Total	$\Sigma\Sigma y_{ij}^2$	IJ	

[a]The factor has I levels, and each cell contains J replications. The value splitting produces $y_{ij} = m + a_i + e_{ij}$.

number of degrees of freedom decreases from three to two. The imposition of a linear constraint (such as forcing the coordinates to sum to zero) reduces the degrees of freedom by one. Walker (1940) has a much more detailed, but elementary, discussion of the geometrical ideas surrounding degrees of freedom.

Each overlay, then, has three summary measures associated with it: sum of squares, degrees of freedom, and mean square. To allow comparison of these measures among overlays and to facilitate the computation of the statistics discussed in Chapter 7 for making inferences, we organize them into an *analysis-of-variance table*. Each row of such a table corresponds to an overlay in the value splitting. The rows are arranged in order of increasing complexity, with the common overlay first and the residual overlay last.

Table 5-7 shows, symbolically, an analysis-of-variance table for a balanced one-way layout. The residual sum of squares, degrees of freedom, and mean square are as described above. The sum of squares for the effects, $\Sigma\Sigma a_i^2$ in equations (3a) and (3b), reduces to $J\Sigma a_i^2$; and the degrees of freedom are $I - 1$, because $\Sigma a_i = 0$. Similarly, the sum of squares for the common term is IJm^2, with one degree of freedom (because m is unconstrained). As we would expect from equation (3a), the total sum of squares is $\Sigma\Sigma y_{ij}^2$. The corresponding number of degrees of freedom is IJ, the number of data values.

Many analyses report the common value separately and use the ANOVA table only for variation about m. Such an ANOVA table omits the line for the common overlay, and its total line gives the corrected total sum of squares and the corresponding degrees of freedom (one fewer than the number of data values).

If we are to compare mean squares from different overlays, we would like some idea of what to expect, at least under reasonable simplifying assumptions. The expected behavior is easiest to derive for the one-way layout, where we routinely compare the mean square for effects and the residual mean square. As above, we assume that all the observations have the same variance, σ^2. We now also assume that the underlying true means μ_i for the groups are all equal, so that differences among the $y_{i\cdot}$ merely represent sampling fluctuations. Then we can estimate the variance of a $y_{i\cdot}$ by treating $y_{1\cdot}, y_{2\cdot}, \ldots, y_{I\cdot}$ as a sample of size I and calculating its sample variance:

$$\text{var}(y_{i\cdot}) = \sum (y_{i\cdot} - y_{\cdot\cdot})^2/(I-1) = \sum a_i^2/(I-1). \qquad (7)$$

Because $y_{i\cdot}$ is the mean of the J observations in its group, we also have $\text{var}(y_{i\cdot}) = \sigma^2/J$, so that $\Sigma a_i^2/(I-1)$ provides an estimate of σ^2/J. Thus, when we assume that the underlying effects are all zero, the mean square, $J\Sigma a_i^2/(I-1)$, estimates σ^2. We have already seen that the residual mean square estimates σ^2; and so, under the simplifying assumptions, the mean square for effects and the residual mean square should not differ greatly. If the mean square for effects is much larger, we would suspect that the underlying effects are not all near zero.

Calculating the residual SS and d.f. can be tedious in large designs having many residuals to square and degrees of freedom to enumerate. A shortcut uses the results that sums of squares add as in equation (3a) and that their d.f. sum to the total number of observations. Then after calculating all SS and d.f. except for the residual overlay, one can obtain the residual SS and d.f. by subtraction.

Example: New-Car Loan Interest Rates

Table 5-8 gives an ANOVA table for the one-factor example of auto interest rates. Each sum of squares in Table 5-8 equals the sum of the squared values in the corresponding overlay in Table 5-5. Because the data involve six cities, the overlay for city effects has 5 d.f., and the residual overlay has 54 − 5 − 1 = 48 d.f.

From Table 5-8 the corrected total sum of squares is 32.71, and hence $R^2 = 10.95/32.71 = .334$. Thus the city effects account for one-third of the squared-scale variation in the data. If we assume that the variance of interest rates is the same in each city (and Figure 5-3 does not appear to contradict this assumption), we can compare MS_{Fit} to MS_{Error}. The ratio is roughly 5:1 and might suggest that the average rate differs from city to city. A formal inference, however, requires the machinery of the F test, discussed in Section 7A.

Table 5-8. ANOVA Table for New-Car Loan Interest Rates

Overlay	SS	d.f.	MS
Common	9140.05	1	9140.05
Cities	10.95	5	2.19
Residual	21.76	48	0.45
Total	9172.76	54	

5E. DESIGNS WITH TWO CROSSED FACTORS

As Section 4B explains, two factors are said to be *crossed* if the data contain observations at each combination of a level of one factor with a level of the other factor. If one factor has I levels and the other has J levels, the two-way crossed design has IJ cells, one for each combination of levels. Arranging the data in a rectangular table, the rows index the levels of one factor and the columns the levels of the other. In this section, we assume that each cell contains only one value. We address replication with two factors in Section 5F.

Extending the one-way description, we add an overlay for the second factor, so that the fit now includes three overlays: the common overlay and two one-factor overlays. As in the one-way layout, a one-factor effect is the deviation of the level mean from the common value. It tells how far above or below the common value the responses for that level of the factor lie on average.

We formulate this additive description as

$$y_{ij} = m + a_i + b_j + e_{ij} \tag{8}$$

for $i = 1, 2, \ldots, I$ and $j = 1, 2, \ldots, J$. Here m is the common value, a_i and b_j are the two sets of one-factor effects, and e_{ij} are the residuals. Sweeping using means produces overlays for the components of equation (8) that satisfy the constraints $\Sigma a_i = 0$, $\Sigma b_j = 0$, $\Sigma_j e_{ij} = 0$ for each i, and $\Sigma_i e_{ij} = 0$ for each j. The common overlay has a single value m in each of its IJ positions; the entries in the overlay for the first factor, which for clarity we call the row factor, are constant by rows with value a_i in row i; the overlay for the column factor is constant by columns with b_j in column j; and the residual overlay contains the e_{ij}. The location of each entry in the data table matches the location of its component pieces in the overlays.

As with the one-factor design, we can form an additive partition of the total sum of squares in the two-factor layout. The partition has four overlays: common, row effects, column effects, and residuals. To calculate the degrees of freedom for the overlays of effects, we note that each set of effects is constrained to sum to zero. Thus each one-factor overlay has degrees of freedom equal to one less than the number of levels for the factor: $I - 1$ for rows and $J - 1$ for columns. Subtracting the sum of the common and the two one-factor d.f. from the total d.f. leaves $IJ - (I - 1) - (J - 1) - 1 = (I - 1)(J - 1)$ residual d.f. All the rows and columns of the residual overlay sum to zero, so each row has $J - 1$ unconstrained entries, except that the last row has none, because its entries are fixed by the summing constraint on the columns.

In constructing overlays for unreplicated layouts with two crossed factors, the sweeping process can follow two different paths. We can sweep on rows and then on columns, or we can sweep on columns and then on rows. If the layout is balanced, the choice does not affect the results. In either way, as with the one-factor layout, sweeping progresses up the ANOVA table, forming more complex overlays before less complex ones.

Sweeping involves one basic step applied to both rows and columns. The step consists simply of computing the mean of each row or column, subtracting the mean from the entries in the row or column, and then placing the mean in a column or row adjoining the table of remainders. Figure 5-4 diagrams the process for a table with two levels of the row factor and three levels of the column factor. Step 1 of the left path sweeps across rows, calculating and subtracting row means, which are placed in a subtable to the left of the rows. The arrows in the figure indicate that we average the values in the area of the table where the arrow originates and put the mean in the cell where the arrow terminates. The remainders after subtracting the means replace the entries in the table. Step 2 diagrams the sweep of the columns, including the column of row means. The arrows show that the basic step of sweeping puts the column means in two subtables above the columns—one subtable above the column of row means and the other above the three columns of the main table. The remainders from the subtraction of these means replace the entries produced by the computations in Step 1. It is

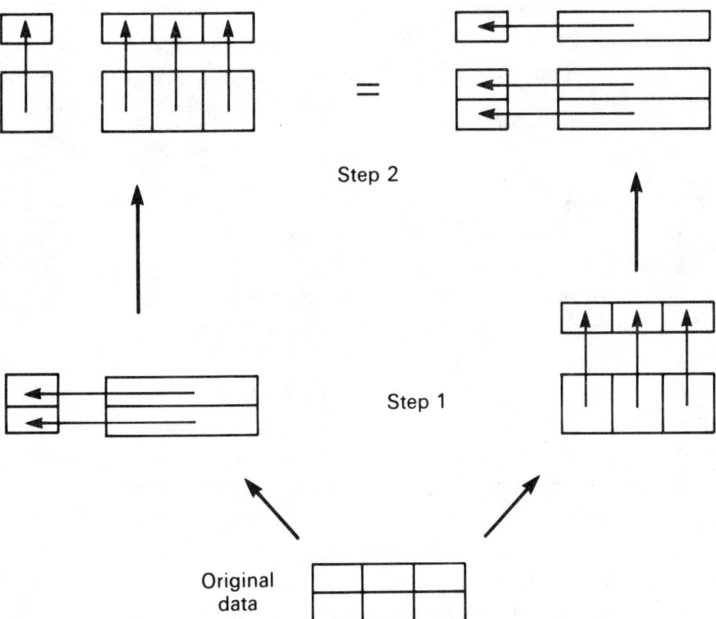

Figure 5-4. Schematic diagram of the sweeping process for an unreplicated two-way layout. The left path corresponds to sweeping on rows first and then on columns. The right path sweeps first on columns, then on rows. When we use means, the order of operations does not affect the final result.

important to understand that the column means computed in Step 2 are not means of the columns of the original data but means of the remainders after subtracting the row means.

Step 2 completes the sweeping. The upper left subtable contains the common value, the upper right subtable the column effects, the lower left subtable the row effects, and the residuals stand in the main table, which is now the residual overlay. This arrangement of the common value, row effects, column effects, and residuals, called a *bordered table*, contains all the principal entries that make up the reduced overlays. The bordered table is thus a concise form of the reduced overlays.

As noted earlier, we could have chosen the right path of Figure 5-4 in sweeping these data. In Step 1 we would sweep the columns, placing the means above; and at Step 2 we would sweep the rows, placing the means to the left. The final result is the same as that obtained by the left path.

Using the decomposition in equation (8) for the layout of two rows and three columns of Figure 5-4, we can dissect the algorithm to see why it works. Consider a single entry y_{ij} in row i and column j. In the left path, Step 1 computes the row means as $y_{i\cdot} = \frac{1}{3}\Sigma_j y_{ij}$ and puts them in the subtable to the left, leaving $y_{ij} - y_{i\cdot}$ for the remainders in the main table. In Step 2, taking means by columns, the mean of the row means, $y_{\cdot\cdot} = \frac{1}{2}\Sigma_i y_{i\cdot} = \frac{1}{6}\Sigma_i\Sigma_j y_{ij}$, is the common value and goes in the upper left subtable; the means of the columns in the subtable of remainders, $\frac{1}{2}\Sigma_i(y_{ij} - y_{i\cdot}) = y_{\cdot j} - y_{\cdot\cdot}$, are the column effects; the entries in the lower left subtable after subtraction of the common value are $y_{i\cdot} - y_{\cdot\cdot}$, the row effects; and the remainders in the main table become the residuals $(y_{ij} - y_{i\cdot}) - (y_{\cdot j} - y_{\cdot\cdot}) = y_{ij} - (y_{i\cdot} - y_{\cdot\cdot}) - (y_{\cdot j} - y_{\cdot\cdot}) - y_{\cdot\cdot}$.

EXAMPLE: SMOKING

Between 1978 and 1980 the Health Interview Survey, conducted by the U.S. National Center for Health Statistics, collected data concerning cigarette smoking in relation to family income and age categories. Table 5-9 presents a part of those data, showing percentages of smokers in certain age and income categories.

The value splitting explores the relationship between age and smoking behavior and the relationship between family income and smoking. It involves four overlays: a common one constant throughout, an income overlay containing the one-factor effects of income and constant by income, an analogous overlay for age group, and a residual overlay. Figure 5-5 details the calculations of the sweeping. Whether we sweep first by rows or by columns, the final result is the same.

Let us focus on the cell corresponding to the combination of people aged 17–30 with yearly family income below \$5000. Table 5-9 tells us that 38% of these people smoked. Sweeping first by rows as in the left path of Figure 5-5, we compute the row mean, 31.3, the average percentage of smokers for people whose families earn less than \$5000. When we subtract this mean

Table 5-9. Percentage of Regular Smokers by Age and Income

Annual Family Income (Dollars)	Age in Years 17–30	31–65	> 65
< 5,000	38	42	14
5,000–9,999	41	41	16
10,000–14,999	36	39	18
15,000–24,999	32	36	15
≥ 25,000	28	33	17

Source: Jeffrey E. Harris (1985). "On the fairness of cigarette excise taxation." In *The Cigarette Excise Tax*. Cambridge, MA: Institute for the Study of Smoking Behavior and Policy, John F. Kennedy School of Government, Harvard University, pp. 106–111 (part of Table 1, p. 107). His source: U.S. National Center for Health Statistics, Health Interview Survey, unpublished data. Reproduced with permission of Jeffrey E. Harris.

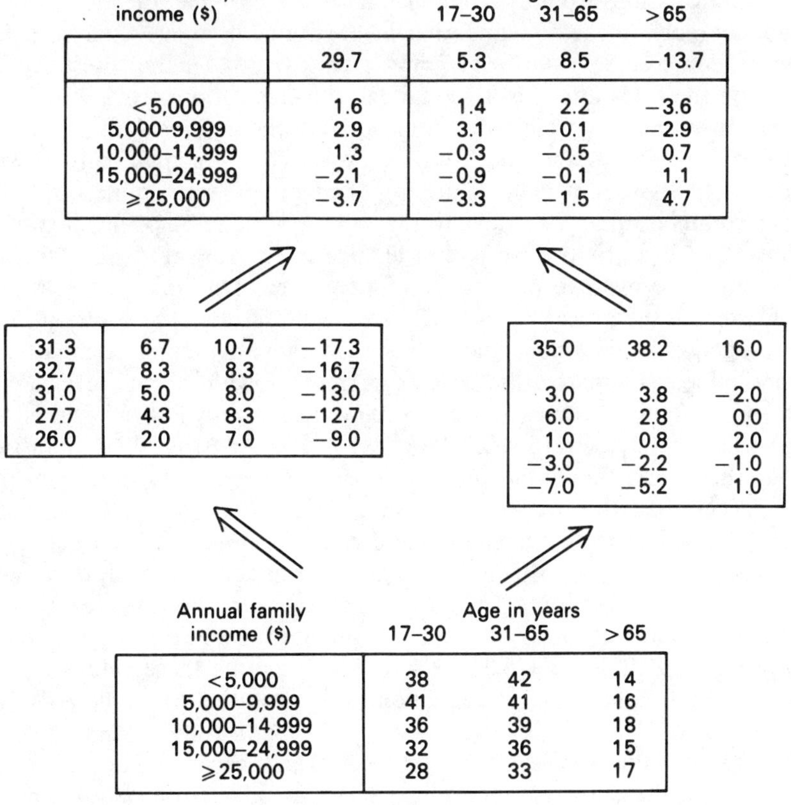

Figure 5-5. Sweeping the data on percentage of regular smokers. The left path and right path give the same value splitting.

from the row, the remainder in the cell of interest is 38 − 31.3 = 6.7. This number appears in the cell in Step 1. We note two features at this point. First, the entries in each row sum to zero. In row 1, the remainders 6.7, 10.7, and −17.3 add to zero (except for rounding error). Second, we can add the row mean to each remainder in each row and recover the original data entry for that cell (e.g., 6.7 + 31.3 = 38). These two characteristics hold throughout our calculations: the remainders after subtracting out a mean add to zero, and we can always get back to the original data by adding up the pieces, as the term "value splitting" implies. This first split shows us that the mean percentage of smokers differs from row to row (i.e., from one income category to another); the higher income categories have lower averages.

Step 2 gives the mean of the remainders from Step 1 for the column of people aged 17–30 as 5.3. Subtracting this from 6.7 leaves a new remainder of 1.4 in the cell after Step 2. Because we have swept out a column mean and a row mean, the remainders sum to zero in each row and in each column. The one-factor effects for age and income also sum to zero because their mean, the common value, has also been removed (as part of Step 2). They represent deviations from the common value. For example, the effects for age group, the column factor, appearing in the subtable above their respective columns, are 5.3, 8.5, and −13.7 and add to zero. Making these internal checks on the calculations is a good idea. The calculations from Step 2 leave a common value, age group effects, income effects, and residuals.

Table 5-10 shows the four full overlays. The principal entries, which appear in the bordered table at the top of Figure 5-5, are emphasized.

The common value, 29.7, tells us that on average 29.7% of this population smokes. The age group effects show change in smoking rates with changing age group. The average percentage of people smoking in the age category 17–30 exceeds the average percentage among all those in the survey by 5.3%. Those aged over 65 smoke much less than those under 65. Similarly, the income effects summarize the change in smoking with income level. People with annual family incomes above $15,000 smoke slightly less on average than those who earn below $15,000. The effect for the first row, 1.6, tells us that among those with family income below $5000 the percentage of smokers is 1.6% larger than the average percentage of smokers in the whole survey. Adding this effect to the common value gives the mean smoking percentage in this income category, 31.3%. Note that this was the row mean from Step 1 for the row corresponding to this income category. Adding together the pieces for the cell of lowest income and youngest age group, we recover the data value as 29.7 + 1.6 + 5.3 + 1.4 = 38. Identifying these values with the pieces in the decomposition of equation (8), we see that the remainder 1.4 is the residual.

To assemble the ANOVA table (Table 5-11), we can calculate each sum of squares from the corresponding full overlay in Table 5-10, or we can work from the reduced overlays (not shown) or from the bordered table at the top of Figure 5-5. If we use the reduced overlays or the bordered table, we

Table 5-10. Full Overlays for the Data on Smoking from the Health Interview Survey[a]

Original data (y_{ij})

38	42	14
41	41	16
36	39	18
32	36	15
28	33	17

= Common (m)

29.7	29.7	29.7
29.7	29.7	29.7
29.7	29.7	29.7
29.7	29.7	29.7
29.7	29.7	29.7

\+ Income (a_i)

1.6	1.6	1.6
2.9	2.9	2.9
1.3	1.3	1.3
−2.1	−2.1	−2.1
−3.7	−3.7	−3.7

\+ Age group (b_j)

5.3	**8.5**	**−13.7**
5.3	8.5	−13.7
5.3	8.5	−13.7
5.3	8.5	−13.7
5.3	8.5	−13.7

\+ Residual (e_{ij})

1.4	**2.2**	**−3.6**
3.1	**−0.1**	**−2.9**
−0.3	**−0.5**	**0.7**
−0.9	**−0.1**	**1.1**
−3.3	**−1.5**	**4.7**

[a]Each overlay has as many entries as the original table. The sum of the entries in any one position over all the overlays equals the entry in that position in the data table.

multiply the square of each principal entry by the number of positions it represents. For example, the sum of squares for the age overlay includes each square of a principal entry five times, and so the overlay sum of squares is the sum of the squares of the principal entries multiplied by 5. Numerically,

$$SS_{age} = 5 \times \left[(5.27)^2 + (8.47)^2 + (-13.73)^2\right] = 1440.13.$$

Table 5-11. ANOVA Table for Smoking Data

Overlay	SS	d.f.	MS
Common	13,258	1	13,258
Age	1,440	2	720
Income	93	4	23.2
Residual	76	8	9.5
Total	14,867	15	

In the ANOVA table, the income overlay has four degrees of freedom, one less than its five levels; the age overlay has two degrees of freedom; and the number of degrees of freedom left for the residual overlay is $(5 - 1) \times (3 - 1) = 8$.

A plot of residuals versus fitted values (Figure 5-6) shows that the residuals are not simply and systematically related to the fitted values. In particular, the data provide no evidence that the assumption of constant residual variance is violated. The large gap in the fitted values occurs because smoking rates differ much more by age than by income and divide into two groups on the basis of age. Among people aged more than 65 years almost 20% fewer smoke than in the other two groups.

In the side-by-side plot of Figure 5-7, all the age group effects are larger than any of the income effects. The effect for those older than 65 is particularly large; this group contains nearly 14% fewer smokers than the population as a whole. The batch of residuals is nearly symmetric, and no outliers appear. The spread of the residuals is about the same as the spread of the income effects.

The large value of R^2, .953, points to a good fit, but examination of the residual overlay in Table 5-10 reveals a pattern. The signs of the residuals in the first and second columns generally agree and are opposite to those in the third column. The only exception occurs in the cell of the second row and second column. The patterned residuals indicate that the additive description does not tell the whole story. When we look back at the data (Table 5-9), we see that income category has essentially no effect on the percentage of smokers in the over-65 age group, whereas it seems to have a noticeable effect in the other two age groups.

This pattern among the residuals apparent from the residual overlay does not show up in the residual boxplot in Figure 5-7. On the other hand, the overlay cannot show distributional patterns as the boxplot can. Viewing residuals both as a batch and as an ordered set with specific positions related to the data layout reveals different aspects of the data. This interplay of residual diagnostics plays a key role in any well-developed data analysis.

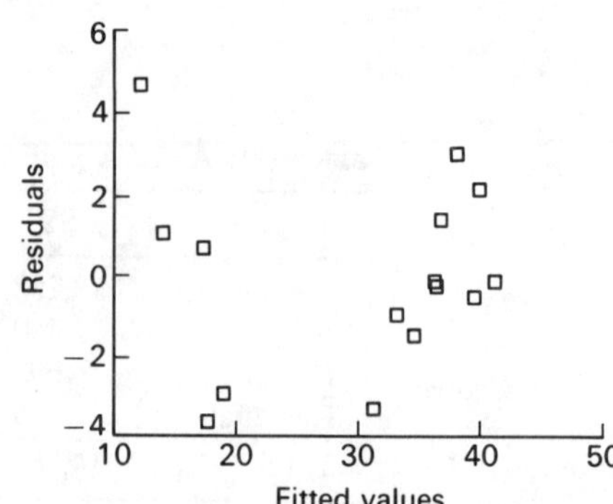

Figure 5-6. Plot of residuals versus fitted values for the smoking data.

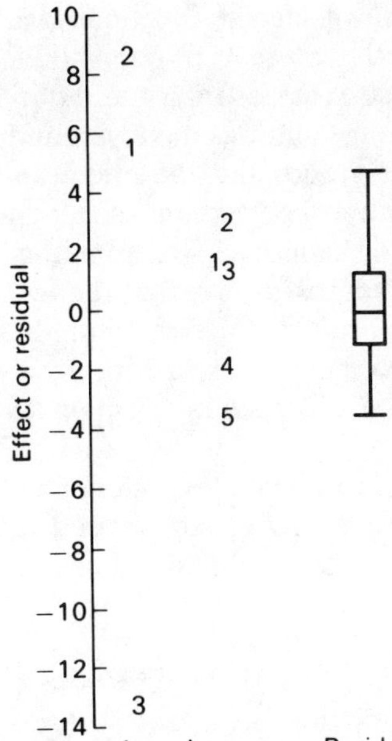

Figure 5-7. Side-by-side plot for the smoking data. (Age: 1 = 17–30; 2 = 31–65; 3 = > 65. Income: 1 = 5000–9999; 2 = 10,000–14,999; 3 = 15,000–19,999; 4 = 20,000–24,999; 5 = 25,000 +.)

5F. INTERACTION AND REPLICATION

Main-effect overlays, constant over all levels of the other factor, cannot capture all the structure in the data when the effects of one factor vary over the levels of the other factor. We require an additional overlay to describe this *interaction* between the two factors. (Section 1A discusses and illustrates the idea of interaction in analysis of variance.)

The *interaction effect* is defined as the difference between the mean response for a cell and the sum of the common term and the two one-factor effects. It is a *two-factor effect* because it involves the levels of two factors simultaneously. Because we interpret the interaction effects as deviations from the lower-order (one-factor) effects, the two-factor interactions are centered about zero also. Indeed, each row is centered, as is each column.

A simple example of interaction occurs in medicine when one drug increases (or reduces) the effect of another drug, so that a patient's response to the combination of drugs differs from the sum of the responses to the two drugs given separately. Another setting involving interaction concerns the joint effect of technological innovation and population increase on the growth of gross national product (GNP). Each factor can spur gains in GNP, but increases might be greater with both factors present and large than the sum of the one-factor effects would indicate.

We can measure the interaction effects in full detail only if the data are replicated. The smoking data lack replication; each cell contains only one value, a percentage of smokers for a combination of income level and age group. If the data were the result of a census of the entire U.S. population, we could interpret the residual overlay as interaction, measuring the difference between the percentage at a combination of age and income levels and the sum of the common term, the effect for age group, and the effect for income category. But a census is a special case; without replication, usually in the form of a sample within each cell, we generally cannot separate interaction from fluctuations in the data. In such circumstances, we say that the two are *confounded*. The residuals may well contain true interactions, but the data do not usually allow us to separate such possible structure from fluctuations within each cell (at least, unless we can limit our attention to specific forms of dependence).

Sometimes, a substantial part of the interaction or nonadditivity follows a simple pattern, in which the size and sign of the residual e_{ij} are related to the product $a_i b_j$ of the corresponding main effects. Chapter 13 considers this pattern further, in part by asking whether the data should be transformed to another scale.

Another common type of nonadditivity occurs when the interaction is concentrated in one cell. Unfortunately, sweeping by means allows such an isolated perturbation to distort the row effects and column effects and hence to leak into the fitted values for other cells. This *leakage* then obscures the location of the disturbed cell, especially when random fluctuation is also present. Daniel (1976, 1978) and Brown (1975) discuss this problem and offer a simple remedy. They suggest picking out the cell with the largest residual and adjusting the data value for this cell so that, for the initial fit, this cell has a residual of zero. Then we refit the data, decreasing the residual d.f. by 1. If single-cell interactions do exist, this adjustment should reduce the residuals in other cells. Generalizations of this procedure apply when the nonadditivity occurs in a small number of cells.

In principle, percentage data offer us another handle on interactions. If we had the sample size for each cell, we could estimate the random variation in each cell and thus determine whether the residual was likely to contain interaction. The cell populations consist of people categorized by age and income level, and the response is a binary random variable describing smoking status. The number of smokers in the cell might behave like a binomial random variable. The variance of a sample proportion $\hat{p}$ when the data follow the binomial distribution is $p(1-p)/n$, and so the standard error of a sample percentage $100\hat{p}$ is $100\sqrt{p(1-p)/n}$, where n is the number of people sampled in the cell.

5G. TWO-FACTOR DESIGNS WITH REPLICATION

When the data have replication within the cells, it is possible to separate the residual and interaction components, cell by cell. Then, in addition to the variation among rows, among columns, and among cells, the replications produce variation within the cells. After sweeping out the variation between cells into the interaction component, a residual component measuring the variation within cells still remains.

The additive decomposition for the two-way replicated design has five overlays. The common overlay is constant over all the data; the one-factor overlays are constant over a column or row of cells; the interaction overlay may have a different value for each cell; and the residual overlay shows the amount that each observation deviates from its cell mean. The additive decomposition

$$y_{ijk} = m + a_i + b_j + (ab)_{ij} + e_{ijk} \tag{9}$$

has m for the common value, a_i for the effect of row i, b_j for the effect of column j, $(ab)_{ij}$ for the interaction effect in row i and column j, and e_{ijk} for the residual corresponding to the kth of the K replications in cell (i, j). The notation $(ab)_{ij}$ does not imply multiplication of the symbols a and b; rather, it symbolizes an effect that applies simultaneously at the ith level of the factor represented by a and at the jth level of the factor represented by b.

The five overlays partition the sum of squares in the usual manner. Degrees of freedom for the main effects and interactions follow simply by considering the cell means as an unreplicated layout. We calculate degrees of freedom for the common and one-factor overlays as in Section 5E and allocate to the interaction overlay the $IJ - 1 - (I - 1) - (J - 1) = (I - 1)(J - 1)$ d.f. remaining from the cell means. The residual degrees of freedom, for within-cell variation, are then $IJK - IJ = IJ(K - 1)$. Each of the IJ cells has $K - 1$ unconstrained values, with the other value fixed by the cell mean, computed in the first step of sweeping.

Figure 5-8 schematically illustrates the first step of sweeping for the two-factor layout with replication. In this first step, we sweep the cells,

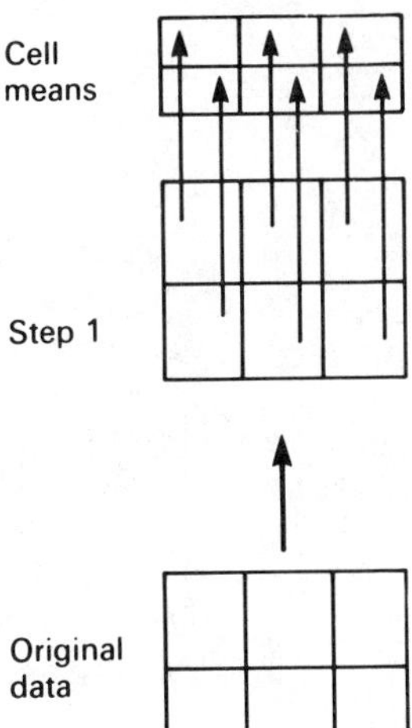

Figure 5-8. Schematic diagram of the first step of sweeping applied to replicated two-factor data. Step 2 and Step 3 work with the table of cell means and reproduce the calculations in Step 1 and Step 2, respectively, of Figure 5-4.

computing the mean of the replications in each cell and subtracting that mean from each replication. The subtable of cell means, placed above the table of remainders, forms the basis for all subsequent sweeping. These means contain all the information we have about the two-factor interaction, the one-factor effects, and the common value. Two further steps of sweeping summarize and separate these components, as illustrated for the unreplicated layout in Figure 5-4. Because the cell mean has been removed, each cell of remainders has zero mean. In the replicated layout, the remainders after Step 1 are the residuals because they measure the variation within cells. This contrasts with sweeping for the unreplicated layout, which produces the residuals only at the last step.

EXAMPLE: RAT FEEDING

As an example of the analysis of a replicated two-way table, consider the data in Table 5-12. They come from an experiment that tested the effectiveness of two factors forming six potential protein feeding treatments for cattle. Seeking to determine which levels gave large weight gains, the experimenters measured the weight gains of 60 rats. Ten rats were randomly assigned to

Table 5-12. Weight Gain (in Grams) of Rats Given Different Diets

	Beef	Pork	Cereal
	118	120	111
	117	108	98
	111	105	95
	107	102	92
High	104	102	88
	102	98	86
	100	96	82
	87	94	77
	81	91	74
	73	79	56
	95	106	107
	90	97	98
	90	86	97
	90	82	95
Low	86	82	89
	78	81	80
	76	73	74
	72	70	74
	64	61	67
	51	49	58

Source: George W. Snedecor and William G. Cochran (1980). *Statistical Methods*, 7th ed. Ames: Iowa State University Press (data from Table 16.6.1, p. 305). ©1980 The Iowa State University Press.

		Beef	Pork	Cereal
	87.9	1.7	1.2	−3.0
High	7.3	3.1	3.1	−6.3
Low	−7.3	−3.1	−3.1	6.3

Step 3

		Beef	Pork	Cereal
High	95.1	4.9	4.4	−9.2
Low	80.6	−1.4	−1.9	3.3

	Beef	Pork	Cereal
	89.6	89.1	84.9
High	10.4	10.4	1.0
Low	−10.4	−10.4	−1.0

Step 2

	Beef	Pork	Cereal
High	100.0	99.5	85.9
Low	79.2	78.7	83.9
High	18.0	20.5	25.1
	17.0	8.5	12.1
	11.0	5.5	9.1
	7.0	2.5	6.1
	4.0	2.5	2.1
	2.0	−1.5	0.1
	0.0	−3.5	−3.9
	−13.0	−5.5	−8.9
	−19.0	−8.5	−11.9
	−27.0	−20.5	−29.9
Low	15.8	27.3	23.1
	10.8	18.3	14.1
	10.8	7.3	13.1
	10.8	3.3	11.1
	6.8	3.3	5.1
	−1.2	2.3	−3.9
	−3.2	−5.7	−9.9
	−7.2	−8.7	−9.9
	−15.2	−17.7	−16.9
	−28.2	−29.7	−25.9

Step 1

	Beef	Pork	Cereal
High	118	120	111
	117	108	98
	111	105	95
	107	102	92
	104	102	88
	102	98	86
	100	96	82
	87	94	77
	81	91	74
	73	79	56
Low	95	106	107
	90	97	98
	90	86	97
	90	82	95
	86	82	89
	78	81	80
	76	73	74
	72	70	74
	64	61	67
	51	49	58

Figure 5-9. Sweeping the rat feeding data.

each of the treatments: beef, cereal, and pork given at high and low amounts of protein.

Sweeping out a common value, row effects, column effects, and interaction effects, as in Figure 5-9, produces a bordered table in Step 3. Its entries, together with the residuals (produced at Step 1), become the principal entries in the value splitting.

To illustrate the value splitting for these data, we focus on the weight gain of 73 grams for the rat with smallest weight gain in the high-protein beef diet group. This value splits into the sum of the common value 87.9, the effect for high protein, 7.3, the effect for beef, 1.7, the interaction of beef with high protein, 3.1, and the residual, −27.0. The residual is the difference between the weight gain for this rat and the mean weight gain for the 10 rats that received the high-protein beef diet. The interaction measures how the protein-level effects differ at each type of protein and how the protein-type effects differ according to the protein level. The subtable of cell means in Figure 5-9 (the result of Step 1) shows that high protein produces larger weight gains for each protein type, but that the excess is 20.8 grams for beef and pork, and only 2.0 grams for cereal. Similarly, beef and pork give larger gains than cereal when used with high protein, but smaller gains with a low level of protein.

The common value tells us that the rats gained 87.9 grams on average. The level effects show that, on average, rats receiving a high level of protein gained 7.3 grams more than this common value, whereas those receiving the low level gained 7.3 grams less than the common value. Similarly, the effect of the beef feed was 1.7 grams, that of pork 1.2 grams, and that of cereal −3.0 grams. Amount of protein clearly had a larger effect than protein type. The interaction effects are larger than the protein-type effects.

Table 5-13 presents an ANOVA table for these data. The reader can verify the sums of squares in the usual manner. The fit describes about one-fourth of the (squared-scale) variation in the data ($R^2 = .285$). The only new term for the replicated data set is the interaction, which has $(I - 1) \times (J - 1) = 2$ degrees of freedom because of the $(I - 1) + (J - 1) + 1$ constraints imposed on the IJ cells by the row, column, and common effects. The residual mean square again estimates σ^2, but now the replications allow us to

Table 5-13. ANOVA Table for Rat Feeding Data

Overlay	SS	d.f.	MS
Common	463,233	1	463,233
Protein level	3,168	1	3,168
Feed type	267	2	133.5
Interaction	1,178	2	589
Residual	11,586	54	215
Total	479,432	60	

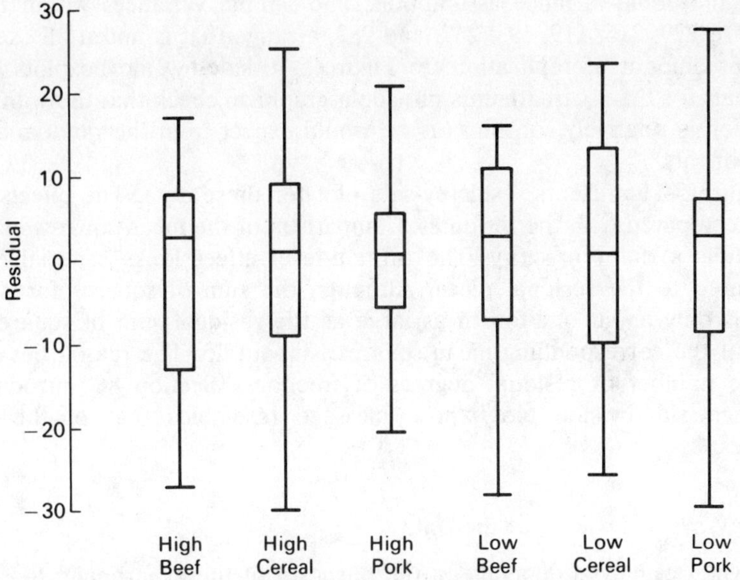

Figure 5-10. Side-by-side boxplots of the residuals by treatment for the rat feeding data. The medians lie near zero, and the variation is similar from group to group.

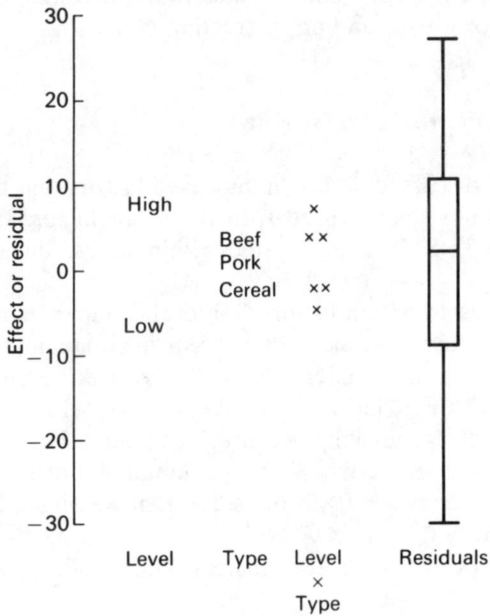

Figure 5-11. Side-by-side plot for the rat feeding data.

check the equal-variance assumption. The sample variances within the six cells are 229, 226, 119, 193, 247, and 282, a range that is not at all excessive for this amount of replication. In Figure 5-10 side-by-side boxplots of the residuals for the six treatments provide a graphical check that the within-cell variation is relatively constant (as we would expect from the random assignment of rats).

Figure 5-11 presents a side-by-side plot for these data. The effects seem small compared with the residuals. Comparison of the mean squares in Table 5-13 tells a different story. The protein-level effect looks especially large compared to the residual piece. Although the sum of squares for protein level is only about one-fourth as large as the residual sum of squares, the ratio of the corresponding mean squares is about 15. The reason lies in the sizable number of residual degrees of freedom. (Section 8C introduces a modified side-by-side plot whose message resembles that of the mean squares.)

Other Ways of Looking at the Data

The type of analysis depends on the purposes of the experiment. Searching for the combination that gives the largest weight gains leads to the selection of the high-protein beef diet. Considering the data this way makes the design a one-way layout with six treatments and ten replications per treatment. A similar design would be appropriate if the researcher were looking for the treatment with best return. With two factors the investigator can determine separate one-factor effects and an interaction effect.

5H. TWO-FACTOR DATA WITH NESTING

As Chapter 4 has discussed, a design may have factors that are *nested* within another factor rather than crossed with the other factor. In the crime rate example (Table 4-3), SMSAs are nested within states, and states are nested within regions.

If factor B is nested within factor A, a level of factor B makes sense only within one level of factor A and cannot occur in combination with any other level of A. Then one-factor effects for factor A are present, but effects for factor B occur within a level of A. We can say either: "no interaction between A and B is possible because we cannot have an effect for B measured at different levels of A," or "a combination of A and B is present without any main effects for B (in the sense that we give a level of A and a level of B to identify this effect)."

For a nested design with two factors and replication, we write the decomposition

$$y_{ijk} = m + a_i + b_{j(i)} + e_{ijk}. \tag{10}$$

As usual, each term corresponds to an overlay. Here m is the common value, a_i is the effect for level i of factor A, and $b_{j(i)}$ represents the effect for the jth level of factor B nested in the ith level of A. Each level of A has J effects of B nested within it, and so B has IJ effects. The fitted value for cell (i, j) is the mean $m + a_i + b_{j(i)}$. The difference $y_{ijk} - (m + a_i + b_{j(i)})$ is the residual for the kth replication in cell (i, j). The term $b_{j(i)}$ represents the difference between the mean for cell (i, j) and the mean for the ith level of factor A because $b_{j(i)} = (m + a_i + b_{j(i)}) - (m + a_i)$. In a crossed two-factor design with replication, this difference $y_{ij\cdot} - y_{i\cdot\cdot}$ equals the sum of the B main effect and the AB interaction.

The ANOVA table for a nested design orders the overlays according to the nesting hierarchy. The overlay for a nested factor follows the overlay of the factor in which it nests. Thus a two-factor nested design having B nested within A has the following ordering from top to bottom: common, A, B within A, and residual. Sums of squares are computed in the usual way with the appropriate multiplier for the results from the principal entries. Degrees of freedom correspond to the number of unconstrained values in an overlay. Factor A has $I - 1$ degrees of freedom. For each level of factor A, the effects for the J nested levels of factor B sum to zero, so that B has $J - 1$ degrees of freedom for each of the I levels of A, or $I(J - 1)$ in all.

Figure 5-12 illustrates sweeping for nested designs. Each row in the large array corresponds to a level of factor B within a level of factor A. The elements of the row are the replications within the level of B, and so each row is a cell. A row mean is $m + a_i + b_{j(i)}$, and the residual e_{ijk} is the difference between the data value y_{ijk} and the row mean. As with value splitting in the presence of replication, we remove the cell means from the cell entries to obtain the residuals and then work with the table of cell means. Step 1 of Figure 5-12 shows how the row means are subtracted from the data values to leave the residual overlay. Step 2 subtracts the mean of a level of A from the levels of B nested within that level of A. The remainders give the effects of B. In Step 3, the common value is subtracted from the A level means, leaving the A effects. Sweeping in nested designs proceeds in only one direction, and each step operates only on the subtable formed in the previous step. Thus each step except the last forms one overlay; the last step forms two overlays: common term and one-factor effects.

As an example of a nested design, we look at some data from a laboratory that tests blood for the presence of antibodies to HIV.

Example: HIV testing

In determining whether a person has been exposed to the human immunodeficiency virus (HIV), which is believed to cause the acquired immune deficiency syndrome (AIDS), medical laboratories subject a sample of the person's blood to a screening test that involves an enzyme-linked immunosorbent assay (ELISA). In addition to the samples of patients' blood, each run of the ELISA test includes positive and negative control samples. The

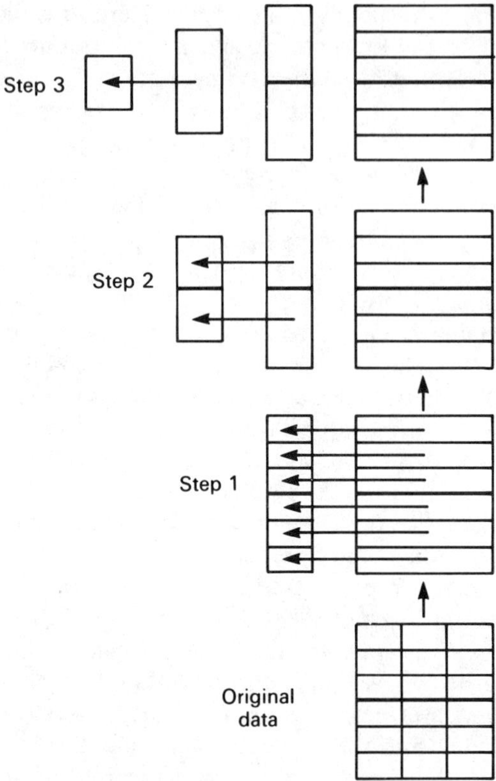

Figure 5-12. Sweeping a nested two-factor design that has a three-level factor nested within a two-level factor.

positive control samples are known to contain antibodies to HIV, and the negative control samples are known to be free of such antibodies. At the final step of the test, a spectrophotometer determines the optical density of each sample (measured as absorbance of light at a particular wavelength). A patient's sample is positive if its absorbance exceeds a cutoff value, which is a function of the absorbance of the positive and negative control samples in the run. The ELISA test involves a number of steps, and so some variation enters into the results, causing the absorbances to differ from sample to sample.

As part of a study of ways to monitor and ensure proficiency in laboratories' performance of the ELISA test, researchers examined variation in control values from run to run. Because some ingredients used in the test deteriorate, the materials for the test come in lots, which contain enough material for several runs. Thus three sources contribute to variation in the results for the control samples: differences among lots, differences among runs within lot, and differences among control samples within a run.

Table 5-14. Measurements of Light Absorbance for Positive Control Samples in an ELISA Test for HIV[a]

Lot	Run	Positive Control Samples		
A	1	1.053	1.708	0.977
	2	0.881	0.788	0.788
	3	0.896	1.038	0.963
	4	0.971	1.234	1.089
	5	0.984	0.986	1.067
B	1	0.996	1.129	1.016
	2	1.019	1.088	1.280
	3	1.120	1.054	1.235
	4	1.327	1.361	1.233
	5	1.079	1.120	0.959
C	1	1.229	1.027	1.109
	2	1.118	1.066	1.146
	3	1.053	1.082	1.113
	4	1.140	1.172	0.966
	5	0.963	1.064	1.086
D	1	0.985	0.894	1.019
	2	0.847	0.799	0.918
	3	1.033	0.943	1.089
	4	0.988	1.169	1.106
	5	1.308	1.498	1.271
E	1	1.128	1.141	1.144
	2	0.990	0.801	0.416
	3	0.929	0.950	0.899
	4	0.873	0.871	0.786
	5	0.930	0.968	0.844

[a]Readings come from 25 runs in 5 lots at one laboratory during May–August 1987.

One manufacturer's version of the ELISA test uses three positive control samples in each run. In 25 runs, involving five lots, one laboratory recorded the absorbances shown in Table 5-14 for the positive control samples. The data thus have a nested structure: positive control samples are nested within runs, which are nested within lots.

Table 5-15 shows the bordered table produced by the three steps of sweeping for the HIV data. Each full overlay would contain 75 values. The residual overlay is created in Step 1. For example, in the first row subtracting the mean for Run 1 in Lot A, 1.246, leaves the residuals −0.193, 0.462, and −0.269. Similarly, Step 2 produces the runs overlay. For example, the effect for Run 1 in Lot A is 1.246 − 1.028 = 0.218. Finally, Step 3 gives the common overlay and the lots overlay. For example, the effect for Lot A is 1.028 − 1.044 = −0.016.

Table 5-15. The Bordered Table Produced by the Three Steps of Sweeping for the HIV Data

Lot	Run	Common	Lots	Runs	Residuals		
A	1			0.218	−0.193	0.462	−0.269
	2			−0.209	0.063	−0.031	−0.031
	3		−0.016	−0.062	−0.070	0.072	−0.003
	4			0.070	−0.127	0.136	−0.009
	5			−0.016	−0.028	−0.026	0.055
B	1			−0.087	−0.051	0.082	−0.031
	2			−0.005	−0.110	−0.041	0.151
	3		0.090	0.002	−0.016	−0.082	0.099
	4			0.173	0.020	0.054	−0.074
	5			−0.081	0.026	0.067	−0.094
C	1			0.038	0.107	−0.095	−0.013
	2			0.026	0.008	−0.044	0.036
	3	1.044	0.045	−0.001	−0.030	−0.001	0.030
	4			0.009	0.047	0.079	−0.127
	5			−0.048	−0.075	0.026	0.048
D	1			−0.092	0.019	−0.072	0.053
	2			−0.203	−0.008	−0.056	0.063
	3		0.014	−0.036	0.011	−0.079	0.067
	4			0.030	−0.100	0.081	0.018
	5			0.301	−0.051	0.139	−0.088
E	1			0.227	−0.010	0.003	0.006
	2			−0.175	0.254	0.065	−0.320
	3		−0.133	0.015	0.003	0.024	−0.027
	4			−0.068	0.030	0.028	−0.057
	5			0.003	0.016	0.054	−0.070

The common absorbance is 1.044 units, and the lot means differ from this common mean by at most 13% (Lot E). As measured by the run-within-lot effects, none of the lots is much more variable than any of the others, although Lot C appears particularly stable, even in individual samples within runs.

The lot and run effects themselves do not have as much importance as the factor effects in previous examples because tests for new patients can never use these lots and runs. The importance of the study lies more in what it reveals about the variability of the lots and runs. Less variable results lead to more confidence that the results of one sample do not depend on the lot used or the run made (equivalently, the day on which the lab ran the test). In effect, the lots and runs used here are a sample from all the lots and runs that could have been used. It is this population variability that concerns us. Formal statistical inference about such variance components must wait for Chapter 9.

Table 5-16. ANOVA Table for Absorbance of Positive Control Samples in an ELISA Test

Overlay	SS	d.f.	MS
Common	81.761	1	81.761
Lots	0.424	4	0.106
Runs (within lots)	1.134	20	0.057
Samples (within runs within lots)	0.782	50	0.016
Total	84.101	75	

To check on the variation among samples within runs, we examine the residuals. A stem-and-leaf display (not shown) suggests that, with the exception of three low values and two high values, the residuals are well-behaved. The interquartile range, .107, suggests a residual standard deviation around .08. Tracing the five extreme residuals back into Table 5-15, we find that they all come from two runs: Run 1 in Lot A and Run 2 in Lot E. If we had an opportunity to investigate the circumstances surrounding these two runs, we would ask questions such as the following. Was the person who carried out these runs someone other than the technician who usually handles the ELISA test (e.g., because of vacation or illness)? Did those runs take place at unusual times (e.g., on overtime or as the first run after a weekend)? Was the air conditioning functioning properly on those days? Even a well-run laboratory has mishaps occasionally. If we found a clue to trouble, we would inquire into other days when the same problem occurred, not just the ones when these runs were unusual. For example, what happened on other days when the regular technician was absent or when the air conditioner was off? Persistent trouble with the air conditioner might turn out not to have any consistent association with unusual results.

Table 5-16 gives an ANOVA table for these data. When calculating the sums of squares using the bordered table or reduced overlays, we must remember the multiplier for the number of values each principal entry represents. For example, each lot contains 15 samples, so the lot SS is the sum of the squares of the five principal entries multiplied by 15. The calculation of the number of degrees of freedom for each overlay proceeds as follows. The five lot effects sum to zero, and so subtracting one degree of freedom for this constraint leaves 4 d.f. for lots. The run effects within each lot sum to zero; these five constraints on the 25 run effects leave 20 d.f. for runs. The residuals within each run also sum to zero, so each run has 2 d.f., leading to $25 \times 2 = 50$ d.f. for samples. Together with the 1 d.f. for the common overlay, the degrees of freedom add to 75, the number of entries in the data. Calculating R^2 gives .67.

If we remove the six samples in the two unusual runs (Run 1 in Lot A and Run 2 in Lot E) from the sum of squares for samples and recalculate the mean square, we get $(.782 - .494)/(50 - 4) = .0063$. This estimate of the variance of an individual sample agrees well with the standard deviation (.08) that we calculated from the interquartile range. The outliers substantially inflated the residual mean square. The close agreement, however, involves a considerable element of chance. As Section 7D explains, sample variances are highly variable, even under ideal conditions.

5I. OTHER DESCRIPTIONS

Much of this chapter has focused on describing a rectangular array of data in terms of its factors. These standard descriptions express variation as deviations from means taken over levels of one or two factors. Sometimes the data are better described without using the framework of main effects and interactions. This section discusses some of the more common nonstandard descriptions.

A simple example of a problem that would profit from a different approach involves data for two factors, each having (say) a high level and a low level, but with a positive effect only when each factor is at its high level. Then three of the four responses are similar, and the fourth is different; the data are best described this way and not in terms of main effects and interactions. Suppose, for example, that we have the data

$$\begin{array}{rr} 12 & 4 \\ 4 & 4 \end{array}$$

We could analyze this 2×2 array in terms of main effects and interactions, finding a common value of 6, row effects and column effects of $+2$ and -2, and residuals of $+2$ and -2. A simpler and possibly better description, however, might be that the common level is 4 and the (1, 1) cell deviates by $+8$.

Sometimes a data set splits naturally into two sections, either by design (as with data for males and females) or for some other less intrinsic reason. The smoking data of Table 5-9, for example, have two dissimilar sets of values: one for the oldest group and the second for the younger two groups. With such data, a separate analysis of each section might be more informative. Exercise 5 explores this option.

In other situations, the components may combine not by addition, but perhaps by multiplication. A transformation or reformulation of the data may then put them in a form for which an additive description becomes appropriate. Taking logs of multiplicative data will transform the scale to additivity. Chapter 13 discusses transformation in some detail.

5J. SUMMARY

Constructing overlays of main effects and interactions promotes a direct, hands-on understanding of the data in a way that the ANOVA table cannot. The overlays provide the quantities from which the sums of squares are calculated and allow the user to study pieces that may contribute excessive amounts of variation. Having this detailed information helps the user to avoid making incorrect or inadequate interpretations based on summing totals. Cuthbert Daniel, a very experienced applied statistician, expressed this idea in *Applications of Statistics to Industrial Experimentation* (1976, p. 38):

> Do not try to interpret sums of squares of quantities whose summands you do not know.

Sweeping facilitates calculation of the overlays for balanced designs. As the amount of data grows, however, more concise summaries than full overlays become necessary.

Finally, value splitting promotes a much more careful approach to residual diagnostics through the residual overlay and various graphical displays. These diagnostics help pinpoint lack of fit, suggest better descriptions, and sometimes promote further investigations.

REFERENCES

Brown, M. B. (1975). "Exploring interaction effects in the ANOVA," *Applied Statistics*, 24, 288–298.

Cobb, G. W. (1984). "An algorithmic approach to elementary ANOVA," *The American Statistician*, 38, 120–123.

Daniel, C. (1976). *Applications of Statistics to Industrial Experimentation*. New York: Wiley.

Daniel, C. (1978). "Patterns in residuals in the two-way layout," *Technometrics*, 20, 385–395.

Snedecor, G. W. and Cochran, W. G. (1980). *Statistical Methods*, 7th ed. Ames: Iowa State University Press.

Walker, H. M. (1940). "Degrees of freedom," *Journal of Educational Psychology*, 31, 253–269.

EXERCISES

1. (a) What are the fitted values for the new-car loans example (Table 5-5)?
   (b) Make and discuss a plot of residuals versus fitted values for these data.
   (c) How does this plot differ from the dotplot of residuals by city in Figure 5-2? What different things do the two plots reveal?

2. Show algebraically that by adding the overlay sums of squares we obtain the sum of the squares of the data values in the two-way layout (a) without replication and (b) with replication. [*Hint:* Write the response as a sum of its components and use the constraints on the main effects and interactions.]

3. **(a)** Verify the sums of squares given in Table 5-11 for the smoking data. Use reduced overlays to simplify calculations.
   (b) Verify that $R^2 = .953$ for these data.

4. **(a)** Verify the sums of squares in Table 5-13 for the rat feeding data.
   (b) Compute R^2 for this example. Is the fit a good one?
   (c) Combine the interaction and residual overlays and recompute R^2. Does the assumption of no interaction appear valid? Why or why not?
   (d) Refer to the right path in Figure 5-9, in which column means are removed first. Note that in the table of remainders after removing column means

	Beef	Pork	Cereal
High	10.4	10.4	1.0
Low	−10.4	−10.4	−1.0

   the values in the cereal column are very small compared to those in the beef and pork columns. What different way does this pattern suggest to summarize the rat feeding data?

5. Reanalyze the smoking data in Table 5-9, treating the percentages of smokers over 65 separately from those under 65.
   (a) Find the means of each group.
   (b) Calculate the age and income effects for the group under 65.
   (c) How do these overlays differ from those in Table 5-10? Does this analysis suggest a different interpretation for the smoking data?

6. After a questionnaire survey of all high school seniors in the state of Wisconsin in 1957, researchers sampled about one-third of the students for a follow-up study in 1964–1965. They obtained data from 87% of the sample—a total of 4386 males and 4621 females. Table 5-17 shows the percentage who attended college in each cell of a cross-classification by socioeconomic status and intelligence, separately by males and females. Analyze the data for males as a two-way layout and interpret the results.

Table 5-17. Percentage of Students in a Wisconsin Cohort Who Attended College, by Socioeconomic Status and Intelligence, Separately for Males and Females

Socioeconomic Status	Intelligence Level: Low	Lower Middle	Upper Middle	High
	(a) Males			
Low	6.3	16.5	28.0	52.4
Lower middle	11.7	27.2	42.6	58.9
Upper middle	18.3	34.3	51.3	72.0
High	38.8	60.8	73.2	90.7
	(b) Females			
Low	3.7	6.3	8.9	27.5
Lower middle	9.3	20.2	24.1	36.7
Upper middle	16.0	25.6	31.0	48.1
High	33.3	44.4	67.0	76.4

Source: William H. Sewell and Vimal P. Shah (1967). "Socioeconomic status, intelligence, and the attainment of higher education," *Sociology of Education*, 40, 1–23 (data from Table 3, p. 13).

7. (Continuation) Analyze the data for females in Table 5-17b as a two-way layout. Interpret these results.

8. (Continuation) Compare the results from the analyses of the data for males and the data for females in Exercises 6 and 7. Interpret similarities and differences. Explain why an analysis that combines the data for males and the data for females into a three-way layout might or might not be appropriate.

9. What complicated relation of residuals to fitted values is suggested by Figure 5-6?

10. (Continuation) What might be the cause of the complicated relation just found?

CHAPTER 6

Value Splitting Involving More Factors

Katherine Taylor Halvorsen
Smith College

To show what happens to the individual data values in analysis of variance, Chapter 5 introduced overlays and the process of value splitting, using data layouts that have at most two factors. Although many sets of data in practice follow such simple structures, more complicated investigations involve larger numbers of factors. Thus the present chapter extends value splitting to situations that involve three and four crossed factors.

As the number of factors increases, the cost of experimentation may make it attractive to use less than a fully crossed design, collecting fewer observations but arranging them to get at sources of variation that seem likely to be important. This chapter includes one such design, the Latin square, which yields main effects for its three factors, but not two-factor interactions.

In all three layouts we use bordered tables instead of full overlays to present the results of value splitting, mainly for economy in displaying the numbers. For graphical display the side-by-side plots, which Chapter 5 uses to show main effects for one or two factors along with the residuals, extend naturally to include all the main effects, two-factor interactions, and higher-order interactions that make up the value splitting.

6A. THREE CROSSED FACTORS

A three-way layout without replication (i.e., one observation per cell) yields eight overlays: a common value, one main effect for each of the three factors, three two-factor interactions, and the residuals. We write the decomposition as follows:

$$y_{ijk} = m + a_i + b_j + c_k + (ab)_{ij} + (ac)_{ik} + (bc)_{jk} + e_{ijk}.$$

If each cell contains n replicate observations, then a full decomposition replaces the single-replication residuals e_{ijk} with the three-factor interaction $(abc)_{ijk}$ and a new set of residuals e_{ijkr}, representing within-cell variation. An additional subscript on y identifies the individual replications:

$$y_{ijkr} = m + a_i + b_j + c_k + (ab)_{ij} + (ac)_{ik} + (bc)_{jk} + (abc)_{ijk} + e_{ijkr}.$$

An unreplicated layout does not allow us to separate the three-factor interaction term from the within-cell variation. We say they are "confounded." (In some situations it might be appropriate to fit a form of three-factor interaction that has a special structure, possibly $a_i b_j + a_i c_k + b_j c_k$, but we do not pursue this specialized topic here.)

Example: variation in measurement of solid tumors

To illustrate three crossed factors, we use data from an experiment in evaluating tumor measurement. Lavin and Flowerdew (1980) asked oncologists to measure simulated solid tumors. Tumor form and texture were included as study factors.

For the study they used three forms (small, oblong, and large) and two materials (cork and rubber—a possible source of different textures). For each combination of form and material they constructed two simulated solid tumors—a total of 12. The large tumors were spherical with a diameter of 38.3 mm, the oblong tumors were ellipsoidal with a longest diameter of 38.3 mm and remaining two diameters of 31.5 mm, and the small tumors were spherical with a diameter of 31.5 mm.

The investigators arranged the 12 tumors at random on top of a folded blanket, covered them with a sheet of foam rubber half an inch (13 mm) thick, and asked oncologists attending the May 1975 meeting of the Eastern Cooperative Oncology Group to measure each simulated tumor just as they would measure a tumor in a patient.

Twenty-six oncologists volunteered to make the measurements. They were unaware of possible differences in texture in the simulated tumors, and they were not told about the replicates among the tumors. The physicians conducted their evaluations independently, using a ruler and calipers.

The investigators expected that their results would underestimate the actual errors of measurement because of the idealized setting. The experiment did not simulate such factors as ascites (an accumulation of fluid in the abdominal cavity), tumor location, and patient discomfort.

To keep this example manageable, we use only the data from the first 13 oncologists. Table 6-1 shows the actual cross-sectional area (in square millimeters) of each tumor form and gives the oncologists' estimates of the area. The entries in Table 6-1 are the average over the two replicates of each tumor form.

Table 6-1. Actual and Estimated Cross-Sectional Area (Average of Two Measurements of Area, in mm²) of Simulated Solid Tumors

	Cork Model			Rubber Model		
Oncologist / Actual Area (mm²):	Small 980.0	Oblong 1199.0	Large 1467.0	Small 980.0	Oblong 1199.0	Large 1467.0
1	598.0	1034.0	1005.0	685.5	892.5	1190.0
2	762.5	987.5	1225.0	762.5	1250.0	1412.5
3	841.5	1085.0	1056.5	650.5	1042.0	1295.0
4	1085.0	1170.0	1460.0	900.0	1067.0	1500.0
5	1400.0	1772.5	1625.0	1207.5	1528.5	1925.0
6	1500.0	1730.0	2368.0	1442.5	1537.5	2205.0
7	1040.0	1300.0	1386.0	992.0	1262.5	1640.0
8	1186.0	1291.0	1485.0	944.5	1232.0	1503.0
9	1062.5	1225.0	1225.0	825.0	1015.0	1522.0
10	1062.5	1325.0	1700.0	1352.5	1800.0	2262.5
11	942.0	1435.0	1512.5	975.0	1372.0	1446.0
12	902.0	1140.0	1440.0	1028.0	1140.0	1741.5
13	855.5	1085.0	1224.0	620.0	1201.0	1369.0

By established convention, oncologists use a streamlined definition of cross-sectional area: the product of the longest dimension and the shortest dimension of the tumor. For the large tumors, for example, this formula gives (38.3 mm)(38.3 mm) = 1467 mm². Geometrically, the spherical form has a circular cross section, whose area would be πr^2 or $\pi d^2/4$ by the usual formula. The conventional definition simply omits the $\pi/4$ from the second of these formulas and uses d^2. In monitoring a tumor over a period of time, however, an oncologist would customarily use the ratio of the present cross-sectional area to a previous value, and so any such multiplicative constant would cancel out. (This practice might lead us to analyze the logarithms of the areas, because the usual comparisons would then be differences.)

Sweeping Upward

For a three-way layout with one observation per cell, four steps of sweeping take us from the data to the eight overlays. We present the process in the specific context of the tumor data and then summarize it more generally.

Step 1 averages the data in each row of the original table over the three tumor forms within each tumor material, cork and rubber. That is, it calculates the mean for each combination of oncologist and material. It places these means in a thirteen-row, two-column bordering table to the left of the original table and subtracts each mean from each observation that contributed to that mean.

For Oncologist 1 and cork material, the mean is (598 + 1034 + 1005)/3 = 879, and the three remainders are −281, 155, and 126.

Step 2 averages the remainders in each row of the partially swept original table over the two materials, giving a bordering table of thirteen rows and three columns (oncologist by form). It also sweeps these means out of the partially swept table, leaving a new set of remainders.

For Oncologist 1 and the small form, Step 2 gives the average (−281 − 237)/2 = −259 and the remainders −22 and 22.

At the completion of Step 2 we have swept out of the original table the joint effects of oncologist by material and the joint effects of oncologist by form. Further operations will separate these sets of joint effects into the common value, main effects for oncologist, main effects for material, main effects for form, two-factor interaction effects for oncologist by material, and two-factor interaction effects for oncologist by form. Thus, among the terms in the decomposition of y_{ijk}, the original table now contains, in addition to residuals, only two-factor interaction effects for material by form.

Step 3 makes a final sweep across columns by computing the mean of each row of the first (oncologist by material) bordering table, and it places these (oncologist) means in a third bordering table to the left of the first two. The new bordering table has thirteen rows and one column.

The mean for Oncologist 1 is (879 + 922.7)/2 = 900.8, leaving the remainders −21.8 and 21.8 in the oncologist by material bordering table.

Step 4, the final sweep, operates on the partially swept original table and, in parallel, on the three bordering tables already computed. In each of these tables it forms the mean in each column and writes these means in bordering tables above their respective source tables. It also subtracts these means out of their source tables.

The calculations for a single entry go as follows, taking the tables from right to left. For cork material and the small form, the mean is (−21.9 + 75.0 + ⋯ + 122.0)/13 = −48.2, and the 13 remainders (now residuals) are −70.2, 26.8, …, 73.8. For cork material the mean (now main effect) is [−21.8 + (−75.0) + ⋯ +(−4.2)]/13 = −15.5, and the 13 remainders (now oncologist by material interactions) are −6.4, −59.5, …, 11.2. For the small form the mean (now main effect) is [−259.1 + (−304.2) + ⋯ +(−321.3)]/13 = −274.3, and the 13 remainders (now oncologist by form interactions) are 15.2, −29.8, …, −47.0. Finally, the common value is (900.8 + 1066.7 + ⋯ + 1059.1)/13 = 1259.8, and the 13 remainders (now oncologist main effects) are −359.0, −193.2, …, −200.7.

Table 6-2 gives the complete bordered table, showing the final cell entries in the original table and in each of the bordering tables. After the final sweep, the original part of the table contains the residuals. Each bordering table condenses the full overlay for the corresponding effects. Each full overlay contains 78 entries; a bordering table contains only the principal entries of the overlay.

Table 6-2. The Bordered Table Containing Effects for the Simulated Tumor Data

	Common	Form			Material		Material × Form					
	Value	S	O	L	C	R	CS	CO	CL	RS	RO	RL
	1259.8	−274.3	6.3	268.0	−15.5	15.5	48.2	24.7	−73.0	−48.2	−24.7	73.0
	Oncologist	Form × Oncologist			Material × Oncologist		Residuals					
1	−359.0	15.2	56.1	−71.3	−6.4	6.4	−70.2	67.9	2.3	70.2	−67.9	−2.3
	−193.2	−29.8	45.8	−15.9	−59.5	59.5	26.8	−81.0	54.2	−26.8	81.0	−54.2
3	−264.7	25.2	62.1	−87.3	14.7	−14.7	48.0	−2.5	−45.5	−48.0	2.5	45.5
	−62.8	69.8	−84.8	15.0	56.8	−56.8	2.9	−14.5	11.6	−2.9	14.5	−11.6
5	316.6	1.7	67.8	−69.4	38.2	−38.2	25.2	74.5	−99.8	−25.2	−74.5	99.8
	537.3	−51.6	−169.7	221.3	84.3	−84.3	−88.3	2.7	85.6	88.3	−2.7	−85.6
7	10.2	20.2	4.8	−25.1	−12.6	12.6	3.8	22.1	−26.0	−3.8	−22.1	26.0
	13.8	66.0	−18.4	−47.6	62.6	−62.6	25.4	−42.3	16.9	−25.4	42.3	−16.9
9	−114.1	72.3	−32.1	−40.2	40.6	−40.6	45.4	55.2	−100.6	−45.4	−55.2	100.6
	323.9	−101.9	−27.6	129.5	−205.8	205.8	28.0	−41.0	13.0	−28.0	41.0	−13.0
11	20.6	−47.6	116.8	−69.2	31.6	−31.6	−80.8	−9.3	90.1	80.8	9.3	−90.1
	−27.9	7.4	−98.2	90.8	−55.8	55.8	−40.0	46.5	−6.5	40.0	−46.5	6.5
13	−200.7	−47.0	77.6	−30.6	11.2	−11.2	73.8	−78.5	4.7	−73.8	78.5	−4.7

Table 6-3. Sweeping Operations and Information Flow in the Process of Upward Sweeping for an Unreplicated Three-Factor Layout[a]

Step 1:	O × M × F → O × M
Step 2:	(O × M × F − O × M) → O × F
Step 3:	O × M → O
Step 4:	(O × M × F − O × M − O × F) → M × F
	(O × M − O) → M
	O × F → F
	O → C

[a]The arrow represents the operation of averaging over the factor that appears only to its left, placing the results in the bordering table on its right, and subtracting them from the table on the left.

Note: O, oncologist; M, material; F, form; C, common value.

Summary of the Sweeping Process

We now summarize the flow of information from the original data array to the final effects. In a shorthand notation Table 6-3 lists the separate sweeping operations that take place at each step. Each arrow represents three calculations: (1) average over one subscript (i.e., factor), (2) record these means in a bordering table, and (3) subtract them from each of the entries on which they are based. Thus in Step 1, O × M × F represents the original table. If the subscript k in y_{ijk} refers to F, then Step 1 records $y_{ij\cdot}$ as the (i, j) entry in the O × M bordering table and leaves $y_{ijk} - y_{ij\cdot}$ as the (i, j, k) entry in the O × M × F table. By referring to the main table as O × M × F − O × M, Step 2 indicates that this partial sweeping has already taken place.

Any effect in the original data that involves only the subscripts i and j (i.e., only the effects involving O and M) moves in Step 1 into the O × M bordering table. Thus the entries in O × M × F − O × M are free of main effects for O, main effects for M, and two-factor interactions for O × M.

To illustrate the flow of information through the steps of the process, we track the calculation of the main effects for O, which we associate with the subscript i in y_{ijk}. Step 1 sweeps these contributions from O × M × F to O × M in the calculation of $y_{ij\cdot}$. Step 3 sweeps them into O when it forms $y_{i\cdot\cdot}$. Finally, Step 4 produces the O main effects when it averages over i and subtracts: $y_{i\cdot\cdot} - y_{\cdot\cdot\cdot}$. The principal entries for the other overlays develop similarly.

Interpreting the Effects

The overall mean for estimated cross-sectional area of the simulated solid tumors is 1259.8. Figure 6-1 displays side-by-side plots of the fitted main effects, two-factor interactions, and residuals. Instead of boxplots for the two-factor interactions and residuals, it uses a vertical line to show the range

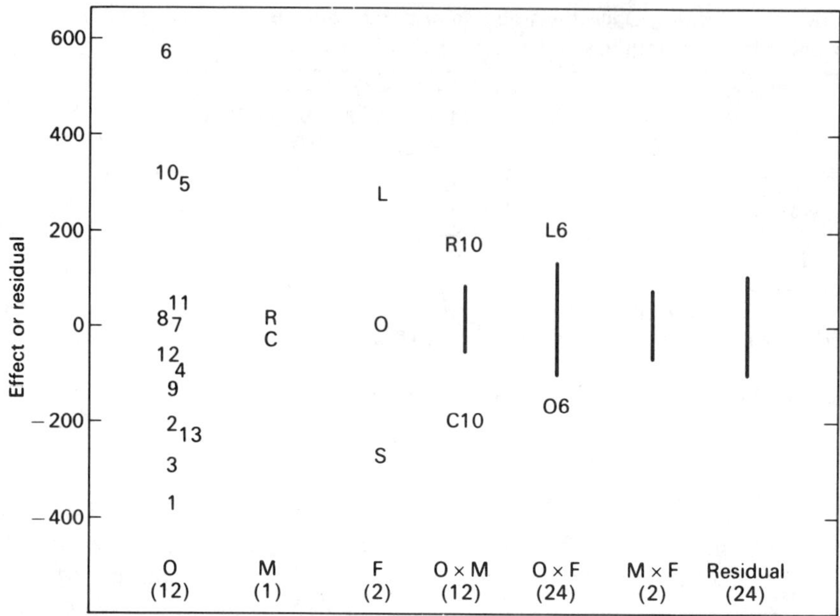

Figure 6-1. Side-by-side plot of effects and residuals for the simulated tumor data. Numbers in parentheses are degrees of freedom.

of the values that are not plotted individually. The two materials have very small effects relative to the form effects or the effects for oncologists. Thus texture seems unimportant in this experiment. Together with the common value, the effects for form show that the areas of the oblong and large tumors tended to be overestimated by about 60 mm^2:

	Small	**Oblong**	**Large**
Actual	980	1199	1467
Estimated ($m + c_k$)	985.5	1266.1	1527.8

This difference corresponds to only 1 mm in each dimension (about $\frac{1}{13}$ of the thickness of the foam rubber), so we are not surprised. Oncologists 6, 10, and 5 tended to overestimate, whereas 1, 3, 13, and 2 tended to underestimate.

Among the two-factor interactions, those for form by oncologist produce the largest effects (in absolute value). The largest values, positive and negative, both come from Oncologist 6, who also has the largest main effect (the greatest overall overestimation). A look back at Table 6-1 confirms that this oncologist generally overestimated by about 500 mm^2. For the large tumors, however, the estimates were high by about 700–900 mm^2. In contrast, the estimate for the oblong rubber tumor was high by only about 340 mm^2, helping to account for the substantial interaction effect.

Similarly, oncologist by material also produces some large effects (in absolute value). Oncologist 10, who overestimated by an average of about 320 mm^2, produced an interaction of 205.8 for rubber tumors and -205.8 for cork tumors. This oncologist's general pattern of overestimation (as summarized in the main effect) involves substantially greater overestimates for the rubber tumors than for the cork tumors. Because the oncologist by material interaction effects must sum to zero over the two materials (for each oncologist), this behavior in the data shows up as a sizable positive interaction effect for rubber and an equally sizable negative interaction effect for cork.

Mean Squares

From the bordered table, we may compute the sums of squares. To compute the residual sum of squares, we square each entry in the residual table and sum these squares. In each of the bordering tables, each squared entry appears in the sum with a repetition factor. This factor is the ratio of the number of entries in the full overlay to the number in the bordering table. For example, the O × M bordering table contains $13 \times 2 = 26$ entries. Thus we multiply the sum of its squared entries by 3 ($= 78/26$) to obtain the sum of squares for the full O × M overlay, 407,342.

Table 6-4 shows the mean squares for the solid tumor data. The mean square for form stands out (in part reflecting its small number of degrees of freedom), followed in decreasing order by those for oncologist and material by form. From Figure 6-1 we might not have expected this result. The repetition factor, however, links the bordering table to the corresponding overlay, and the full overlays provide the basis for partitioning the variation in the squared scale (analysis of variance). Section 8C introduces a modified type of side-by-side plot that is better attuned to the mean squares.

Three-Way with Replication

As mentioned earlier, each entry in Table 6-1 is the average of two replicates. Because Lavin and Flowerdew gave the details, their data allow us to

Table 6-4. Table of Mean Squares for Simulated Tumor Data

Source	SS	d.f.	MS
Oncologist	4,733,531	12	394,461
Material	18,693	1	18,693
Form	3,825,101	2	1,912,551
O × M	407,342	12	33,945
O × F	435,237	24	18,135
M × F	214,815	2	107,407
Error (O × M × F)	215,418	24	8,976

Table 6-5. The Original Simulated Tumor Data with Replicate Observations: Actual and Estimated Cross-Sectional Areas (in mm²) of Simulated Solid Tumors, Estimated by 13 Oncologists

	Cork Model						Rubber Model					
Actual Area	Small		Oblong		Large		Small		Oblong		Large	
Oncologist (mm²):	980	980	1199	1199	1467	1467	980	980	1199	1199	1467	1467
1	667	529	1258	810	1050	960	696	675	875	910	1225	1155
2	625	900	750	1225	1225	1225	625	900	900	1600	1225	1600
3	783	900	986	1184	1089	1024	676	625	884	1200	1330	1260
4	1120	1050	1050	1290	1400	1520	960	840	950	1184	1400	1600
5	1575	1225	1520	2025	1225	2025	1365	1050	1482	1575	2250	1600
6	1600	1400	1710	1750	2236	2500	1485	1400	1500	1575	2100	2310
7	1056	1024	1280	1320	1404	1368	1023	961	1365	1160	1640	1640
8	1188	1184	1287	1295	1332	1638	961	928	1344	1120	1600	1406
9	1225	900	1225	1225	1225	1225	900	750	1050	980	1600	1444
10	1225	900	1600	1050	1600	1800	1225	1480	1600	2000	2025	2500
11	960	924	1470	1400	1800	1225	900	1050	1344	1400	1692	1200
12	1020	784	1160	1120	1480	1400	900	1156	1160	1120	1849	1634
13	928	783	1050	1120	1224	1224	840	400	1110	1292	1369	1369

Source: Philip T. Lavin and Gordon Flowerdew (1980). "Studies in variation associated with the measurement of solid tumors," *Cancer*, 46, 1286–1290 (part of Table 2, p. 1288). Reproduced by permission.

illustrate the analysis of a three-way layout with replication. Table 6-5 shows the detailed data, and Table 6-6 gives the corresponding overlay of residuals and all the bordering tables. In each cell the two residuals are simply the differences between the data values and the cell mean, tabulated in Table 6-1. The top half of Table 6-6 differs from Table 6-2 in only one respect: the bordering table at the lower right is now labeled "Material × Form × Oncologist," rather than "Residual," to indicate that it contains the three-factor interaction effects.

Compared to the residuals (which tell us about the oncologists' variation in measuring two identical tumors under the conditions of the experiment), the three-factor interactions seem quite modest. Figure 6-2 supports this impression by including, at the right, a summary of the range of the residuals. Only the main effects for the oncologists have a wider range than the residuals. Thus the measurements do not seem to be highly reproducible, either by the individual oncologist or among oncologists.

For the analysis of variance, Table 6-7 gives the sums of squares and the mean squares. The SSs and MSs in this table are twice the corresponding entries in Table 6-4, because introducing the replicates has produced overlays (if we wished to show them in full) that have exactly twice as many entries as in the unreplicated analysis. As a result, each repetition factor doubles as

Table 6-6. The Bordered Table Containing Effects for the Simulated Tumor Data with Replicate Observations

	Common Value	Form			Material		Material × Form					
		S	O	L	C	R	CS	CO	CL	RS	RO	RL
	1259.8	−274.3	6.3	268.0	−15.5	15.5	48.2	24.7	−73.0	−48.2	−24.7	73.0
	Oncologist	Form × Oncologist			Material × Oncologist		Material × Form × Oncologist					
1	−359.0	15.2	56.1	−71.3	−6.4	6.4	−70.2	67.9	2.3	70.2	−67.9	−2.3
	−193.2	−29.8	45.8	−15.9	−59.5	59.5	26.8	−81.0	54.2	−26.8	81.0	−54.2
3	−264.7	25.2	62.1	−87.3	14.7	−14.7	48.0	−2.5	−45.5	−48.0	2.5	45.5
	−62.8	69.8	−84.8	15.0	56.8	−56.8	2.9	−14.5	11.6	−2.9	14.5	−11.6
5	316.6	1.7	67.8	−69.4	38.2	−38.2	25.2	74.5	−99.8	−25.2	−74.5	99.8
	537.3	−51.6	−169.7	221.3	84.3	−84.3	−88.3	2.7	85.6	88.3	−2.7	−85.6
7	10.2	20.2	4.8	−25.1	−12.6	12.6	3.8	22.1	−26.0	−3.8	−22.1	26.0
	13.8	66.0	−18.4	−47.6	62.6	−62.6	25.4	−42.3	16.9	−25.4	42.3	−16.9
9	−114.1	72.3	−32.1	−40.2	40.6	−40.6	45.4	55.2	−100.6	−45.4	−55.2	100.6
	323.9	−101.9	−27.6	129.5	−205.8	205.8	28.0	−41.0	13.0	−28.0	41.0	−13.0
11	20.6	−47.6	116.8	−69.2	31.6	−31.6	−80.8	−9.3	90.1	80.8	9.3	−90.1
	−27.9	7.4	−98.2	90.8	−55.8	55.8	−40.0	46.5	−6.5	40.0	−46.5	6.5
13	−200.7	−47.0	77.6	−30.6	11.2	−11.2	73.8	−78.5	4.7	−73.8	78.5	−4.7

Table 6-6. (*Continued*)

	Residuals (Repetition within Material × Form × Oncologist)											
Oncologist	CS		CO		CL		RS		RO		RL	
1	69.0	−69.0	224.0	−224.0	45.0	−45.0	10.5	−10.5	−17.5	17.5	35.0	−35.0
	−137.5	137.5	−237.5	237.5	0.0	0.0	−137.5	137.5	−350.0	350.0	−187.5	187.5
3	−58.5	58.5	−99.0	99.0	32.5	−32.5	25.5	−25.5	−158.0	158.0	35.0	−35.0
	35.0	−35.0	−120.0	120.0	−60.0	60.0	60.0	−60.0	−117.0	117.0	−100.0	100.0
5	175.0	−175.0	−252.5	252.5	−400.0	400.0	157.5	−157.5	−46.5	46.5	325.0	−325.0
	100.0	−100.0	−20.0	20.0	−132.0	132.0	42.5	−42.5	−37.5	37.5	−105.0	105.0
7	16.0	−16.0	−20.0	20.0	18.0	−18.0	31.0	−31.0	102.5	−102.5	0.0	0.0
	2.0	−2.0	−4.0	4.0	−153.0	153.0	16.5	−16.5	112.0	−112.0	97.0	−97.0
9	162.5	−162.5	0.0	0.0	0.0	0.0	75.0	−75.0	35.0	−35.0	78.0	−78.0
	162.5	−162.5	275.0	−275.0	−100.0	100.0	−127.5	127.5	−200.0	200.0	−237.5	237.5
11	18.0	−18.0	35.0	−35.0	287.5	−287.5	−75.0	75.0	−28.0	28.0	246.0	−246.0
	118.0	−118.0	20.0	−20.0	40.0	−40.0	−128.0	128.0	20.0	−20.0	107.5	−107.5
13	72.5	−72.5	−35.0	35.0	0.0	0.0	220.0	−220.0	−91.0	91.0	0.0	0.0

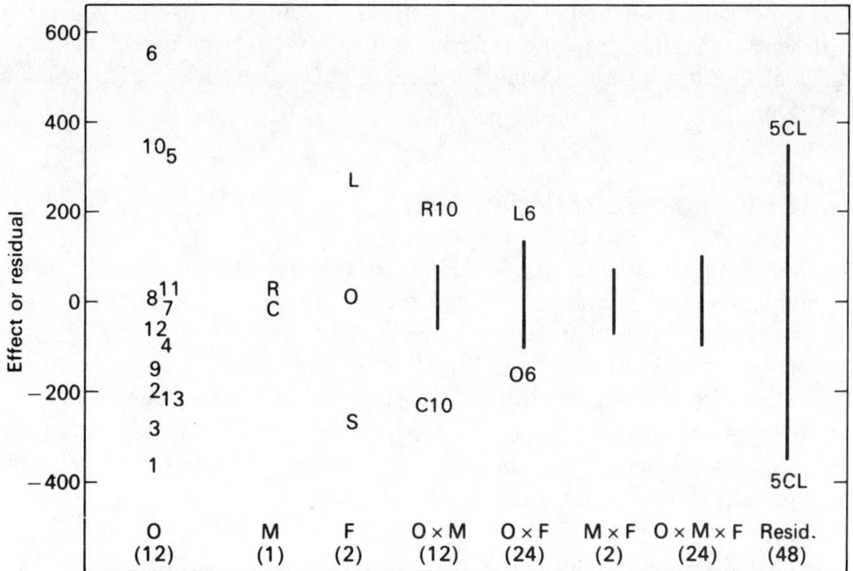

Figure 6-2. Side-by-side plot of effects and residuals for the simulated tumor data with replicate observations. Numbers in parentheses are degrees of freedom.

well. Our comparisons among SSs and MSs are relative, so this doubling does not change any of our impressions.

By having a new residual mean square, however, based solely on the replications, we are led to regard several of the other mean squares (M, O × M, O × F, and, of course, O × M × F) as relatively small. In Table 6-4, O × M did not seem so small, relative to that table's residual mean

Table 6-7. Table of Mean Squares for Simulated Tumor Data with Replicate Observations

Source	SS	d.f.	MS[a]
Oncologist	9,467,062	12	788,922
Material	37,386	1	37,386
Form	7,650,202	2	3,825,101
O × M	814,684	12	67,890
O × F	870,474	24	36,270
M × F	429,630	2	214,815
O × M × F	430,836	24	17,951
Error	2,815,680	78	36,098

[a]Each mean square other than Error is exactly twice the corresponding mean square in Table 6-4. The change arises because the individual observation is now a single replicate measurement instead of a *mean* of *two* replicate measurements.

square (which we now label O × M × F). Thus the experimenters' decision to design replication into the experiment can give us a modified picture of the contributions of the various sources of variation, relative to residual variation.

6B. FOUR CROSSED FACTORS

The most complex of the completely crossed models that we present in this chapter is the four-way model. A full four-factor model without replication has 16 overlays: a common value, one main effect for each of the four factors, six two-factor interactions, four three-factor interactions, and residuals. In general, n completely crossed factors produce 2^n overlays, and we can count the number of overlays for each type of interaction by computing the combinations $\binom{n}{2}$ for the number of two-factor interactions, $\binom{n}{3}$ for the number of three-factor interactions, and so on.

We plan to decompose a four-way unreplicated layout according to

$$\begin{aligned} y_{ijkl} = m & \\ & + a_i + b_j + c_k + d_l \\ & + (ab)_{ij} + (ac)_{ik} + (ad)_{il} + (bc)_{jk} + (bd)_{jl} + (cd)_{kl} \\ & + (abc)_{ijk} + (abd)_{ijl} + (acd)_{ikl} + (bcd)_{jkl} \\ & + e_{ijkl}. \end{aligned} \tag{1}$$

If each cell contains n replicate observations, then a full decomposition contains two terms—a term for the four-factor interaction $(abcd)_{ijkl}$ and a term for the residuals e_{ijklr}, representing within-cell variation—in place of e_{ijkl}. An additional subscript r on y then denotes the replication: y_{ijklr}. In unreplicated experiments we cannot make the corresponding separation.

Example: hardness of dental golds

To illustrate the four-factor model, we use data from an experiment in dentistry that investigated possible factors affecting the hardness of gold fillings. Xhonga (1970, 1971) described the original experiment, and Brown (1975) presented the data and used them to illustrate a statistical technique. This example uses the data that deal with the feasibility of using gold alloys in place of pure gold. (The full data table includes two other types of gold. See Exercise 13.)

Pure gold, one of the first restorative materials used in dentistry, serves well because it welds at room temperatures, does not corrode, and has high elasticity and ductility. It works well in cavities close to the nerve or in small cavities near the gum line, because it does not irritate these tissues. One problem with using pure gold is the softness of the metal, but this can be substantially overcome by using well-condensed gold foil. Addition of 0.1–5.0% by weight of other metals will further increase its strength. Another disadvantage of gold comes from the difficulty of the technique. Gold work, today generally done in dental schools, requires much patience, careful observation, and good hand–eye coordination. A single gold filling can take several hours to complete.

In general, the finer the crystal structure of the gold material, the harder it is. Thus powdered gold alloy is placed in a gold foil envelope to make it convenient to use as a filling. The packet is heated to temperatures below the melting point. This process, called sintering, causes the individual particles to stick together. Sintering strengthens the mass and allows it to be handled more easily. This experiment used two powdered gold alloys, called 97-1-1-1 and AuCa, sintered at three temperatures: 1500, 1600, and 1700°F.

In preparing the material for use in a filling, the dentist further processes the material to increase its hardness. This process, known as condensation, may be carried out by hand using a special instrument called a condenser to press the gold into the cavity. The dentist needs special knowledge and skill to apply the proper amount of force in the right direction to achieve the necessary compaction. An alternate method of condensation known as malleting may use either a hand mallet or a mechanical mallet. In hand malleting the dentist places the condenser in the proper position, and the dental assistant hits the condenser with a hammer weighing about 50 grams. Mechanical mallets accommodate various shapes of condenser points and contain a small mallet in the handle of the instrument that may be activated when the dentist has the condenser point in the proper position. Malleting allows the dentist to focus on positioning the condenser while the dental assistant or the mechanical mallet provides uniform taps on the instrument when required.

Condensation causes dislocations in the crystal structure of the mass, and these dislocations obstruct slippage between the crystals, thereby increasing the hardness of the metal. Xhonga's study used three methods of condensation: hand malleting, electric malleting, and hand condensation. At the time of this experiment (1970), hand condensation was thought to be an advantage for these gold materials. It takes less time, and more complicated fillings can be done more easily. Hammering, either with a hand mallet or with an electromallet, may cause damage to the tooth pulp if the hammering takes too long or uses too much force.

The investigator asked five dentists associated with the Operative Division of the School of Dentistry at the University of California, Los Angeles, to prepare the six types of gold filling material (the two alloys 97-1-1-1 and AuCa, sintered at the three temperatures) using each of the three methods of condensation. Thus the four factors are dentist (D), method (M), alloy (A), and temperature (T). A single experiment consisted of one dentist preparing

Table 6-8. Diamond Pyramid Hardness Number (Sum of 10 Measurements) of Dental Fillings Made from Two Gold Alloys Sintered at Three Temperatures then Prepared by Five Dentists Using Three Methods of Condensation

		Alloy 97-1-1-1			Alloy AuCa		
Dentist	Method	1500°F	1600°F	1700°F	1500°F	1600°F	1700°F
1	1	813	792	792	907	792	835
	2	782	698	665	1115	835	870
	3	752	620	835	847	560	585
2	1	715	803	813	858	907	882
	2	772	782	743	933	792	824
	3	835	715	673	698	734	681
3	1	743	627	752	858	762	724
	2	813	743	613	824	847	782
	3	743	681	743	715	824	681
4	1	792	743	762	894	792	649
	2	690	882	772	813	870	858
	3	493	707	289	715	813	312
5	1	707	698	715	772	1048	870
	2	803	665	752	824	933	835
	3	421	483	405	536	405	312

Sources: Morton B. Brown (1975). "Exploring interaction effects in the ANOVA," *Applied Statistics*, 24, 288–298 (data from Table 1, p. 290). Frida A. Xhonga (1971). "Direct gold alloys—part II," *Journal of the American Academy of Gold Foil Operators*, 14, 5–15 (description of gold types in Table 1, p. 6). Data reproduced by permission of the Royal Statistical Society.

one alloy, sintered at one of the temperatures, using one of the methods of condensation to fill a 2 × 2 mm hole in an ivory block.

Hardness is the resistance of one material to indentation, scratching, cutting, abrasion, or wear produced by contact with another material. The Diamond Pyramid Hardness test presses a standard pyramid-shaped diamond against a test material and measures the area of the indentation. The Diamond Pyramid Hardness Number (also known as the Vickers Hardness Number) is a constant multiple of the ratio of the force applied in the test to the area of the indentation. In this experiment, the investigator measured the surface hardness of the test specimen and recorded the Diamond Pyramid Hardness Number (DPHN). The data shown in Table 6-8 represent the sum of 10 DPHN readings from each sample, so our response is in units of 10 × DPHN.

Because the 10 replicate hardness measures are unavailable, we analyze these sums as though they were unreplicated observations. We might divide the sums by 10 and present the mean DPHN values, but for ease of reading we avoid the decimal point.

Sweeping Upward

For a four-way layout, six steps of sweeping take us from the data to the 16 overlays. To make them concrete, we focus on the dental-gold example. Figure 6-3 depicts the bordered table obtained by sweeping upward. Each numbered arrow indicates a step in the sweeping process. An arrow originates in the table whose entries we average to obtain the means we sweep out. Each arrow terminates in the table where we write these means. Arrow 1 indicates the first sweep, arrow 2, the second sweep, and so on.

The large box in the lower right of Figure 6-3 represents an array of 90 numbers arranged in 15 rows (combining dentist and method) and 6 columns (combining alloy and temperature), as in Table 6-8. Before we begin sweeping, this box contains the original data. After we complete the sweeping, the box contains the residuals. We label each box in Figure 6-3 by the array of effects it represents, and we label the 1 × 1 box for the common effect "C."

Both the large box representing the original data table and the bordering boxes omit lines that separate individual entries. The bordering boxes,

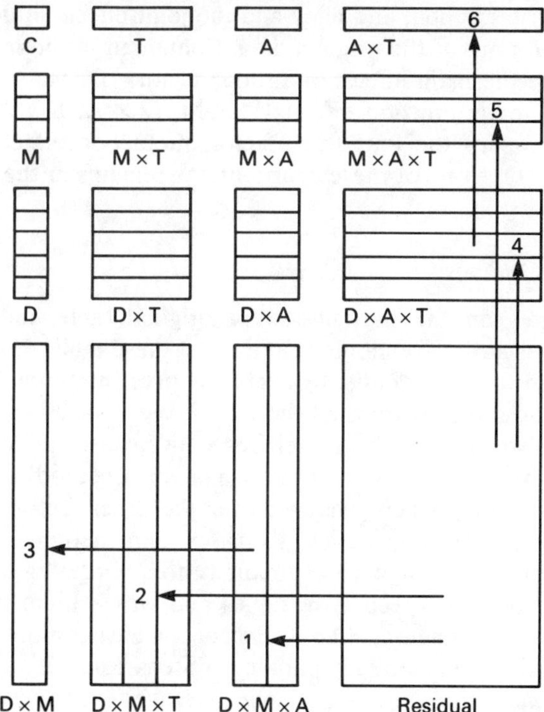

Figure 6-3. The table of residuals and the bordering tables that result from sweeping the four-way layout of dental-gold data. Each numbered arrow indicates a step in the sweeping process. An arrow originates in the table whose entries are averaged to obtain the means that are then swept out and written in the table where the arrow terminates.

however, do show those lines if their structure differs in one dimension from that of the data table or from an adjacent box that has more structure. Specifically, the bordering boxes that result from sweeps across columns show the number of columns the sweep produces. Similarly, boxes resulting from sweeps across rows show the number of rows we obtain.

Sweeps Along Rows

Step 1 averages the data in the original table, on each row, over the three sintering temperatures within each alloy. It yields the first bordering table, which contains two columns (one for each alloy) and 15 rows (as in the original data table). The process of sweeping upward transfers these means from the data table into the bordering table to the left of the original table.

Step 2 returns to the now partially swept original table and sweeps out means, on each row, over the alloys within each level of temperature. This box appears to the left of the first bordering box.

The final sweep across columns, at Step 3, computes means over the two alloys in the first bordering box. This single column appears to the left of the second box. At the completion of these three steps, we have swept the joint effects of dentist, method, and alloy and the joint effects of dentist, method, and temperature out of the original data. Contained in those two three-variable effects are the main effects of all four factors (D, M, A, and T), five of the six two-factor interaction effects (D × M, D × A, D × T, M × A, and M × T), and two of the four three-factor interaction effects (D × M × A and D × M × T). None of these contributions remains in the partially swept original table.

Sweeps Along Columns

Step 4 operates on the partially swept original table and on the three bordering tables already computed. In each of these tables it forms the mean in each column for each dentist by averaging over the three methods, and it then sweeps these means from all the tables. We obtain four new bordering tables, and we write these above their respective source tables.

Step 4 has averaged over one of the two factors (method) whose composite labels the rows. Step 5 now averages over the other factor (dentist). More specifically, Step 5 averages over dentists within methods in the partially swept original table and in each of the first three bordering tables. It sweeps these means from their respective tables and writes them in boxes placed above the tables obtained at Step 4. Except for having more pieces to work on, Steps 4 and 5 are essentially parallel to Steps 1 and 2.

The final sweep across rows, at Step 6, averages over dentists in the set of four bordering tables obtained at Step 4 and writes these means in boxes above the tables produced by Step 5.

Table 6-9 gives the complete bordered table, showing the final cell entries in the primary table and in each of the bordering tables. The primary table,

Table 6-9. The Bordered Table for the Dental-Gold Data, Showing Common Value, Main Effects, Two-Factor Interaction Effects, Three-Factor Interaction Effects, and Residuals

	Common Value	Temperature			Alloy		A × T					
		5	6	7	9	A	95	96	97	A5	A6	A7
	742	31	10	−41	−34	34	−14	−8	22	14	8	−22
	Method	M × T			M × A		M × A × T					
M1	52	−19	−7	26	−8	8	4	−13	8	−4	13	−8
M2	63	2	−10	8	−25	25	8	17	−25	−8	−17	25
M3	−115	17	17	−35	34	−34	−12	−4	16	12	4	−16
	Dentist	D × T			D × A		D × A × T					
D1	41	55	−77	22	1	−1	−40	29	12	40	−29	−12
	45	−16	−8	24	9	−9	11	11	−23	−11	−11	23
D3	7	3	−11	8	3	−3	28	−24	−4	−28	24	4
	−28	−12	78	−66	2	−2	−28	17	11	28	−17	−11
D5	−65	−31	18	12	−15	15	29	−33	3	−29	33	−3
	D × M	D × M × T			D × M × A		Residual					
	−13	−29	45	−15	19	−19	25	15	−41	−25	−15	41
	−18	33	16	−49	−54	54	−8	7	1	8	−7	−1
	31	−4	−60	64	35	−35	−17	−22	39	17	22	−39
	−9	−39	31	9	−19	19	−21	10	12	21	−10	−11
	−42	28	−13	−15	8	−8	−44	17	27	44	−17	−27
	51	11	−17	6	10	−10	65	−27	−38	−65	27	38
	−56	41	−41	0	3	−3	−40	15	25	40	−15	−25
	−41	13	36	−48	9	−9	19	11	−30	−19	−11	30
	97	−54	6	48	−11	11	21	−26	6	−21	26	−6
	6	71	−85	14	35	−35	−7	−14	21	7	14	−21
	38	−83	−16	99	25	−25	5	13	−19	−5	−13	19
	−44	13	101	−113	−59	59	1	1	−3	−2	−1	3
	73	−43	50	−7	−37	37	42	−26	−16	−42	26	16
	63	10	−22	12	12	−12	28	−48	20	−28	48	−20
	−135	34	−29	−5	25	−25	−70	74	−4	70	−74	4

after the final sweep, contains the residuals. Each bordering table represents a shorthand way of presenting the full overlay for the effects shown in the table. Each full overlay contains 90 entries; a bordering table contains only the principal entries in the overlay.

When we look at the results, we see relatively few entries, even in the residual table, that are between −10 and +10. Thus there can be no appreciable information in the fractional parts of any of these entries, and we have rounded them to the nearest integer.

Summary of the Sweeping Process

Because the six steps of sweeping for the four-way layout have produced a total of 16 overlays, we now briefly summarize the flow of information from the original data array to the final effects. In a shorthand notation Table 6-10 lists the separate sweeping operations that take place at each step. Each arrow represents three calculations: (1) average over one subscript (i.e., factor), (2) record these means in a bordering table, and (3) subtract them from each of the entries on which they are based. Thus in Step 1 D × M × A × T represents the original data. If the subscript l in y_{ijkl} refers to T, then Step 1 records $y_{ijk\cdot}$ as the (i, j, k) entry in the D × M × A bordering table and leaves $y_{ijkl} - y_{ijk\cdot}$ as the (i, j, k, l) entry in the D × M × A × T table. By referring to the main table as D × M × A × T − D × M × A, Step 2 indicates that this partial sweeping has taken place.

Any contribution in the original data that involves only the subscripts i, j, and k (i.e., only the versions of D, M, and A) moves in Step 1 into the D × M × A bordering table. Thus the entries in D × M × A × T − D ×

Table 6-10. Sweeping Operations and Information Flow in the Process of Upward Sweeping for an Unreplicated Four-Factor Layout[a]

Step 1:	D × M × A × T → D × M × A
Step 2:	(D × M × A × T − D × M × A) → D × M × T
Step 3:	D × M × A → D × M
Step 4:	(D × M × A × T − D × M × A − D × M × T) → D × A × T (D × M × A − D × M) → D × A D × M × T → D × T D × M → D
Step 5:	(D × M × A × T − D × M × A − D × M × T − D × A × T) → M × A × T (D × M × A − D × M − D × A) → M × A (D × M × T − D × T) → M × T (D × M − D) → M
Step 6:	D × A × T → A × T D × A → A D × T → T D → C

[a]The arrow represents the operation of averaging over the factor that appears only to its left, placing the results in the bordering table indicated on its right, and subtracting them from the table indicated on the left.

M × A are free of main effects for D, main effects for M, main effects for A, two-factor interactions for D × M, two-factor interactions for D × A, two-factor interactions for M × A, and three-factor interactions for D × M × A.

To illustrate the flow of information through the steps of the process, we track the calculation of the main effects for A, which we associate with the subscript k in y_{ijkl}. Step 1 sweeps these contributions from D × M × A × T to D × M × A in the calculation of $y_{ijk\cdot}$. Then Step 3 leaves them in D × M × A − D × M when it forms $y_{ijk\cdot} - y_{ij\cdot\cdot}$. Step 4 sweeps them into D × A by calculating $y_{i\cdot k\cdot} - y_{i\cdot\cdot\cdot}$. Finally, Step 6 produces the A main effects when it averages over i (i.e., D): $y_{\cdot\cdot k\cdot} - y_{\cdot\cdot\cdot\cdot}$. The principal entries for the other overlays develop similarly. For example, the D × M two-factor interaction goes through Steps 1, 3, 4, and 5 to emerge as

$$y_{ij\cdot\cdot} - y_{i\cdot\cdot\cdot} - y_{\cdot j\cdot\cdot} + y_{\cdot\cdot\cdot\cdot}$$

from the subtraction phase of (D × M − D) → M.

Interpreting the Effects

The overall mean for 10 × Diamond Pyramid Hardness Number is 742. Figure 6-4 displays dotplots of the main effects and two-factor interactions, side by side with boxplots of the three-factor interactions and residuals. The gold alloy 97-1-1-1 has an average 10 × DPHN of 708, 34 less than the

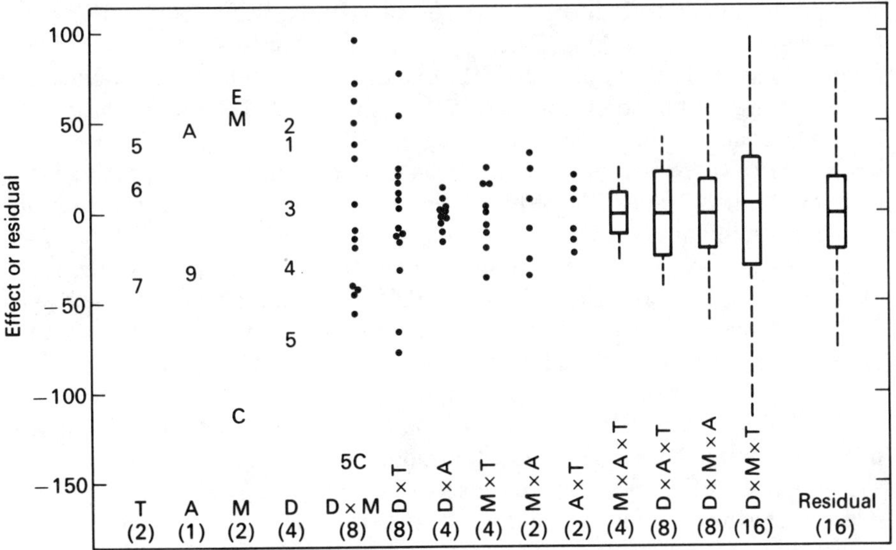

Figure 6-4. Side-by-side plot of effects and residuals for the dental-gold example.

overall average; thus the AuCa average has to be 776, 34 greater than the overall average, because we have constrained each set of main effects to sum to zero. Hand malleting gives a 10 × DPHN above the overall mean by 52, and electric malleting produces a 10 × DPHN 63 above the mean, but hand condensation leaves hardness 115 below the mean. Overall, malleting appears to give greater hardness than hand condensation.

The dentists also have a clear effect on results. Dentists 1, 2, and 3 have positive effects. They produce harder fillings than Dentists 4 and 5, who each have negative effects (i.e., below the mean for the five dentists) on hardness. Dentist 5 appears to produce fillings of particularly low hardness. This dentist's fillings average 65 less than the overall mean.

The temperature effects decrease with increasing temperature. The lowest temperature, 1500°F, makes 10 × DPHN 31 above the common mean; the 1600°F temperature has a small positive effect of 10; and the highest temperature, 1700°F, balances these with a negative effect of 41. Although this relationship between effect size and temperature seems clear, the magnitudes are smaller than the effects for dentists and methods.

Among the six two-factor interactions, that for dentist by method produces the largest effects (in absolute value). Method 3, hand condensation, combines with Dentist 5 to produce an interaction of −135. We interpret this to mean that Dentist 5 using the hand condensation technique produces fillings that appear to have a 10 × DPHN 135 lower than would have been expected from the overall mean of 742, the Dentist 5 effect of −65, and the hand condensation effect of −115.

The dentist by temperature interaction also contains some large effects. Dentists 1 and 4 have the largest interactions with temperature, and these effects work in opposite directions for the two dentists. The 1500 and 1700°F temperatures appear to help Dentist 1 and harm Dentist 4, and the reverse occurs at the 1600°F temperature. Hardness may not behave monotonically over the range of temperatures used in this experiment. The intent was to get hard material, so the investigator may have tried to choose a range of temperature where hardness would reach a maximum. (The inverted curvature for Dentists 1 and 4, however, deserves comparison with an estimate of their measurement error.)

The three-way interaction effects larger than 100 involve Method 2, Dentist 5, and Temperatures 1600 and 1700°F. Thus, after removing the main effect and the two-factor effects, Method 2 shows some large three-factor interactions with dentist and temperature. (Dentist 5's results are more irregular than the others.)

Figure 6-4 has shown us the sizes of the fitted effects. For some purposes, this is the preferred plot. For other purposes we want to look at how much each effect contributes to the corresponding sum of squares. Because the sizes of the fitted effects in different bordering tables do not necessarily reflect the effects' comparative contributions to the respective sums of squares, we offer a different plot in Section 8C.

Table 6-11. Table of Mean Squares for Hardness of Dental Gold

Source	SS	d.f.	MS
Dentist	157,795	4	39,449
Method	593,427	2	296,714
Alloy	105,816	1	105,816
Temperature	82,178	2	41,089
D × M	306,472	8	38,309
D × A	5,687	4	1,422
D × T	134,424	8	16,803
M × A	54,685	2	27,343
M × T	30,652	4	7,663
A × T	21,725	2	10,863
D × M × A	76,929	8	9,616
D × M × T	181,518	16	11,345
D × A × T	47,443	8	5,930
M × A × T	16,571	4	4,143
Error	75,772	16	4,736
Total	1,891,096	89	

Mean Squares

With the bordered table in hand we now compute the sums of squares. To compute the residual sum of squares, we square each entry and sum these squares. However, in each of the bordering tables, each squared entry appears in the sum with a repetition factor. This factor is the ratio of the number of entries in the full overlay to the number of entries in the bordering table.

For example, the D × M bordering table contains 15 entries. Thus we multiply its sum of squared entries by 6 (= 90/15) to obtain the sum of squares for the full D × M overlay.

Table 6-11 shows the mean squares for the dental-gold data.

6C. LATIN SQUARE DESIGNS

As the number of factors that we wish to consider as possible sources of variation in an experimental outcome increases, the cost of carrying out a completely crossed factorial design increases rapidly. For a three-way completely crossed factorial design with p levels for each of the three factors, we need p^3 experimental units. And n factors will call for at least 2^n experimental units, requiring this minimum number only when each factor has exactly two levels.

Constraints on cost, space, and time led researchers to develop several types of experimental design that we call "incomplete n-way layouts." Researchers needed ways to measure the main effects of many factors without using as many experimental units as a complete n-way design would require. Fractional factorials, Latin squares, Graeco-Latin squares, lattice designs, Youden squares, and split-plot designs name a few of the incomplete designs available.

Often the amount of experimental material available is not large enough to conduct fully crossed designs, or the fully crossed design may be infeasible for some other reason. For example, a study tested the comfort and convenience of automatic safety belt systems in automobiles equipped with motorized or mechanical passive safety belt systems. Suppose we have 20 such automobiles. We hire 20 testers and ask each tester to evaluate each restraint system. A completely crossed two-factor design (20×20) could handle this question, but suppose, in addition, that fatigue may affect a tester's evaluation of a restraint system. We could consider a fully crossed three-factor experiment in which the third factor represents the order of evaluating the 20 restraint systems. In this design each tester would evaluate each system 20 times. If we assume that a tester could complete 20 evaluations in one day, then the entire experiment would require 20 days to complete. For a particular tester, a particular restraint system would be the first one evaluated on one of the 20 days. It would be the second one evaluated on another day, and so on, so that on some day it would be the last one evaluated by this tester. It is unlikely that the testers would be willing to make 20 separate judgments of the same restraint system, and thus this design is not practical.

The Latin square design offers a solution to this design problem if we believe that only the main effects of restraint system, tester, and order are important. In this design the entire experiment can be completed in one day, and each tester evaluates each restraint system only once. Each restraint system is the first one evaluated by some tester. Each system is the second one evaluated by some other tester, and so on, so that each one is also evaluated last by some tester. This kind of balance allows us to measure the main effects of each factor, but because the design is not fully crossed, we cannot separate interaction effects from one another.

The Latin square design involves three main factors. All factors have the same number of levels, say p levels, and the investigator takes observations on p^2 of the p^3 possible treatment combinations. In this design every pair of factors is fully crossed. We display the design by drawing a p by p square, letting the p rows represent the p levels of one factor, and letting the p columns represent the p levels of another factor. The third factor is arranged so that each of its p levels appears exactly once in combination with each level of the first factor and exactly once in combination with each level of the second factor. Table 6-12 gives an illustration of a 5 by 5 Latin square design. The letters A, B, C, D, and E arranged in the square represent the five levels of the third factor. Each letter appears exactly once in each row and exactly once in each column.

Table 6-12. An Illustration of the Treatment Combinations in the 5 by 5 Latin Square[a]

	Factor 1				
Factor 2	1	2	3	4	5
1	A	B	C	D	E
2	B	C	D	E	A
3	C	D	E	A	B
4	D	E	A	B	C
5	E	A	B	C	D

[a]Each level of the third factor (denoted by the letters) appears exactly once in each row and exactly once in each column.

The decomposition for the Latin square design may be written

$$y_{ijk} = m + a_i + b_j + c_k + e_{ijk}$$

for $i = 1, 2, \ldots, p$, $j = 1, 2, \ldots, p$, and $k = 1, 2, \ldots, p$. In this model y_{ijk} represents the observed response at the ith level of factor 1, the jth level of factor 2, and the kth level of factor 3. The common value is given by m; a_i, b_j, and c_k represent the three main effects, and e_{ijk} represents the residual.

The primary advantage of the Latin square design is the savings in observations, a factor of $1/p$ observations over the complete three-way layout. That only the main effects of the three factors can be estimated and the model's heavy dependence on the assumption of additivity constitute the main disadvantages of the Latin square design.

Value splitting often simplifies the interpretation of incomplete n-way designs and provides a clearer picture of the structure of the data. We illustrate the method of sweeping out means to split the data values. We use the example of an experiment in making sour cream.

EXAMPLE: HOMEMADE SOUR CREAM

Dr. John W. McDonald, a statistician and amateur cook, likes to cook "from scratch" whenever possible. One of his from-scratch products is homemade sour cream, made in the same way yogurt is made except that sweet cream is substituted for milk. He adds a small amount of yogurt culture (the starter) to a batch of sweet cream, places the mixture in an electric yogurt maker, and allows it to incubate for 22 to 23 hours.

Dr. McDonald compared five brands of yogurt as starters in order to find the best starter. He used brands available in grocery stores or in a food cooperative in Seattle, Washington: Continental, Healthglow, Maya, Nancy's, and Yami. To test which yogurt produced the best end product,

Dr. McDonald used a Latin square design. He reasoned that the best starter would have the largest amount of bacterial growth. As a proxy for bacterial growth, he measured acidity, because yogurt bacteria produce lactic acid as a by-product of their growth. He assumed that the more lactobacilli present, the higher the acidity of the sour cream. In these data the response variable is based on acidity; we explain further below.

The sour cream is made in a thermostatically controlled yogurt maker, which has five recessed holes, arranged in a line, for the five covered containers of cream. Each container holds approximately 6 ounces of cream. After the five containers are placed in the five positions, a plastic cover is placed over the whole yogurt maker to prevent heat loss. The yogurt maker maintains a temperature between 80 and 100°F so that the bacteria grow rapidly. Dr. McDonald thought that the heating system in the yogurt maker might heat the five positions differently. The end positions might lose heat faster and thereby maintain a lower temperature than the inside positions. Thus he planned his experiment to try each starter at each position. This plan required five days to complete the experiment, and he used a fresh batch of sweet cream on each day.

Because bacterial growth is affected by the amount of milk solids and butterfat content of the sweet cream, the batches might affect the end product. To minimize any batch effects on the sour cream, Dr. McDonald purchased one brand of sweet cream at the same store for the entire experiment. The experiment required almost two pints of sweet cream for each batch, mixed thoroughly before using. He added two level teaspoons of starter to one of the yogurt containers, half filled the container with sweet cream, stirred the mixture, topped off the container with sweet cream, placed the cover on the container, and put it in the yogurt maker. He used a different brand of starter for each container; a single carton of each brand of yogurt provided the starter on all five days.

This Latin square design uses five brands of yogurt as starters in each of five positions in the yogurt maker, and it requires five days to carry out, producing one batch of sour cream each day. By using a Latin square rather than a fully crossed three-factor design, Dr. McDonald needed only 25 observations rather than the 125 observations required for the fully crossed design. A fully crossed experiment would have taken 25 days. Table 6-13 gives the raw data and shows the Latin square pattern used to assign yogurt starters to positions and batches. The labels indicating the Latin square pattern represent the different yogurt starters: C, Continental; H, Healthglow; M, Maya; N, Nancy's; and Y, Yami. The response variable is the amount of titrant (in milliliters) used to neutralize the acidity of 9 ml of sour cream: the more titrant, the greater the acidity.

Table 6-14 shows the bordered table that results from sweeping out means. The sweeping procedure can be completed in three steps. First, we average each row of the original data table and write these means in a bordering table to the left of the original table. We sweep these means out of

Table 6-13. The Latin Square Design and the Raw Data Used by John McDonald in His Experiment in Making Sour Cream[a]

	Position				
Batch	1	2	3	4	5
1	8.04 M	6.61 H	11.99 N	7.78 Y	8.40 C
2	9.58 N	6.58 Y	6.66 C	5.34 M	7.92 H
3	7.98 Y	7.98 M	8.98 H	7.94 C	11.32 N
4	9.74 H	9.46 C	9.14 M	12.00 N	9.32 Y
5	9.66 C	11.28 N	8.04 Y	8.12 H	6.72 M

[a]The letter beneath each data entry in the table represents the brand of yogurt used as a starter. The brands are C = Continental, H = Healthglow, M = Maya, N = Nancy's, and Y = Yami.

Source: John W. McDonald III, unpublished data. Used with permission.

Table 6-14. The Bordered Table Containing the Value Splitting for the Sour Cream Experiment

	Common	Starter				
	8.66	−1.22 M	2.57 N	−0.72 Y	−0.39 H	−0.24 C
		Position				
		0.34 1	−0.28 2	0.30 3	−0.43 4	0.07 5
	Batch	Residual				
1	−0.10	0.36 M	−1.28 H	0.56 N	0.37 Y	0.01 C
2	−1.45	−0.55 N	0.36 Y	−0.62 C	−0.23 M	1.02 H
3	0.18	−0.48 Y	0.64 M	0.23 H	−0.23 C	−0.16 N
4	1.27	−0.14 H	0.05 C	0.13 M	−0.07 N	0.04 Y
5	0.10	0.80 C	0.23 N	−0.30 Y	0.18 H	−0.89 M

the original data, leaving a remainder in each entry in the table. Second, we sweep column means out of the data table and out of the bordering table of row means. The average of the row means gives the grand mean, here called the common value. The remainders left in the row means are the batch effects, and the column averages we obtain on the second sweep are the position effects. The third sweep averages the five values for each starter and subtracts these means from their respective table entries. These means represent the starter effects. The remainders in the 5 by 5 table are the residuals.

Figure 6-5 gives the side-by-side plot of the effects and the residuals obtained in Table 6-14. This plot allows us to see at a glance the large effects. The starter effects show the widest spread, mainly a large difference between the effect of Nancy's yogurt starter and the other starters. Nancy's yogurt produced the highest acidity and presumably the highest bacterial growth. Batch 4 appears to have had somewhat more acidity overall than the other batches; maybe this batch was left to incubate slightly longer than the others. Batch 2 had the lowest acidity. The positions show little difference in acidity. If the end positions in the yogurt maker were slightly cooler than the middle positions, then we might expect the end positions to have less bacterial growth and hence lower acidity. This does not appear to have happened.

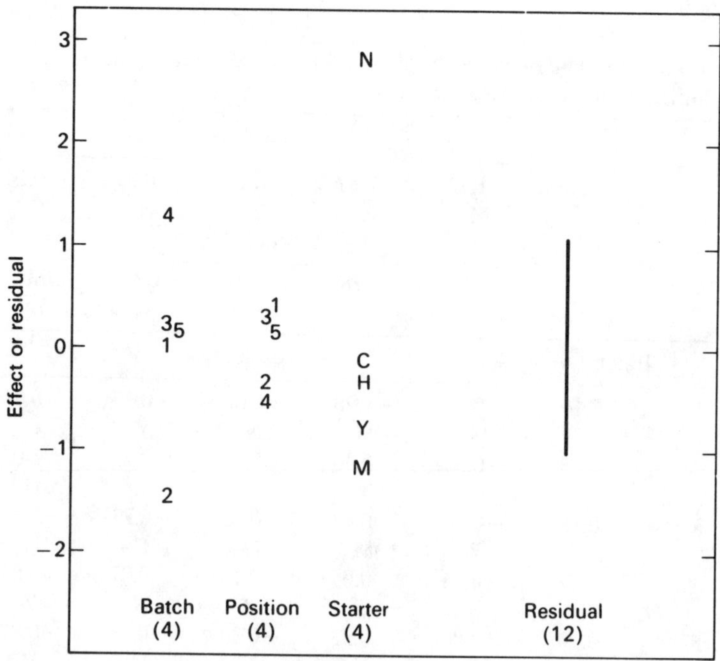

Figure 6-5. Side-by-side plot of effects and residuals for the data on homemade sour cream.

Table 6-15. Table of Mean Squares for the Sour Cream Experiment

Source	SS	d.f.	MS
Batch	18.84	4	4.71
Position	2.37	4	0.59
Starter	44.11	4	11.03
Residual	6.55	12	0.55
Total	71.87	24	

Position 1 had the highest average acidity, and position 5 fell in the middle. The residuals show a tight clustering about zero with a few values in the ranges of +1 and −1.

Table 6-15 gives the sums of squares, the degrees of freedom, and the mean squares for the analysis-of-variance table. Each of the main-effect mean squares in a Latin square design has $p - 1$ degrees of freedom. In this example p is 5, and each factor has 4 degrees of freedom.

Because the mean square for position has almost the same magnitude as the residual mean square, we believe that the variation in position effect is no greater than we would expect by chance. Thus position seems not to have affected the acidity of the sour cream.

However, batch and starter have mean squares much greater than the residual mean square, suggesting that both factors may affect acidity. The main finding is that Nancy's starter gave the highest acidity compared with the other starters.

Although we see some effect from batches, it does not hold primary interest for us. We do not know whether Dr. McDonald purchased all the sweet cream used in the experiment on the same day or bought fresh cream daily. If he purchased it on one day, the batch effect may represent an aging effect. Because batches 3, 4, and 5 were more acidic than batches 1 and 2, we might guess that older cream tends to produce more acidity. If he bought new sweet cream daily, the batch effect could still represent an aging effect if the store held the same batch of cream over several days, and he bought cream from that batch each day. Alternate explanations of the batch effect might be that his procedure differed slightly from day to day, that ambient air temperature varied daily, and the length of time the cream was incubated in the yogurt maker differed on different days, or that the bottles of cream he bought came from different batches at the factory.

6D. SUMMARY

Together, this chapter and the preceding one have examined many of the basic data layouts that arise in situations involving one, two, three, or four factors. The discussion has emphasized the overlays (in practice, usually reduced to bordering tables) that make up the value splitting of the data, along with graphical techniques for displaying and comparing the overlays and numerical ways of summarizing them. The basic approaches extend readily to other types of data layout, including those with more than four factors.

A number of techniques discussed in subsequent chapters usually serve to round out an exploratory analysis within the ANOVA framework. The F test (Chapter 7) uses mean squares to assess the overall significance of sets of main effects and interactions. An adjusted side-by-side plot (Chapter 8) displays all the components of a value splitting in a way that takes into account their mean squares, and other graphical techniques complement the analysis. Chapter 11 presents a systematic approach for considering whether some overlays contain sufficiently much noise that they should be recombined with other related overlays—a process sometimes called "downsweeping" to show its inverse relation to sweeping upward, as described in Chapter 5 and the present chapter. A number of procedures provide confidence statements for comparing individual effects and interactions (Chapter 12). Finally, Chapter 13 discusses the use of transformations to promote additivity and make variability more nearly constant across the cells of the data layout.

REFERENCES

Brown, M. B. (1975). "Exploring interaction effects in the ANOVA," *Applied Statistics*, 24, 288–298.

Lavin, P. T. and Flowerdew, G. (1980). "Studies in variation associated with the measurement of solid tumors," *Cancer*, 46, 1286–1290.

Xhonga, F. A. (1970). "Direct golds—part 1," *Journal of the American Academy of Gold Foil Operators*, 13, 17–22.

Xhonga, F. A. (1971). "Direct gold alloys—part II," *Journal of the American Academy of Gold Foil Operators*, 14, 5–15.

EXERCISES

1. Transform the estimated cross-sectional areas of Table 6-1 into the logarithmic scale and obtain the bordered table corresponding to Table 6-2. Then make a side-by-side plot and compare it to Figure 6-1.

2. (Continuation) Obtain the mean squares for the analysis in the log scale. Compare their relative sizes to the relative sizes of the mean squares in Table 6-4. (To simplify matters, express each set of mean squares as multiples of the error mean square in its table.)

Table 6-16. Cross-Sectional Area (in mm^2) of Simulated Solid Tumors, as Estimated by the Other 13 Oncologists

	Cork Model						Rubber Model					
	Small		Oblong		Large		Small		Oblong		Large	
Oncologist / Actual Area (mm^2):	980	980	1199	1199	1467	1467	980	980	1199	1199	1467	1467
14	1225	900	1225	1360	1365	1406	992	960	1188	1240	1369	1681
15	1023	1120	1110	1280	1368	1330	1056	1122	1209	1360	1482	1722
16	784	754	1050	980	1225	1155	900	900	1140	1050	1330	1444
17	1400	1225	1400	1400	1800	1400	1225	1225	1350	1350	1600	1400
18	900	1350	1184	1200	1369	1225	900	1225	1120	1200	1600	1444
19	780	638	864	875	1225	1188	840	780	875	665	900	1184
20	900	840	1369	1050	1600	1400	900	960	1050	1000	1260	1225
21	1054	900	1330	1116	1600	1440	1224	1088	1260	1394	1520	2025
22	1080	1089	1295	1575	1800	1400	1225	1326	1440	1280	1813	1936
23	1296	1330	992	1428	1764	1600	1122	1156	1470	1240	1681	1680
24	900	784	784	900	1225	1024	725	960	840	1050	1225	960
25	960	930	1200	1254	1600	1295	1024	1225	1287	1400	1600	1600
26	784	729	1295	1053	1295	1680	1054	1024	899	918	1147	1443

Source: Philip T. Lavin and Gordon Flowerdew (1980). "Studies in variation associated with the measurement of solid tumors," *Cancer*, 46, 1286–1290 (part of Table 2, p. 1288). Reproduced by permission.

3. Transform the more detailed data of Table 6-5 into the log scale and obtain the bordered table corresponding to Table 6-6. Add the boxplot of the residuals to the side-by-side plot constructed in Exercise 1. Compare the resulting plot to Figure 6-2.

4. (Continuation) Obtain the mean squares for the analysis of Table 6-5 in the log scale. Compare their relative sizes to the relative sizes of the mean squares in Table 6-7. How has transforming the data into the log scale affected your conclusions from the analysis?

5. (Continuation) Compare the table of mean squares obtained in Exercise 4 to that obtained in Exercise 2. Review the impact that including the replications had on the original analysis (Figure 6-2 and Table 6-7 versus Figure 6-1 and Table 6-4). Discuss the impact that including the replications has had on the analysis in the log scale.

6. Subtract the actual area of each simulated tumor in Table 6-5 from the estimated area given by each oncologist. Analyze the resulting measurement errors.

7. Divide each estimated area in Table 6-5 by the corresponding actual area and transform the resulting ratios into the log scale. Analyze these (logarithmic) measurement errors.

8. Lavin and Flowerdew concluded that the material (cork or rubber) had no effect on the oncologist's ability to measure the tumors. Reanalyze the data of Table 6-5 as a two-way layout (oncologist by form) with four observations per cell. Compare the results to Figure 6-2 and Table 6-7. Discuss the appropriateness of this simplification.

9. A total of 26 oncologists estimated the cross-sectional area of the solid tumors. Table 6-16 gives the data from the other 13 oncologists. Analyze these data and compare the results to those for the first half of the data (Tables 6-6 and 6-7 and Figure 6-2). Discuss similarities and differences.

10. Analyze the data of Table 6-16 in the log scale. Compare the results to those obtained in Exercises 3 and 4.

11. Reconstruct the full solid tumor data set by combining Tables 6-5 and 6-16. Analyze the resulting data set and compare the results to those obtained for the two halves.

12. From the dental-gold data in Table 6-8, extract the entries for Alloy AuCa and Methods 1 and 2 (hand malleting and electric malleting). Analyze the resulting three-way layout (5 dentists, 2 methods, and 3 temperatures). How do the main effects and two-factor interactions compare with corresponding values in Table 6-9 (take into account, for

Table 6-17. Diamond Pyramid Hardness Number (Sum of 10 Measurements) of Dental Fillings Made from Two Additional Types of Gold (see Table 6-8)

		Type	
Dentist	Method	Gold Foil[a]	Goldent[b]
1	1	792	824
	2	772	772
	3	782	803
2	1	803	803
	2	752	772
	3	715	707
3	1	715	724
	2	792	715
	3	762	606
4	1	673	946
	2	657	743
	3	690	245
5	1	634	715
	2	649	724
	3	724	627

[a] Pure gold cylinders, size $\frac{1}{2}$ gr.

[b] Pellets of powdered gold, in gold foil envelope, assorted sizes.

Sources: Morton B. Brown (1975). "Exploring interaction effects in the ANOVA," *Applied Statistics*, 24, 288–298 (data from Table 1, p. 290). Frida A. Xhonga (1971). "Direct gold alloys—part II," *Journal of the American Academy of Gold Foil Operators*, 14, 5–15 (description of gold types in Table 1, p. 6). Data reproduced by permission of the Royal Statistical Society.

example, the fact that the main effects in the present analysis are specific to Alloy AuCa). What temperature tends to give the hardest fillings with AuCa? Does this result depend on the method used?

13. Reconstruct the full dental-gold data set by combining the data in Table 6-17 (gold types 1 and 2) with the data in Table 6-8 (gold types 3 through 8). Analyze this three-way table (5 dentists, 3 methods, and 8 gold types). Compare the various effects with corresponding values in Table 6-9.

14. In Table 6-14 recombine the main effects for position with the residuals. Then rearrange the entries in the resulting table so that the rows correspond to batches and the columns correspond to starters. Does this new table show evidence of interaction?

15. Analyze the data on shelf space and sales volume, given in Table 4-7, as a Latin square. The three factors are supermarket, number of shelves, and week.

CHAPTER 7

Mean Squares, *F* Tests, and Estimates of Variance

Frederick Mosteller
Harvard University

Anita Parunak

John W. Tukey
Princeton University

Chapters 5 and 6 have introduced sums of squares and mean squares as one way to summarize the sizes of the elements in an overlay. A tabulation of the sums of squares, degrees of freedom, and mean squares associates each of these with the corresponding overlay in a value splitting of a factorial data table. Classical ANOVA gives far more emphasis to these summaries than we have so far in this volume. The present chapter discusses the two main customary roles of mean squares: as the basis for the simplest significance tests and as estimates of population variances.

7A. FORMAL INFERENCES: THE *F* TEST

In Chapters 5 and 6, we made rough comparisons among the sizes of various pieces of a breakdown. The simplest design, the one-way layout, invites us to compare the overlay for the effects of a factor to the overlay for the residuals. A dotplot of the one-factor effects side by side with a boxplot of the residuals (as in Figure 5-3) presents this comparison graphically. When it is adequate to summarize away the bulk of the detail from these two overlays and reduce each to a single number, we can use the corresponding mean squares.

Making this comparison, either by graphs or by numbers, can lead us to ask about differences among the effects for the factor. The leading questions are likely to be whether we know the direction of a difference and then how large the difference is.

The side-by-side plot may show very clearly that the effects differ by more than chance fluctuations or that their differences could easily have arisen by chance. Between these two extremes we may use our judgment, but the plot offers no firm rule for deciding whether two effects have been shown to be different in a specified direction. Classical techniques for ANOVA address the question of overall inequality of the effects by referring the ratio of the mean squares to a suitable F distribution. Some form of this standard procedure often applies when we need to compare two mean squares, with a view to asking, "Are these effects demonstrably different?"

Setting for the F Distribution

In theoretical developments we usually start by assuming that the data come from normal distributions, one distribution for each group in the one-way layout. These distributions are usually assumed to be equally variable, but with an unknown common variance, σ^2. The observations for any one level of the factor (one group) come from the same distribution, but the distributions for the levels of the factor may have different means. Before we see the data, we do not know whether and how the means differ, and that is what we want to assess.

To make the classical test of this kind, we need to disentangle the unknown means from the unknown variance, σ^2, so that we do not have to deal with two kinds of unknowns at the same time. The ratio

$$\frac{\text{mean square for the deviations among the groups}}{\text{mean square for the deviations within the groups}} \qquad (1)$$

provides a starting point for doing this. We refer to the numerator as the *mean square for the effects* and to the denominator as the *residual mean square*. With the assumptions we have adopted, the expected (or average) value of the residual mean square—the denominator in expression (1)—is σ^2. When the population means for the groups actually are the same, the mean square for the effects also averages to and estimates σ^2. The distribution of the ratio (1) does not depend on σ^2 if there are no treatment effects. When the population means for the groups differ, those differences cause the average value of the numerator to be larger than σ^2. If we think of the true effects as multiples of σ, however, we can regard the unknown σ^2 as canceling out.

When the true effects are all zero and the data are Gaussian (i.e., normal), the ratio (1) of the mean square among groups to the mean square within groups follows an F distribution. To specify the particular F distribution, we need to give the number of degrees of freedom for the numerator, ν_1, and the number of degrees of freedom for the denominator, ν_2. As Section 5D explains, when a one-way layout has I groups and J replications in each group, these degrees of freedom are

$$\nu_1 = I - 1 \quad \text{and} \quad \nu_2 = I(J - 1).$$

If n denotes the total number of observations in the one-way layout (i.e., $n = IJ$), we may write $\nu_2 = n - I$.

As we discuss in Section 7B, the F distribution is a good approximation to the distribution of the ratio in expression (1) in a wide variety of circumstances. Thus we can use the F distribution without relying heavily on the assumption that the data are Gaussian.

EXAMPLE: INTEREST RATES FOR NEW-CAR LOANS

Table 5-3 gives the annual percentage rate of interest charged on new-car loans at nine of the largest banks in six U.S. cities. In this one-way layout the estimated effects for the cities are .19, −.40, .30, .24, .48, −.81, as displayed in the full overlay of Table 5-5 and the reduced overlay of Table 5-6.

We would like to consider whether the data show the associated true effects to be unequal. (It is highly unlikely that the true effects are all equal to zero to many decimal places.) Figure 5-3, which shows the dotplot of city effects and the boxplot of residuals, suggests that the effects for Philadelphia and Chicago may be significantly lower than those for the other four cities. In overall terms, then, the six observed effects may suggest that differences among the average rates for the six cities are real, and not the result of chance fluctuation.

The F statistic leads more formally to the same conclusion of overall inequality. The mean square for cities (from Table 5-8) is 2.19 with 5 degrees of freedom, and the residual mean square is 0.45 with 48 degrees of freedom. The F statistic is $2.19/0.45 = 4.83$. A table of the F distribution gives the upper 5% critical value for F on 5 and 48 degrees of freedom as 2.41 and the upper 1% critical value as 3.42. Furthermore, the probability of an observation larger than 4.83 from F on 5 and 48 degrees of freedom is .0012. Thus it is very unlikely that we would get the observed effects for the cities if the true effects were all equal (i.e., all zero). The F statistic offers a formal way to infer that the true effects are not all equal—to confirm what we saw in the plots of their estimates.

The Analysis-of-Variance Table

An analysis-of-variance table often shows the ratio of the mean square among groups to the residual mean square, and the corresponding P value. We summarize the calculations, using basic formulas from Section 5D.

Table 7-1. Formal Analysis-of-Variance Table for a One-Way Layout with I Groups and J Observations in Each Group[a]

Source	Sum of Squares	Degrees of Freedom	Mean Square	F
Groups	$(I-1)\text{MS}_B = J\sum_{i=1}^{I} a_i^2$	$I-1$	MS_B	$\dfrac{\text{MS}_B}{\text{MS}_e}$
Residuals	$(n-I)\text{MS}_e = \sum_{i=1}^{I}\sum_{j=1}^{J} e_{ij}^2$	$n-I$	MS_e	
Total	$(n-1)s^2 = \sum_{i=1}^{I}\sum_{j=1}^{J}(y_{ij}-y_{..})^2$	$n-1$		

[a]The effect for group i is $a_i = y_{i.} - y_{..}$, and observation j in group i leaves the residual $e_{ij} = y_{ij} - y_{i.}$.

In a one-way layout with I groups and J observations per group, we let a_i denote the effect for group i, and e_{ij} the residual for observation j in group i. Each overlay contains IJ entries. The squares of the entries in the overlay for the group effects sum to

$$\sum_i \sum_j a_i^2 = J\left(\sum_i a_i^2\right).$$

(The right-hand side comes from the reduced overlay, and the left-hand side comes from the full overlay, which displays all J replications.) The mean square for the effects is MS_B ("B" for differences "between" groups):

$$\text{MS}_B = J\left(\sum_i a_i^2\right)\Big/(I-1).$$

This mean square appears in the first line of the analysis-of-variance table (Table 7-1).

From the residuals e_{ij} we obtain the residual mean square,

$$\text{MS}_e = \sum_i \sum_j e_{ij}^2 / I(J-1).$$

The residual mean square and its degrees of freedom appear in the second line of Table 7-1. The total sum of squares and degrees of freedom appear in the last line.

The sums of squares $(I-1)\text{MS}_B$ and $(n-I)\text{MS}_e$ add up to $(n-1)s^2$, and the degrees of freedom for groups and for residuals add up to $n-1$.

Recall that the missing degree of freedom in $n - 1$ is associated with the grand mean. The ratio of MS_B to MS_e is the F statistic

$$F = \frac{J\left(\sum_i a_i^2\right) \Big/ (I - 1)}{\left(\sum_i \sum_j e_{ij}^2\right) \Big/ (n - I)},$$

to be referred to the F distribution on $I - 1$ and $n - I$ degrees of freedom.

For an example we turn to a more complicated design, a complete three-way layout with replication.

Example: simulated tumors

For the analysis of the simulated tumor data with replication, Table 6-7 gives sums of squares, degrees of freedom, and mean squares. If we use F to compare each of the mean squares for main effects, two-factor interactions, and three-factor interaction to the error mean square, we get the analysis-of-variance table in Table 7-2. The P values point to substantial main effects for oncologists and the three forms of tumor, but the difference between the two materials appears to be mostly noise. Among the interactions only the two-factor interactions for material by form and oncologist by material stand out. Against the absence of any real main effects for material, these interactions suggest that we examine the data in detail, to see whether some isolated anomalous observations may be responsible. If these two-factor interactions involving M could be neglected, we would consider a simpler breakdown that contained only the common value, main effects for oncologist, main effects for form, and a corresponding set of residuals.

Table 7-2. Analysis-of-Variance Table for the Simulated Tumor Data with Replication

Source	SS	d.f.	MS	F ratio	P
Oncologist	9,467,062	12	788,922	21.85	.0001
Material	37,386	1	37,386	1.04	.3120
Form	7,650,202	2	3,825,101	105.96	.0001
O × M	814,684	12	67,890	1.88	.0497
O × F	870,474	24	36,270	1.00	.4710
M × F	429,630	2	214,815	5.95	.0039
O × M × F	430,836	24	17,951	0.50	.9721
Residual	2,815,680	78	36,098		

Note: Each F ratio in this table has the residual mean square as its denominator. Other choices may be considered later.

Theoretical Background Using Gaussian Assumptions

From the assumption that the observations are Gaussian, one can show that, when the true group effects are zero, $(I - 1)\text{MS}_B$ is distributed as σ^2 times a chi-squared variable with $I - 1$ degrees of freedom, that $(n - I)\text{MS}_e$ is distributed as σ^2 times a chi-squared variable with $n - I$ degrees of freedom, and that MS_B and MS_e are statistically independent. (The unknown σ^2 is the same for the two chi-squared variables.) The statistic $F = \text{MS}_B/\text{MS}_e$ is the ratio of two independent chi-squared variables, each divided by its degrees of freedom. Thus the statistic follows the F distribution. Conveniently, the distribution of the ratio MS_B/MS_e does not depend on the common value of σ^2.

Because the expected value of a chi-squared random variable is equal to the degrees of freedom, both the numerator and the denominator of the F ratio have expected value σ^2. Thus, when the data involve only fluctuations (and no true effects), we expect the F statistic to be in the vicinity of 1. More precisely, the expected value of F on ν_1 and ν_2 degrees of freedom is $\nu_2/(\nu_2 - 2)$, as long as $\nu_2 > 2$. Thus the rather small departures of the average value of F from 1 arise only because of variability in the denominator (i.e., failure of ν_2 to be very large).

(For those familiar with conditional expectations, to derive this expression, we condition on MS_e and then find the expected value of the resulting conditional expectation. That is,

$$
\begin{aligned}
E(F) &= E[E(F|MS_e)] \\
&= E[E(MS_B|MS_e)/MS_e] \\
&= E[\sigma^2/MS_e] \\
&= \nu_2/(\nu_2 - 2),
\end{aligned}
$$

using the independence of MS_B and MS_e and the expected value of the reciprocal of a chi-squared variable.)

As we mentioned earlier, when the true means differ (i.e., when the true effects are not all zero), MS_B will tend to be larger than MS_e. (Technically, $(I - 1)\text{MS}_B/\sigma^2$ follows a noncentral chi-squared distribution, whose expected value reflects the extent of the differences among the true means.) Thus a large value of F encourages us to think that the true means differ. We can formalize this step by comparing the observed value of the F ratio to a percentage point (often the 5% point) in the upper tail of the F distribution on ν_1 and ν_2 degrees of freedom, with the help of either a table or a function in a statistical computing environment. In the present book, however, we use mean squares more as summaries in various steps of exploration than in formal F tests.

7B. BROADENING THE BASE FOR THE F TEST

The preceding section discusses the use of the F test to decide whether the effects for a factor are unequal. That approach compares the mean square for those effects to the residual mean square. When the observations follow Gaussian distributions with a common variance, the F distribution provides the standard of comparison for the ratio of the two mean squares.

What if the distribution of the observations is not Gaussian? What can we say, or do? One fruitful approach of long standing takes the observed responses as given and then considers all appropriate rearrangements (permutations) of those observations over the given layout. By calculating the F ratio for each rearrangement and then comparing the observed F ratio to the resulting distribution of F ratios, one judges whether the observed F ratio is larger than one would expect by chance if the effects were all zero. In this approach, we regard the chance mechanism as having equally likely rearrangements, given the observed responses, instead of coming from repeated sampling from Gaussian distributions. A so-called permutation distribution replaces the F distribution as a basis for calculating probabilities associated with the observed F ratio.

We now examine this approach more concretely in the setting of a two-way layout in which the data come from a randomized block design with J blocks, each block containing I experimental units. In some situations the experimental units form natural blocks. Then the investigator would use a randomized block design, which involves assigning the I treatments randomly within each block. The block effects allow for the possibility that the general level of response may vary from block to block, in a way that is unrelated to the effects of the treatments. For example, an experiment with two mosquito repellents might recruit a number of male volunteers and put one repellent on a subject's left arm and the other repellent on his right arm, assigning repellents to arms at random. Under controlled conditions, the response would be the number of bites during a specified time of exposure to mosquitoes. The blocking handles the possibility that men vary in their attractiveness to mosquitoes; it allows us to compare the repellents within the blocks (here male subjects), whose experimental units (here left arm and right arm) are relatively homogeneous. (A separate part of the experiment would recruit only women, to allow for a general difference between men and women.) The use of blocking generalizes the familiar design from a matched-pairs t test to more than two treatments per block.

In what follows we use the notation MS_A for the treatment mean square (on $I - 1$ d.f.), MS_B for the block mean square (on $J - 1$ d.f.), and MS_e for the residual or error mean square (on $(I - 1)(J - 1)$ d.f.). If the treatments have no effect, and the observations in each block are either (1) a random sample from some distribution, or (2) the result of randomized assignment of treatments to the cells in that block (independently from block to block), we can usefully work conditionally on the values of observed responses. Under either (1) or (2), every permutation of the observed responses within a block is exactly as likely as any other. Indeed, the equality still holds when we permute all the blocks separately and simultaneously. Conditional inference on the set of all simultaneous permutations is thus justified by either (1) or (2). Assumption (1) is sometimes called a *population model*, and (2) is called a *randomization model*.

In principle, we could generate the $(I!)^J$ simultaneous permutations and do exact inference based on the simultaneous randomization. This could, however, be a lot of work, because $(I!)^J$ becomes large for even modest I and J. For example, when $I = 2$ and $J = 6$—a smaller layout than one would expect in practice—$(I!)^J = 2^6 = 64$. When $I = 3$ and $J = 4$, $(I!)^J = 6^4 = 1296$; and the more realistic combination $I = 3$ and $J = 10$ yields $(I!)^J = 60{,}466{,}176$.

To illustrate the layout for the mosquito repellent example, if we have $J = 3$ men with $I = 2$ repellents, there are $(2!)^3 = 8$ permutations. If entries for rows give the number of bites with repellent 1 and 2, respectively, and entries for columns give results for man 1, 2, and 3, respectively, then the pattern and the permutations might be as shown in Table 7-3. We take the upper left panel as "obtained," and repellent 1 looks somewhat more effective in reducing bites than repellent 2.

The between-blocks sum of squares is 31, the same for each permutation, as, of course, is the total sum of squares, 122. The between-repellents sum of squares differs from one permutation to another (actually four different values, each occurring twice), as does the residual sum of squares. The calculations for the upper left-hand panel give residual sum of squares as $122 - 31 - 24 = 67$. Thus the repellent mean square (with 1 d.f.) is 24, the residual mean square (with 2 d.f.) is 33.5, and the F ratio is $24/33.5 = .72$ with 1 and 2 d.f. If we were looking this value up in an F table, we would go no further because an F less than 1 would not reach the usual levels of significance.

If we want to refer the observed value of F to the permutation distribution, we see at the bottom of Table 7-3 that 4 of the 8 permutations have values as large as or larger than the .72 observed. One other permutation has repellent MS = 24, and two have repellent MS = $242/3$ and hence $F = 15.61$. Roughly speaking, then, we are at about the 50% level (perhaps a bit smaller because we are in the top half of a discrete distribution). We would need a larger study to get an impressive result because even the F value of 15.61 is only in the top quarter. The point of this example is to illustrate the permutation approach in this very small randomized block experiment.

It may seem that large I and J would render the permutation approach impractical, but that is not the situation today. In many problems, it is possible to take a large sample of the permutations, say several thousand, and carry out the calculations for each and thus get an excellent approximation to the distribution. This approach has attractions, but our current discussion aims not so much to provide a computational method as to

Table 7-3. Eight Permutations Associated with the Hypothetical Experiment of $I = 2$ Repellents and $J = 3$ Men in the Mosquito Repellent Trial[a]

	M1	M2	M3	Total	Rows SS	M1	M2	M3	Total	Rows SS
R1	8	14	14	36		8	14	9	31	
R2	19	20	9	48		19	20	14	53	
Total	27	34	23	84	24	27	34	23	84	242/3
R1	8	20	14	42		8	20	9	37	
R2	19	14	9	42		19	14	14	47	
Total	27	34	23	84	0	27	34	23	84	50/3
R1	19	14	14	47		19	14	9	42	
R2	8	20	9	37		8	20	14	42	
Total	27	34	23	84	50/3	27	34	23	84	0
R1	19	20	14	53		19	20	9	48	
R2	8	14	9	31		8	14	14	36	
Total	27	34	23	84	242/3	27	34	23	84	24

Total SS = 122   Between-blocks SS = 31

F ratios (2 and 2 d.f.)

0.72	15.61
0.00	0.45
0.45	0.00
15.61	0.72

[a]The upper left panel is regarded as the "actual" outcome (bites received by the three men).

indicate that the F distribution has wider application than to data with Gaussian distributions. Therefore we do not pursue this any further except to say that some computer packages make this approach to the permutation distribution very practical.

A useful approach to the permutation distribution was found fifty-odd years ago by Pitman (1937) and Welch (1937). They use the mean and the variance, not of the ratio of mean squares $F = \mathrm{MS}_A/\mathrm{MS}_e$, but of a monotone function of F, specifically the ratio W given by

$$W = \frac{(I-1)\mathrm{MS}_A}{(I-1)\mathrm{MS}_A + (I-1)(J-1)\mathrm{MS}_e}.$$

The advantage of W is that its denominator is unaffected by randomization within the blocks. So only the moments for the numerator are needed, simplifying the algebra greatly. Because W is a monotonic increasing function of F $(= \mathrm{MS}_A/\mathrm{MS}_e)$, specifically, $W = F/(F + J - 1)$, inferences based on W will be equivalent to inferences based on F.

The following discussion explains that the F distribution we ordinarily use often gives results close to those of the permutation distribution. The work of Pitman and Welch relates the ordinary F distribution to the distribution derived from the permutation approach. For any permutation distribution based on I treatments and J blocks, we can find an F distribution based on normal theory with I^* treatments and J blocks that matches the permutation distribution fairly closely. Furthermore, in many applications I^* is close to I, so that our usual F distribution, even with I treatments and J blocks, will often fit the permutation distribution closely enough. Thus the ordinary F tables have wider application than normal theory would suggest because they give an approximation to the exact inference coming from the permutation distribution. For those who may wish to employ the randomization test in practical applications, Edgington (1987) may be helpful. We turn now to a more technical explanation of this idea. Some may prefer to skip the next section.

7C. (OPTIONAL) NOTE ON THE RELATION OF THE PITMAN–WELCH WORK TO AN ORDINARY F DISTRIBUTION

The distribution of W (or F) based on randomization of the actual data does not coincide with the distribution based on sampling from the Gaussian, but the two distributions often are not greatly different, even though the randomization-based distribution is discrete and the Gaussian-based distribution is continuous. We relate them by asking how we should modify I and J for the Gaussian-based distribution (obtaining I^* and J^*), so that it then has the same mean and variance as the randomization-based distribution (a standard approach to approximating one distribution by another).

If the data are Gaussian with common variance, W follows a beta distribution $B\{\frac{1}{2}(I-1), \frac{1}{2}(I-1)(J-1)\}$, whose mean is

$$\text{mean} = E_{\text{Gau}}(W) = 1/J$$

(the reciprocal of the number of blocks or the number of replications for each treatment) and whose variance is

$$\text{variance} = \text{var}_{\text{Gau}}(W) = \frac{2(J-1)}{J^2(IJ-J+2)}. \qquad (2)$$

Note that the mean and variance of W for Gaussian data depend only on I and J.

The Pitman–Welch randomization moments, based on equally likely permutations, work even for data that are not Gaussian. But now the randomization variance must incorporate a summary of the observations:

$$E_{\text{rand}}(W) = 1/J$$
$$\text{var}_{\text{rand}}(W) = \frac{2(1-A)}{J^2(I-1)}, \tag{3}$$

where the data enter through

$$A = \frac{\sum_j \left(S_j^2\right)^2}{\left(\sum_j S_j^2\right)^2}.$$

Here $S_j^2 = \Sigma_i (y_{ij} - y_{\cdot j})^2$ is the sum of squared deviations within block j (treatments *not* allowed for). The quantity A lies between 1 (when only one S_j^2 differs from zero) and $1/J$ (when the S_j^2 are all equal).

In both the Gaussian distribution of W and the permutation distribution, the number of treatments, I, and the number of replications, J, are the same. Nevertheless, the distribution of the Gaussian-based W and that of the permutation-based W differ, as we have seen above, because of the discreteness and because they may have differing variances. If we want to approximate a permutation-based distribution by a Gaussian-based one, for which tables are widely available (for F rather than W), we wish to find parameters I^* and J^* to replace I and J, respectively, so that the significance of an observed W from the permutation-based distribution can be approximated from the Gaussian-based distribution. The change from I to I^* and J to J^* refers only to the F table we use. What we calculate from the data is still the ratio of the conventional mean squares. To make the mean of W the same for the two cases, we do not change the quantity for number of replications, and so $J^* = J$.

Next we equate the randomization variance of W (for I and J) with the Gaussian variance (for I^* and J), and we find

$$\frac{2(1-A)}{I-1} = \frac{2(J-1)}{I^*J - J + 2}.$$

After some algebra, we solve for $I^* - 1$ and find

$$I^* - 1 = (I-1)\frac{J-1}{J(1-A)} - \frac{2}{J}. \tag{4}$$

Thus if $A = 1/J$ (all S_j^2 equal), then

$$\text{approximate d.f.} = I^* - 1 = (I - 1) - \frac{2}{J} = \text{nominal d.f.} - \frac{2}{J}.$$

That is, in this extreme case the randomization distribution is somewhat more spread out than the nominal Gaussian-based distribution. If $A > 1/J$, the approximate d.f. will be larger than this. Once $I^* - 1$ is greater than $I - 1$, the nominal Gaussian-based distribution becomes conservative. Let us review two examples.

For $J = 3$ and $S_j^2 = 1, 2$, and 3, for instance, we have

$$A = \frac{(1 + 4 + 9)}{(1 + 2 + 3)^2} = \frac{14}{36} = \frac{7}{18}$$

and

$$I^* - 1 = (I - 1)\frac{3 - 1}{3 - (7/6)} - \frac{2}{3} = (I - 1)\frac{12}{11} - \frac{2}{3};$$

whereas for the S_j^2 a little more divergent, say, $S_j^2 = 1, 4$, and 9, we have

$$A = \frac{(1 + 16 + 81)}{(1 + 4 + 9)^2} = \frac{98}{196} = \frac{1}{2}$$

and

$$I^* - 1 = (I - 1)\tfrac{4}{3} - \tfrac{2}{3}.$$

Thus for $A = \frac{1}{2}$ and $I = 2$ we would have $I^* = \frac{5}{3} < I$, but for $I \geq 3$ we would have $I^* - 1 \geq I - 1$.

If the S_j^2 are all equal, then, we lose $2/J$ degrees of freedom in the numerator (and an equal fraction of the nominal degrees of freedom in the denominator). As the S_j^2 diverge, we lose less and less; and soon we tend to have $I^* - 1 \geq I - 1$ and to gain degrees of freedom for the randomized statistic as compared to the Gaussian-based one.

Thus we can proceed in several ways: (1) use the Gaussian-based critical values of F (on $I - 1$ and $(I - 1)(J - 1)$ degrees of freedom) for all randomization analyses; (2) for each randomization analysis separately, find the values of the S_j^2 and hence the values of A and $I^* - 1$, and then use Gaussian-based critical values for $I^* - 1$ and $(I^* - 1)(J - 1)$ degrees of freedom; (3) use the exact randomization distributions; or (4) use sampling from the random permutations. The conventional analysis is always a first approximation, and we can always have second approximations or better.

Pitman and Welch also looked at the third moment of W under randomization and found it reasonably close to that given by these approximations.

In practice, in this book we get along with an ordinary F table and do not use the permutation distribution explicitly. We have only occasional examples of randomized-block investigations. We can use the F-table approximation when the population model seems appropriate.

7D. CONFIDENCE INTERVALS FOR σ^2 UNDER IDEAL CONDITIONS

We have examined the mean squares associated with the overlays in the value splitting of a factorial data table. Each of the mean squares gives an estimate of a component of variance, perhaps after multiplying by a suitable constant. We have learned that ratios of these estimates—the F statistics—offer a way to determine whether a particular factor or a particular interaction of two or more factors contributes substantially to the overall variability in the measurements being analyzed. We have ample reason to believe that estimates of variance are as central to assessing the sources of variability as are sample means in ascertaining the location or levels of measurements. We turn next to the *precision* of our estimates of variability.

We examine the variability of sample variances and sample standard deviations mainly under the ideal conditions of sampling from a Gaussian (normal) distribution. The main finding is that, even under these pleasant conditions and even with large numbers of measurements, our estimates of the true variance σ^2 and of the true standard deviation σ will not be very close to the true values when measured in terms of percentage error. When the number of degrees of freedom is modest, say less than 20—as it often is in practical problems—we can expect errors of hundreds of percent. In exploring this issue, we exploit the confidence limits for σ^2 and for σ, paying special attention to the ratio of the upper confidence limit to the lower confidence limit.

Partly because we are more used to the variation of linear measures like means, the high variability of sample variances gives us an unwelcome surprise—first stemming from the squared deviations and second from the sensitivity to assumptions. In beginning texts on statistics we learn that when the measurements, x_i, are drawn from a normal distribution with variance σ^2,

$$\frac{(n-1)s^2}{\sigma^2} = \frac{\sum_{i=1}^{n}(x_i - \bar{x})^2}{\sigma^2}$$

is distributed according to the chi-squared distribution with $n - 1$ d.f. This result seems to let us set 95% probability limits $\chi^2_{L,n-1}$ and $\chi^2_{U,n-1}$ on this ratio and thus derive the 95% confidence limits on σ^2 given in inequality (5).

CONFIDENCE INTERVAL FOR σ^2 IN GAUSSIAN DATA

$$\sum (x_i - \bar{x})^2/\chi^2_{U,n-1} \le \sigma^2 \le \sum (x_i - \bar{x})^2/\chi^2_{L,n-1}. \tag{5}$$

The ratio of the upper limit on the right to the lower limit on the left is $\chi^2_{U,n-1}/\chi^2_{L,n-1}$. Such a ratio gives us a way of thinking about the uncertainty of our estimate of σ^2 under ideal conditions. When the ratio is large, we have broad limits; when small, relatively narrow ones. Variances are measures of scale and are naturally compared by using factors and ratios—that is, in the multiplicative framework—whereas means, describing location, are ordinarily compared in an additive scale using differences. As we go along, we also take square roots to obtain results that describe σ instead of σ^2. In either scale, the ratio of an upper limit to a lower limit is often larger than one might anticipate.

The graph in Figure 7-1 shows that for the ratio of variances to be less than 50, we need at least 3 d.f.; for it to be less than 10, we need at least 7 d.f.; and for it to be less than 3, we need at least 30 d.f. Note that a ratio of 3 means the limits include a percentage error of at least 100%.

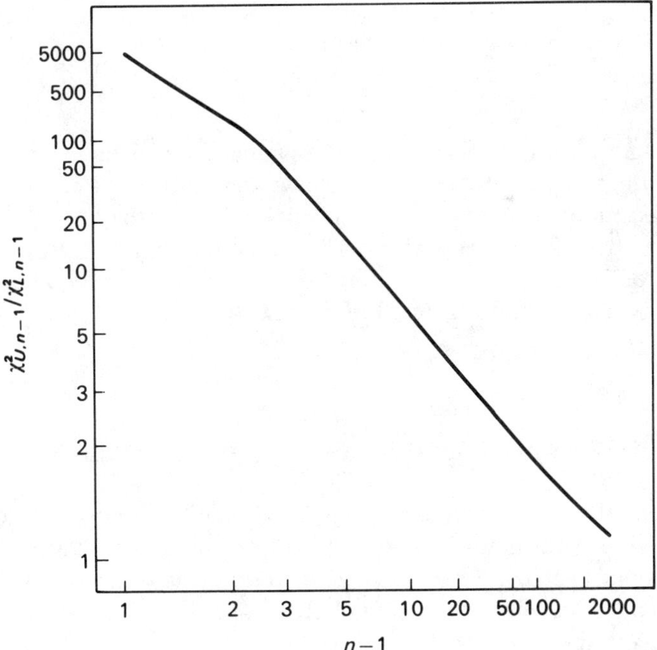

Figure 7-1. Ratio of the chi-squared-based 95% confidence limits for σ^2 for $n - 1$ degrees of freedom. Both scales have been transformed to bring the graph nearer to a straight line. The limits depend heavily on the assumption of normality.

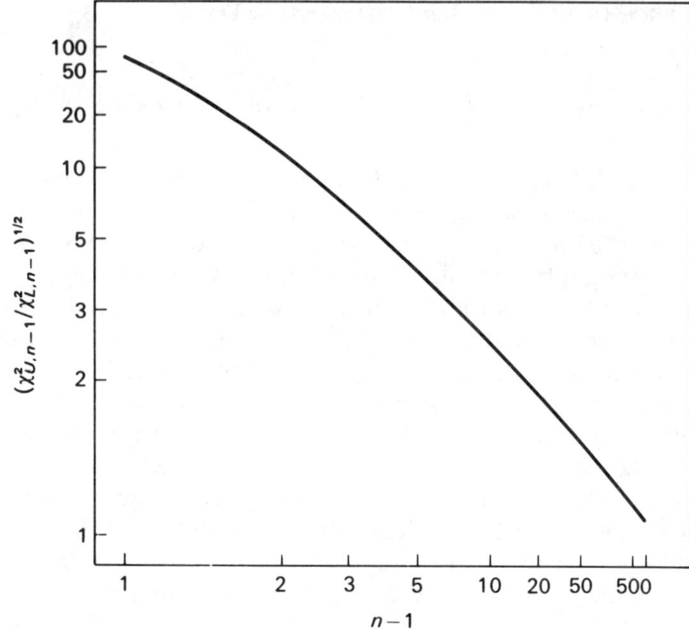

Figure 7-2. Ratio of the chi-squared-based 95% confidence limits for σ for $n - 1$ degrees of freedom. Both scales have been transformed to bring the graph nearer to a straight line. The limits depend heavily on the assumption of normality.

This same information applied to confidence limits for σ instead of σ^2 produces the more comforting graph shown in Figure 7-2. For the 95% limits to have a ratio of less than 13, we need 2 d.f., less than 5 we need about 4 d.f., less than 2 we need about 18 d.f. Our intuitive guesses about the precision of estimates of spread may be more nearly correct when we think about standard deviations instead of variances.

7E. SENSITIVITY TO THE ASSUMPTION OF NORMALITY

It turns out that the confidence limits in inequality (5) are very sensitive to the normality assumption, so sensitive that practitioners rarely use these limits in data analysis. That point only strengthens our current exposition, which emphasizes that even under ideal conditions sample variances are extremely variable.

For samples of size n, the expectation of the sample variance, $s^2 = \Sigma(x_i - \bar{x})^2/(n - 1)$, is σ^2 for distributions with finite variance, and the variance of s^2 for normal data is $2\sigma^4/(n - 1)$. More generally, the variance of the sample variance is of the sample variance is

VARIANCE OF SAMPLE VARIANCE

$$\operatorname{var}(s^2) = \sigma^4\left(\frac{2}{n-1} + \frac{\gamma_2}{n}\right), \tag{6}$$

where γ_2 is an index of kurtosis:

KURTOSIS INDEX

$$\gamma_2 = \frac{E(X-\mu)^4}{\sigma^4} - 3. \tag{7}$$

The 3 being subtracted is the fourth moment of the standard normal distribution. For the normal, then, $\gamma_2 = 0$, but even modest departures from normality can produce substantial γ_2. Because n and $n - 1$ will be close to each other for fair-sized n, the issue is how γ_2 compares with 2 (the numerator of $n - 1$) in expression (6). If $\gamma_2 = 6$, instead of 0, then the variance of the sample variance has roughly quadrupled. Equivalently, the standard deviation of the sample variance has doubled. Roughly speaking, such a value of γ_2 would double the lengths of the confidence intervals for the variance if we were to maintain the confidence level.

Let us consider a contaminated distribution composed of $100(1 - \epsilon)\%$ of a normal with mean μ and variance σ^2 and $(100\epsilon)\%$ of a normal with mean μ and variance $k\sigma^2$. Then its kurtosis index is

KURTOSIS INDEX FOR CONTAMINATED NORMAL

$$\gamma_2 = 3\left[\frac{1 - \epsilon + \epsilon k^2}{(1 - \epsilon + \epsilon k)^2} - 1\right],$$

and if $\epsilon = .2$ (20% contamination) and $k = 6$, then $\gamma_2 \approx 3$. Formula (6) shows that the variance of s^2 is about 150% bigger than it would have been for an uncontaminated normal distribution. It is not hard to imagine situations where, as here, 20% of the observations have standard deviations 2.4 times as large as the basic 80%. These calculations all help make the point that s^2 is a rather unstable quantity, though we cannot readily evade its use. (It is surprising that Student's t and F ratios work as well as they do for non-Gaussian situations.)

REFERENCES

Edgington, E. S. (1987). *Randomization Tests*, 2nd ed. New York: Marcel Dekker.

Pitman, E. J. G. (1937). "Significance tests which may be applied to samples from any populations: III. The analysis of variance test," *Biometrika*, 29, 322–335.

Welch, B. L. (1937). "On the z-test in randomized blocks and Latin squares," *Biometrika*, 29, 21–52.

ADDITIONAL REFERENCE

Snedecor, G. W. and Cochran, W. G. (1980). *Statistical Methods*, 7th ed. Ames, Iowa: Iowa State University Press, pp. 79–81. (pp. 80–81 in 8th ed., 1989)

EXERCISES

1. When samples of the same size are drawn from many populations with identical means, what does the mean square for effects estimate?

2. In Exercise 1, if the populations have different true means, how will that affect the ratio of the mean square for effects to the mean square for residuals?

3. In Exercise 1, if the populations are all Gaussian, what will be the average ratio of the mean square for effects to the mean square for residuals?

4. An F ratio has associated with it two numbers representing degrees of freedom. What determines these numbers? (Note that this does not have anything to do with the Gaussian assumption.)

5. (Continuation) If we have samples of 4 from each of 6 populations, determine the degrees of freedom associated with the F ratio.

6. Find the upper 5% level for F with 5 and 10 degrees of freedom.

7. Examine an F table, such as Table A-3 in the Appendix. For a fixed number of degrees of freedom for the numerator, what happens to the upper 5% critical value as the number of degrees of freedom for the *denominator* increases. Explain why this might be.

8. Examine an F table. For a fixed number of degrees of freedom for the denominator, what happens to the upper 5% critical value as the number of degrees of freedom for the *numerator* increases. Explain why this might be.

9. Three treatments for fleas are tried on 10 breeds of dogs in a large kennel, 3 dogs of a breed to a randomized block. When dogs are assessed for fleas, how many degrees of freedom are associated with the total sum of squares, the sum of squares for blocks, the sum of squares for treatments, and the residual sum of squares.

10. If, in Exercise 9, a permutation analysis were planned, how many possible permutations are there?

11. In a randomized-block study with two treatments, the permutations produce each analysis-of-variance table twice, so that every F value occurs twice. Explain why.

12. Check the analysis of variance leading to the F ratio for the upper left-hand permutation in Table 7-3.

13. If $A = 0.3$, $I = 8$, and $J = 10$, compare I^* with I in equation (4).

14. When $A = 1/J$, discuss I^*.

15. Even when the number of permutations is huge, you may be able to use the permutation approach as a practical matter. How?

16. When the degrees of freedom are 10, find from chi-squared tables (such as Table A-2) the ratio of the 95% confidence limits $\chi^2_{U,10}/\chi^2_{L,10}$ for σ^2 in normally distributed data.

17. Use Figure 7-1 to find how many degrees of freedom are required to bring the chi-squared ratio down to 5. Check your result using chi-squared tables.

18. **(a)** When n is large so that $n/(n-1)$ is nearly 1, find the standard deviation of the sample variance.

(b) Compare it to the standard deviation of the sample variance for the normal distribution.

(c) Use the result to argue that the confidence interval for s^2 for a nonnormal distribution will be approximately $(1 + \gamma_2/2)^{1/2}$ times as long as that for a normal distribution. (Chapter 9 discusses this further.)

19. The principal difference between a t distribution and a Gaussian distribution is that a t distribution has heavier tails that give positive kurtosis. This is especially true when the number of degrees of freedom, m, for t is modest. Suppose that data thought to come from a normal distribution actually come from a t distribution with 10 degrees of freedom; the index of kurtosis of this distribution for $m > 4$ is known to be $\gamma_2 = 3[(m - 2)/(m - 4) - 1]$. Describe the impact of the difference in distribution on a 95% confidence interval for the variance of the distribution of the data.

CHAPTER 8

Graphical Display as an Aid to Analysis

John D. Emerson
Middlebury College

Visual display of information—whether in the raw data, effects, residuals, comparisons, or relationships among variables—helps discovery, understanding, and effective communication. This chapter focuses on graphical methods in ANOVA. It emphasizes the interpretation of graphical information, and it illustrates how graphical displays can aid our understanding of variability, its sources, and its meanings.

Section 8A recalls methods presented in Chapter 3 for displaying raw data effectively; it also gives additional perspective on techniques, some already illustrated, for displaying the overlays produced by one-factor ANOVA. Section 8B extends these graphical techniques to two-factor ANOVA; it begins to show how suitably chosen plots can aid in detecting, describing, and interpreting nonadditivity, or interaction, when it is present. Section 8C develops a new plot, analogous to the parallel dotplot but better suited for use with mean squares, that invites comparison of the magnitudes of effects found in different overlays. Section 8D examines a four-factor data set on the percentage of Americans never married; these data illustrate the adjusted parallel plot, and they provide further illustrations throughout the remainder of the chapter.

Because much of our work examines many comparisons, Section 8E begins to explore challenges posed when we face many overlays, each with many effects.

The careful examination of residuals aids in deciding whether we have done too little or too much value splitting. Section 8F focuses on graphical analysis of residuals, and a concluding section summarizes the major messages of this chapter.

8A. AN OVERVIEW OF GRAPHICAL METHODS FOR ONE-WAY ANOVA

One-way analysis of variance provides a useful setting for introducing both purposes and techniques of graphical display. Earlier chapters—especially Chapters 3 and 5—have used graphical display to convey important messages about tables of data. Chapter 3 discusses two main graphical techniques for displaying the raw data: the dotplot and the boxplot. To display the results of an analysis, we also plot means and standard deviations, combine a dotplot of effects with a boxplot of residuals, and plot residuals against fitted values.

Plots of Means and Standard Deviations

Classical analysis of variance relies on means for location summaries and on standard deviations for summaries of variability. One popular graphical display uses parallel line segments. The segment for group i is centered at $\bar{x}_i$ and has endpoints at $\bar{x}_i \pm s_i$. Although this display can be more compact than the corresponding side-by-side boxplots, it tells nothing about skewness, extreme spread, and outliers. It does have the advantages of simplicity and of displaying some of the ingredients for making classical inferences about differences among the groups, but in other ways it seems less effective than parallel boxplots. A variant of the plot, often used in medical research literature, displays "error bars"—the sample mean plus and minus the standard error of the sample mean—or only half such a bar.

Plots for Fits and Residuals

Following an analysis by value splitting, a side-by-side display of the effects and the residuals helps us to learn more from the analysis. Figure 5-3 presents such a plot of effects and residuals for the interest rates on auto loans at large banks in six U.S. cities. That plot shows clearly that the variation among cities does not dwarf the variation within cities. Differences among rates in a particular city (say, Chicago, with rates of 11.9% and 14.25%) seem somewhat larger than differences among average rates for cities (say, New York, whose average is 13.5%, and Philadelphia, whose average is 12.2%). Although the means of the effects and of the residuals are zero, the medians are not zero.

One should habitually explore residuals graphically to search for patterns. A plot of residuals against fitted values offers a good opportunity to:

- Detect some kinds of patterns not yet discovered.
- Identify residuals (and their associated factor levels) that have unusually large magnitudes.
- Observe whether the spread of the residuals is constant across groups, or whether it increases with the magnitude of the effects in different groups.

Figure 8-1 plots the residuals against the fitted values for the six cities.

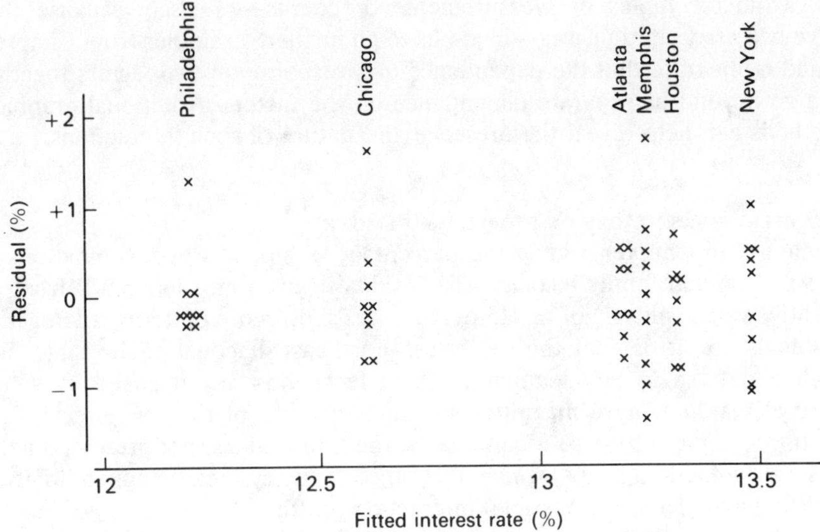

Figure 8-1. Plot of residuals from fit by city average rates against the fitted value.

The plot shows that all rates for a city are within 1.8% of the average for the city. It shows that eight rates for Philadelphia are very close together, and that the rate at a ninth bank is more than 1.2% higher than these. This display shows that Atlanta and Memphis have similar average rates, but those of Memphis are more spread out. For this example with a single factor, cities, the side-by-side boxplots may be more effective for assessing a relationship between spread and level than the plot of residuals against fitted values. Section 13D discusses another type of plot that can help to determine whether a transformation of the data would make spread more nearly constant among groups. The data on interest rates, however, do not appear to need transformation.

8B. GRAPHICAL DISPLAY FOR TWO-FACTOR DATA

Each graphical method already described is also useful for displaying responses that have two associated factors. We can make dotplots and side-by-side boxplots of the response against each of the factors separately. We can make parallel displays of means and standard deviations (or of means and their standard errors). We can plot values from the overlays (for effects of each factor, and for interaction when we have replication) beside each other, and put a boxplot of the residuals alongside these. Indeed, Chapter 6 makes good use of these displays for higher-way analysis—for example, Figures 6-1 and 6-2. In search of further patterns and structure in the residuals, we plot residuals from a two-way fit against the fitted values.

Two-factor tables of measurements or counts—precisely because they have two associated factors—invite us to go further. Examples from Chapters 5 and 6 illustrate that the dependence of a response on two factors together can go beyond its separate dependence on the factors. Additional graphical methods can help reveal the presence and nature of such interactions.

EXAMPLE: PERCENTAGES OF PEOPLE WHO SMOKE

Table 5-9 in Chapter 5 gives the percentage of smokers by combinations of age and annual family income. The residuals (given in Table 5-10) from an additive value splitting of the 15 percentages show more patterns: all negative residuals are at or near the southwest–northeast diagonal of the table. For such a pattern to be meaningful, the effects must be arranged by value. Several residuals have magnitudes similar to those of the row effects. Furthermore, a plot of the residuals against the fitted values (not presented here) has a prominent U shape, which also suggests the systematic nonadditivity.

We have detected possible interaction in the data. Age and income contribute together, beyond the sum of their separate contributions, to the smoking percentages. Lower income status corresponds to much less smoking for older people than for others. The residuals of −3.6 and −2.9 (shown in Table 5-10) for older people at lower income levels suggest that these groups smoke even less than predicted by the additive value splitting. One possible explanation is that heavy smokers among older people with lower incomes tend to die. Another explanation is that poorer elderly people experience more health problems than their more affluent counterparts and are therefore encouraged by their diseases and their concerned acquaintances to stop smoking. But, whatever the reasons, the message is clear: the smoking habits of older people are less strongly related to income status than are those of younger people.

A Display for Two-Factor Interaction

Before leaving the data on smoking, we return to the original data in Table 5-9 and graph the percentages of adults who smoke. Figure 8-2 presents a plot, with one trace for each of the three age groups, of the percentages against the ordered income categories. Although the horizontal scale is arbitrary (separations of income categories are not equal, and we could readily use midpoints only for the middle three categories), this plot does convey visually the structure in this data table. Figure 8-2 displays interaction that we partially described previously. Graphical techniques for learning about the nature and extent of interaction—similar to, but more general than, those of Figure 8-2—are especially helpful.

If the data gave the percentages of smokers separately for men and women, then each trace in Figure 8-2 could be separated into two traces, one

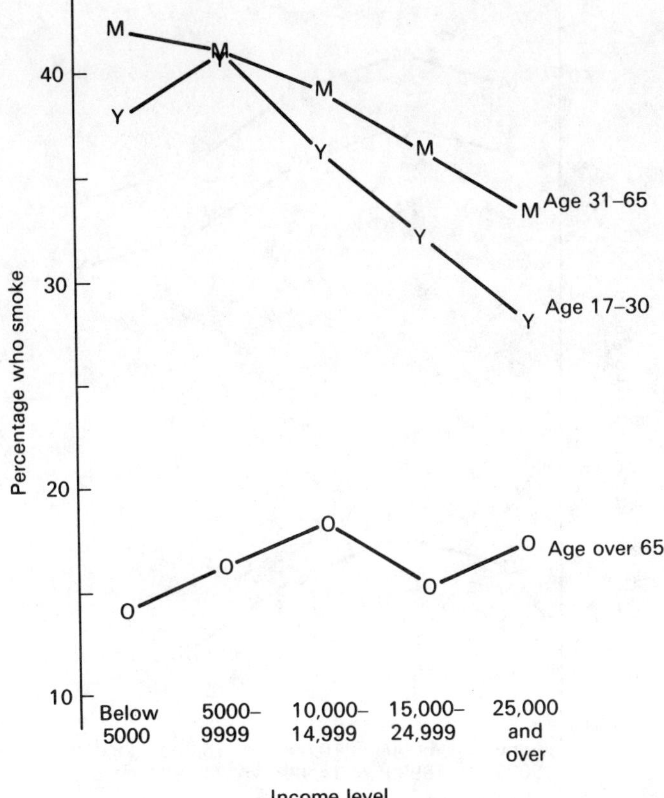

Figure 8-2. Plot of percentage of people who smoke against family income category for each of three age categories: 17–30 (Y), 31–65 (M), and over 65 (O).

for each gender. Information about ethnicity (say, Caucasian, Black, and Hispanic) would further multiply the number of traces by 3, giving $3 \times 2 \times 3 = 18$ connected traces in place of the three displayed in Figure 8-2. We would now face the difficulty of sorting out the effects for race, those for gender, and those for age, as well as those for income category.

Let us suppose that a main-effects-only fit had described perfectly the data about smokers. How would Figure 8-2 have represented this noninteraction? The main-effects-only fit to the data, presented in Table 5-10, has no interaction by definition. Figure 8-3 displays this fit. The important message is that "no interaction," or perfect additivity, in a two-factor table of values y_{ij} is equivalent to the graphs of the y_{ij} against the levels of either factor being *parallel* traces, one for each level of the other factor. Two traces are parallel when their separation, measured in the vertical direction, is constant everywhere.

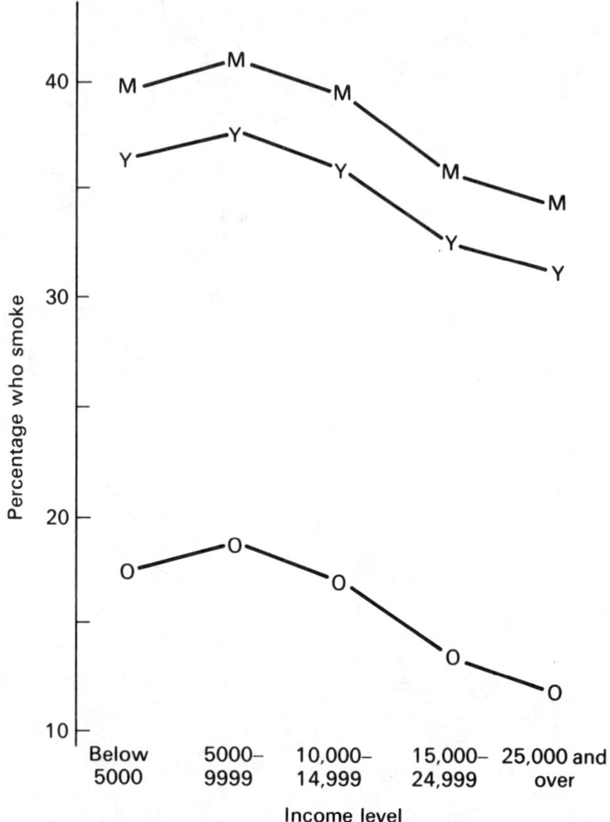

Figure 8-3. Plot of main-effects-only fit to percentage of people who smoke. Parallel traces indicate perfect additivity. Plotting symbols identify age group: 17–30 (Y), 31–65 (M), and over 65 (O).

8C. A SIDE-BY-SIDE PLOT ATTUNED TO MEAN SQUARES

Chapters 5 and 6 display the results of value splitting in a side-by-side plot that shows one dotplot for each bordering table. For an overall numerical summary, they also reduce the value splitting to a table of mean squares—one mean square for each bordering table—and the corresponding degrees of freedom. Although the side-by-side plot and the table of mean squares both provide valuable information, they are not ideal companions. That is, bordering tables can yield the same mean square, although the side-by-side plot shows that their effects have substantially different sizes. Conversely, two bordering tables whose effects are the same size lead to substantially different mean squares when they have substantially different degrees of freedom associated with them. We now consider how to rescale the entries in each bordering table separately so as to produce a side-by-side plot better attuned to the mean squares. David Hoaglin and John Tukey have developed a suitable adjustment, and they suggest using the adjusted effects in place of ordinary effects in side-by-side displays. The present section describes and expands their work.

In pursuing this change, we emphasize that we are not abandoning the basic side-by-side plot. Instead, we recognize different purposes. When we want to examine the effects in the same scale as the original data, we prefer the side-by-side dotplot. For example, we might hope to choose a combination of effects that tends to produce a higher value of the response. When we focus on mean squares, we prefer instead a plot that suggests how clearly the effects have been delineated. A larger mean square, relative to the residual mean square, corresponds to clearer delineation. The F test (Section 7A) formalizes the comparison of mean squares for effects to the residual mean square.

Because the adjusted side-by-side plot is new and rests on a somewhat technical foundation, we first develop it for the special situation of a two-way layout. We show how the row effects and the column effects can be adjusted to make them comparable when working with mean squares.

A second development handles any bordering table for a layout associated with any number of factors. This development is partly heuristic; we provide motivation and insight for the algebraic steps. In particular, we show how the development moves toward sets of adjusted effects whose extents are about the same size when the mean squares are the same for the different bordering tables.

Development for a Two-Way Layout

We consider a two-way layout with I rows, J columns, and a single observation per cell. To place the row effects a_i and the column effects b_j in comparable scales, we divide each effect by its standard deviation to give what are sometimes called "standardized effects." This step needs formulas for the variances of a_i and of b_j, which we now derive.

Chapter 5 describes in detail how to calculate the effects for a two-way layout. The main effect for row i is

$$\begin{aligned} a_i &= \frac{1}{J}\sum_{j=1}^{J} y_{ij} - \frac{1}{IJ}\sum_{i=1}^{I}\sum_{j=1}^{J} y_{ij} \\ &= y_{i\cdot} - y_{\cdot\cdot}\,. \end{aligned}$$

The usual model for analysis of variance assumes that each y_{ij} has the same variance, σ^2. By writing a_i as a linear combination of the individual y_{ij} and then doing some algebra, we obtain

$$\operatorname{var}(a_i) = \frac{1}{J}\sigma^2 - \frac{1}{IJ}\sigma^2 = \frac{I-1}{IJ}\sigma^2.$$

A similar calculation gives

$$\operatorname{var}(b_j) = \frac{J-1}{IJ}\sigma^2.$$

Thus the standardized effects are

$$\frac{a_i}{\sqrt{\operatorname{var} a_i}} = \sqrt{\frac{IJ}{I-1}}\,\frac{a_i}{\sigma} \quad \text{and} \quad \frac{b_j}{\sqrt{\operatorname{var} b_j}} = \sqrt{\frac{IJ}{J-1}}\,\frac{b_j}{\sigma}.$$

Apart from σ, an unknown factor that is common to the variances of all effects and residuals, the right-hand side of these equations involves the total number of data values (IJ) and the numbers of degrees of freedom associated with the row effects and column effects ($I - 1$ and $J - 1$). Its form is

$$(\text{effect})\left[\frac{\text{number of observations}}{\text{d.f.}}\right]^{1/2},$$

and the subsection below shows how this expression generalizes.

The adjusted effects give plots better attuned to the mean squares, because adjusted effects from two overlays have the same general size when they lead to the same mean squares and because similar mean squares give similar adjusted effects. We now turn to the more general development of the adjusted side-by-side plot attuned to mean squares. The plot displays effects (for *any* bordering table) and residuals in a scale where all have comparable variability.

A More General Development

For a given bordering table we can calculate the corresponding sum of squares by squaring each entry, summing these, and multiplying the result by the number of times that the bordering table appears in the full overlay. For the previous example, in an $I \times J$ two-way layout with one observation per cell, the overlay for the effects of the row factor is constant by rows. That is, a copy of the bordering table for the row effects appears in each of the J columns of the overlay. Thus $\text{SS}_{\text{rows}} = J\Sigma a_i^2$.

The mean square for a bordering table is just the sum of squares divided by the corresponding degrees of freedom, d.f. We can think of each entry in the bordering table as contributing a squared effect, multiplied by the number of appearances of the bordering table and divided by d.f. If we take square roots to move from the squared scale to the scale of the data, we obtain the adjusted effect:

$$(\text{effect})\left[\frac{\text{number of appearances of bordering table}}{\text{d.f.}}\right]^{1/2}. \tag{1}$$

For the row effects of a two-way layout, this expression will be

$$a_i\left[\frac{J}{I-1}\right]^{1/2};$$

note that summing the square of this expression over i gives the mean square:

$$\sum_{i=1}^{I} a_i^2\left[\frac{J}{I-1}\right] = J\frac{\sum_{i=1}^{I} a_i^2}{I-1} = Js_a^2.$$

This calculation illustrates why we say that the adjusted effect is "attuned to mean squares."

For the plot we want to go further, however, and arrange things so that, when plotted, sets of effects that give the same value of the mean square come out about the same size. From expression (1), a given value of the mean square corresponds to smaller effects when I is larger, and to larger effects when I is smaller. To further adjust, we multiply expression (1) by

$$(\text{number of entries in bordering table})^{1/2}.$$

To motivate this choice, we regard the mean square as a sum of squared terms of the form given by expression (1), and we consider a situation in which all these terms have the same magnitude. Then the individual squared term equals

$$\text{MS}/(\text{number of entries in bordering table}).$$

Thus the further rescaling factor places on the same footing two bordering tables that have different numbers of entries.

We get a useful simplification in the final expression for the rescaled effects by observing that

$$\begin{aligned}&(\text{number of appearances of bordering table})\\ &\quad\times(\text{number of entries in bordering table})\\ &\quad=(\text{number of observations}).\end{aligned}$$

Thus the side-by-side dotplot attuned to mean squares uses

$$(\text{effect})\left[\frac{\text{number of observations}}{\text{d.f.}}\right]^{1/2} \tag{2}$$

as the quantity to be plotted against the vertical axis within each bordering table. For the example of row effects for the two-way layout, expression (2) becomes

$$a_i\left[\frac{IJ}{I-1}\right]^{1/2},$$

and this agrees with our earlier development.

Example: feeding protein to rats

Figure 5-9 gave the raw data and the effects for the weight gain of rats on six forms of protein. The study used 60 rats, 10 at each combination of two levels of protein and three types of protein. Figure 5-11 gave a side-by-side display of the two sets of main effects, the level × type interaction, and the residuals from a full-effects fit. We now compare the adjusted side-by-side display to this original display.

The factors for adjusting the effects presented in Figure 5-9 are $\sqrt{\frac{60}{1}} = 7.7$ for protein level, $\sqrt{\frac{60}{2}} = 5.5$ for both protein type and the (two-factor) interaction, and $\sqrt{\frac{60}{54}} = 1.05$ for the residuals. We omit giving a new table of the adjusted effects. Figure 8-4 reproduces the side-by-side display of Figure

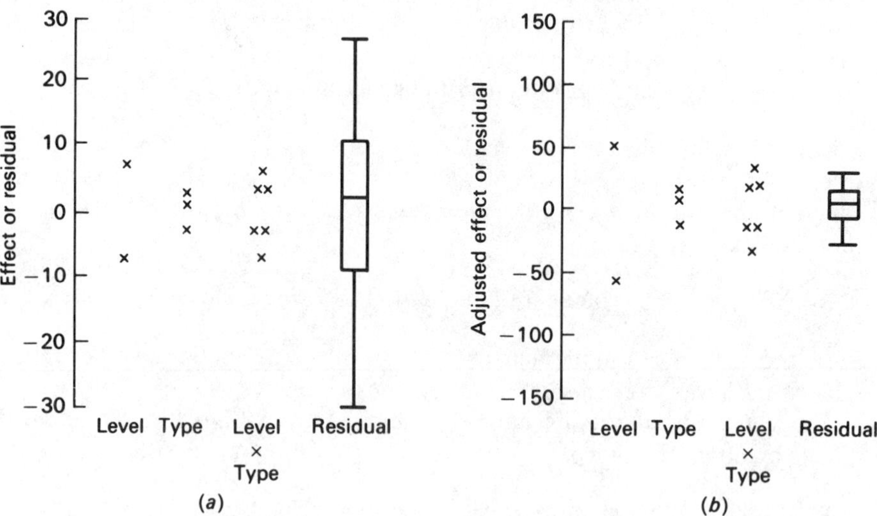

Figure 8-4. Two side-by-side plots of effects and residuals for the weight gains of rats treated with two protein levels and three protein types. (*a*) Side-by-side plot of unadjusted effects. (*b*) Side-by-side plot of effects adjusted to compare mean squares.

5-11 and adds, at the right, the new adjusted display. Although the scales for the plots differ by a factor of 5, we have chosen the visual size for the new plot so that the effects for protein type and the interactions are the same in both side-by-side plots; this choice makes the comparison easy, although we could have chosen instead to match the plots on some other set of effects.

In this illustration, the adjustment has substantial impact. The effect for level increases by a factor of $\sqrt{2} = 1.4$. Relative to the type effects and the interactions, the magnitudes of the residuals decrease more dramatically, by a factor of $\sqrt{\frac{2}{54}} = 0.2$. The display in Figure 8-4b is better suited than Figure 8-4a to interpreting the mean squares presented in Table 5-13.

8D. A DETAILED EXAMPLE: PERCENTAGE OF AMERICANS WHO HAVE NEVER MARRIED

The U.S. Census gathers data on martial status in the population. Bianchi and Spain (1986, p. 12) present the percentage of Americans who have never married. They give the percentages for race, age, gender, and year. Table 8-1 presents the data in a $2 \times 5 \times 2 \times 5$ layout, where age and year both have five ordered categories. Before we examine a side-by-side plot attuned to mean squares, we explore these data briefly using graphical methods in order to become acquainted with some of the principal messages.

Inspection of the data suggests that the age category will have the largest effects. Whereas fewer than 10% of people of age 35 and over have never married, more than 80% of those at ages 15–19 have not married. To begin to examine the role of other factors, we present five plots of the percentages, one for each age group, by race and gender, against year. Figure 8-5 presents these plots, which display all 100 of the percentages given in Table 8-1.

Comparison among the five panels (i.e., the age categories) requires care. Although the vertical units are the same, the values differ across the panels: 50–100% in the first, 30–80% in the second, 10–60% in the third, and 0–50% in the last two. These choices spread out the percentages more for easier examination of each plot.

What can we learn from these graphical displays of the data? The percentage never married generally increased over the 33-year period, except for those over 35. Blacks are more likely never to have married, especially in recent decades. For each race, men are more likely never to have married than women, and the gender difference is more pronounced in recent decades, with the exception of teenagers. About 90% of all Americans do eventually marry, but the clear trend in recent decades is toward marriage later in life.

Each of the statements above describes main effects or two-factor interactions (e.g., age-by-year interaction). For higher-way data sets like this one, we need other plots (not discussed in this volume) to explore more carefully the interaction effects—especially three-factor or higher-order interactions.

Table 8-1. Percentage of Americans Who Have Never Married

	Women					Men				
Age (years)	1950	1960	1970	1980	1983	1950	1960	1970	1980	1983
					White					
15–19	83.5	83.9	88.0	90.7	92.8	96.8	96.1	95.9	97.2	97.9
20–24	32.4	27.4	35.1	48.6	52.2	59.5	52.6	54.9	66.9	71.1
25–29	13.1	9.8	10.8	19.2	21.8	23.6	20.0	18.7	30.5	35.5
30–34	9.3	6.6	6.7	9.1	10.9	13.1	11.3	10.0	13.8	18.3
35 +	8.5	7.4	6.4	5.3	4.8	8.9	7.6	6.8	5.8	5.8
					Black					
15–19	78.9	83.8	88.6	95.1	96.9	95.6	96.2	95.5	98.0	99.6
20–24	31.2	35.4	43.6	67.5	75.3	54.7	57.1	58.4	77.8	85.2
25–29	14.1	15.7	21.3	37.0	42.6	25.2	27.6	25.4	42.6	56.6
30–34	8.9	9.6	12.8	21.5	27.2	14.4	17.2	16.1	23.1	29.3
35 +	5.2	6.1	7.1	8.4	9.0	7.9	9.5	9.5	9.8	8.8

Note: Data for 1983 are from the Current Population Survey and not strictly comparable to census data for earlier years; data for 1950 and 1960 are for nonwhites.

Source: U.S. Bureau of the Census, *Census of Population*: *1950*, Vol. 2, Pt. 1, U.S. Summary, Table 104; *Census of Population*: *1960*, Vol. 1, Pt. 1, U.S. Summary, Table 176; *Census of Population*: *1970*, Vol. 1, Pt. 1, U.S. Summary, Table 203; *Census of Population*: *1980*, Vol. 1, Chap. D, U.S. Summary, Table 264; "Marital Status and Living Arrangements: March 1983," *Current Population Reports*, Series P-20, No. 389, Table 1. Taken from *American Women in Transition* by Suzanne M. Bianchi and Daphne Spain. © 1986 The Russell Sage Foundation. Used with permission of the Russell Sage Foundation.

Side-by-Side Plots for Data About Marriage

The raw material for either version of a side-by-side plot for effects is the collection of effects for each factor, for all interactions among factors, and for residuals (Table 8-2).

To examine the impact of adjustments for degrees of freedom, we begin by presenting Figure 8-6 , a version of the usual side-by-side plot of the four sets of main effects, the six sets of effects for two-factor interactions, the four sets of effects for three-factor interactions, and the residuals from a full-effects analysis.

To construct the side-by-side plot attuned to mean squares, we rescale each set of effects in Table 8-2 according to expression (2) with the appropriate degrees of freedom from Table 8-2 (these data have 100 observations). Thus, for example, the main effects for age are 53.7, 15.5, -13.3, -24.4, and -31.4, and rescaling them by $\sqrt{100/4} = 5$ gives 268.4, 77.4, -66.6, -122.0, and -157.2. Figure 8-7 shows a side-by-side plot attuned to mean squares. To facilitate comparisons of the two side-by-side plots, we (arbitrarily) match the plots visually on the age effects.

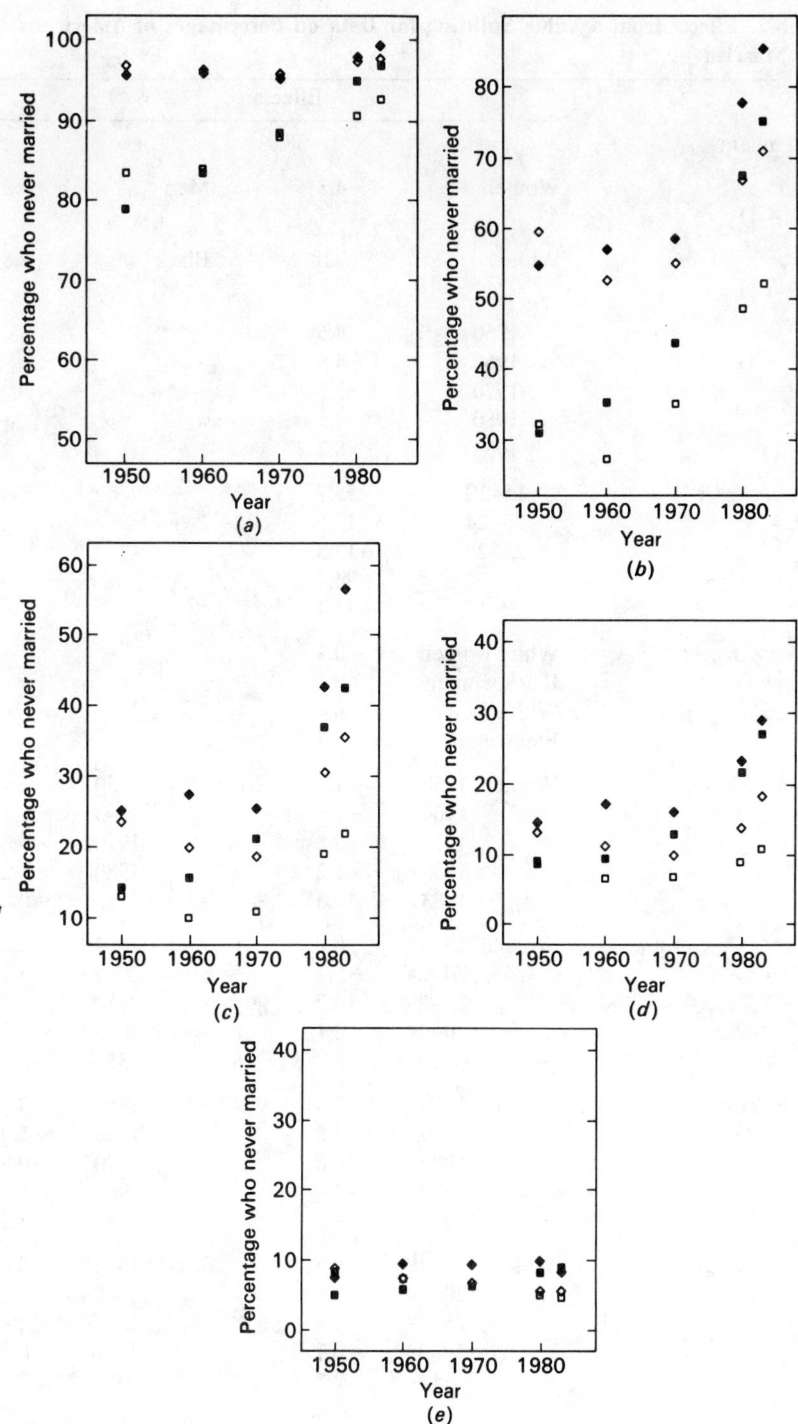

Figure 8-5. Plots of percentage of Americans never married against year for each of five age groups. Plotting symbols identify gender and race: □ white women, ■ black women, ◇ white men, ◆ black men. (*a*) Age 15–19 years. (*b*) Age 20–24 years. (*c*) Age 25–29 years. (*d*) Age 30–34 years. (*e*) Age 35 years and over.

Table 8-2. Effects from a Value Splitting for Data on Percentages of Americans Never Married

	Effects					
Grand mean:			38.9			
Gender: (d.f. = 1)	Women		−4.3	Men		4.3
Race: (d.f. = 1)	White		−2.8	Black		2.8
Year: (d.f. = 4)		1950	−4.6			
		1960	−4.8			
		1970	−3.3			
		1980	4.5			
		1983	8.2			
Age group: (d.f. = 4)		15–19	53.7			
		20–24	15.5			
		25–29	−13.3			
		30–34	−24.4			
		35 +	−31.4			
Gender × Race: (d.f. = 1)	White women		−0.4			
	Black women		0.4			
	White men		0.4			
	Black men		−0.4			
Gender × Year: (d.f. = 4)	Women	1950	−1.4	Men	1950	1.4
		1960	−1.1		1960	1.1
		1970	0.8		1970	−0.8
		1980	1.2		1980	−1.2
		1983	0.6		1983	−0.6
Gender × Age group: (d.f. = 4)	Women	15–19	−0.0	Men	15–19	0.0
		20–24	−5.1		20–24	5.1
		25–29	−0.7		25–29	0.7
		30–34	2.1		30–34	−2.1
		35 +	3.7		35 +	−3.7
Race × Year: (d.f. = 4)	White	1950	3.4	Black	1950	−3.4
		1960	1.0		1960	−1.0
		1970	0.6		1970	−0.6
		1980	−1.9		1980	1.9
		1983	−3.2		1983	3.2
Race × Age group: (d.f. = 4)	White	15–19	2.5	Black	15–19	−2.5
		20–24	−1.5		20–24	1.5
		25–29	−2.4		25–29	2.4
		30–34	−0.7		30–34	0.7
		35 +	2.1		35 +	−2.1

Table 8-2. (*Continued*)

		Effects				
Year × Age group: (d.f. = 16)						
		15–19	20–24	25–29	30–34	35 +
	1950	0.8	−5.3	−1.9	1.6	4.8
	1960	2.3	−6.4	−2.5	1.5	5.0
	1970	2.7	−3.1	−3.2	0.2	3.3
	1980	−1.8	6.3	2.2	−2.1	−4.6
	1983	−4.0	8.4	5.4	−1.2	−8.5
Gender × Race × Year: (d.f. = 4)						
		1950	1960	1970	1980	1983
White women		0.6	0.6	−0.0	−0.6	−0.5
Black women		−0.6	−0.6	0.0	0.6	0.5
White men		−0.6	−0.6	0.0	0.6	0.5
Black men		0.6	0.6	−0.0	−0.6	−0.5
Gender × Race × Age group: (d.f. = 4)						
		15–19	20–24	25–29	30–34	35 +
White women		0.2	−1.1	0.0	0.2	0.7
Black women		−0.2	1.1	−0.0	−0.2	−0.7
White men		−0.2	1.1	−0.0	−0.2	−0.7
Black men		0.2	−1.1	0.0	0.2	0.7
Gender × Year × Age group: (d.f. = 16)						
		15–19	20–24	25–29	30–34	35 +
Women	1950	−1.8	−1.8	1.0	1.3	1.2
	1960	−0.7	−1.1	0.6	0.3	0.9
	1970	−0.2	0.0	1.2	−0.2	−0.9
	1980	0.8	1.2	−0.4	−0.5	−1.0
	1983	1.8	1.7	−2.5	−0.8	−0.2
Men	1950	1.8	1.8	−1.0	−1.3	−1.2
	1960	0.7	1.1	−0.6	−0.3	−0.9
	1970	0.2	−0.0	−1.2	0.2	0.9
	1980	−0.8	−1.1	0.4	0.5	1.0
	1983	−1.8	−1.7	2.5	0.8	0.2
Race × Year × Age group: (d.f. = 16)						
		15–19	20–24	25–29	30–34	35 +
White	1950	−1.7	2.3	1.2	−0.1	−1.7
	1960	−0.8	0.1	0.8	0.3	−0.5
	1970	−0.3	0.7	0.4	−0.1	−0.7
	1980	0.8	−1.3	−0.3	0.0	0.8
	1983	2.0	−1.9	−2.1	−0.1	2.1

Table 8-2. (*Continued*)

		Effects				
Race × Year × Age group: (d.f. = 16)						
		15–19	20–24	25–29	30–34	35 +
Black	1950	1.7	−2.3	−1.2	0.1	1.7
	1960	0.8	−0.1	−0.8	−0.3	0.5
	1970	0.3	−0.7	−0.4	0.1	0.7
	1980	−0.8	1.3	0.3	0.0	−0.8
	1983	−2.0	1.9	2.1	0.1	−2.1
Residuals: (d.f. = 16)						
		1950	1960	1970	1980	1983
White women	15–19	.4	−.4	−.1	−.1	.1
	20–24	−.0	−.0	.2	.1	−.3
	25–29	−.1	.2	−.6	−.5	.9
	30–34	.0	.3	.2	.0	−.6
	35 +	−.4	−.1	.1	.5	−.1
Black women	15–19	−.4	.4	.1	.1	−.1
	20–24	.0	.0	−.2	−.1	.3
	25–29	.1	−.2	.6	.5	−.9
	30–34	−.0	−.3	−.2	−.0	.6
	35 +	.4	.1	−.2	−.5	.1
White men	15–19	−.4	.4	.1	.1	−.1
	20–24	.0	.0	−.2	−.1	.3
	25–29	.1	−.2	.6	.5	−.9
	30–34	−.0	−.3	−.2	−.0	.6
	35 +	.4	.1	−.2	−.5	.1
Black men	15–19	.4	−.4	−.1	−.1	.1
	20–24	−.0	−.0	.2	.1	−.3
	25–29	−.1	.2	−.6	−.5	.9
	30–34	.0	.3	.2	.0	−.6
	35 +	−.4	−.1	.2	.5	−.1

Within each bordering table, the messages in Figures 8-6 and 8-7 are the same, because relative scaling of bordering tables does not matter. A close comparison of the effects reveals that the adjustment for mean squares leads to relatively larger effects for gender and for race; without the adjustment, these two factors seem to have effects with substantially smaller range than the effects for year. Among interactions, the most pronounced impact of the adjustment is to reduce the relative sizes of effects for year-by-age, so that the range of these effects is now smaller than that for gender-by-age. All three-factor interaction effects and the residuals appear slightly less important, following the adjustment.

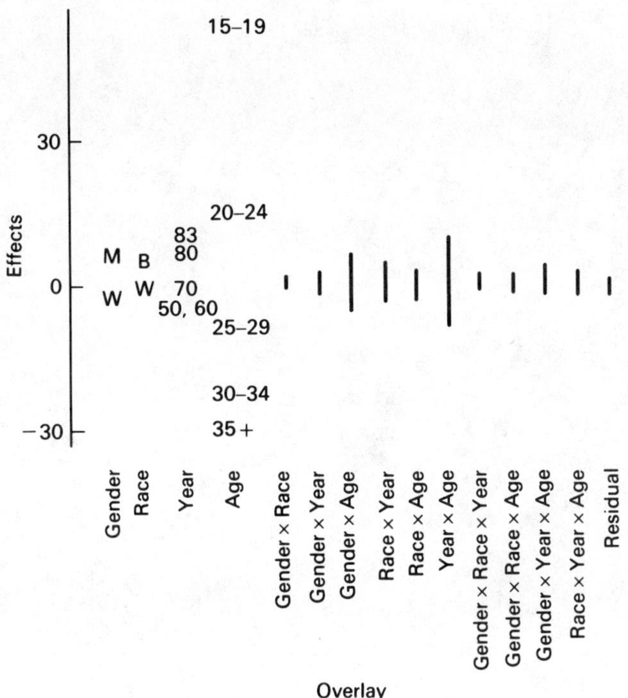

Figure 8-6. Unadjusted side-by-side plot of effects and residuals from full-effect fit to data on percentage never married.

We can learn about the data on marriages from either version of the side-by-side plot. We are not surprised that the age effects are very substantial; by definition, people are more likely to have married at least once as they grow older. Gender effects are somewhat greater than race effects; men are more likely to remain unmarried than women, and blacks are somewhat more likely to remain unmarried than whites. The five effects for year are modest but show a trend toward fewer first marriages in recent decades. The gender-by-age interactions are the largest among the interaction effects, and this may be associated in part with the relatively large values of the main effects for age and gender.

8E. PATTERNS OR NOISE?

An important theme in the analysis of variance is a desire to recognize systematic variation in a data layout, and to distinguish it from variation that is random or, at least, not accounted for by information at hand. Plots for checking interaction, such as Figure 8-2, illustrate the challenges in trying to make this distinction. How close to parallel must two or more traces be in order to indicate that interaction is absent? When is a departure from

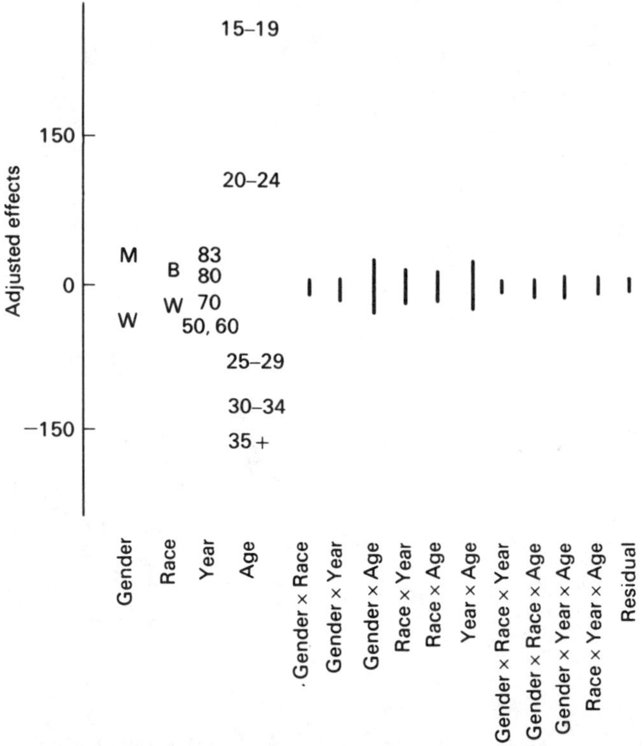

Figure 8-7. Side-by-side plot attuned to mean squares, for effects and residuals from full-effects fit to data on percentage never married.

parallelism simply a representation of (random) noise? Having a sizable number of comparisons thrust upon us in a display makes it almost essential that we begin to face the questions of multiplicity.

Answers to such questions require that we make simultaneous statements about many effects or about many comparisons of effects; therefore the statistical concepts of multiple comparisons and of contrasts have relevance. Entire books discuss these and related types of questions; see, for example, the book on contrasts by Rosenthal and Rosnow (1985) and the more advanced texts by Miller (1981) and by Hochberg and Tamhane (1987) on multiple-comparison procedures.

When we try to judge whether two or more traces are nearly parallel, we make judgments about many effects, or comparisons of effects, at the same time. If, for example, we focus on whites aged 20–24 years in Figure 8-5b, a judgment about the absence of interaction amounts to deciding whether the differences between women and men, at the five dates, are attributable to chance fluctuations.

The percentages of people never married arise from data compiled by the U.S. Bureau of the Census. This special situation gives us greater confidence that the percentages shown in Table 8-1 have high precision, both because of care in gathering and organizing the data and because of the large numbers. Thus we believe that the differences are real instead of being an artifact of chance fluctuation; of course, we cannot know whether the differences reflect underlying differences in the population or whether they may also reflect differences in reporting for the U.S. Census. However, in many situations the framing of multiple questions about the validity and nature of interactions leads to the difficulties inherent in multiple tests. We now briefly outline two elementary techniques for beginning to respond to the challenge of making multiple comparisons. Chapter 12 describes and illustrates other techniques for addressing this challenge.

A Role for *F* Tests

F tests for effects and interactions, as introduced for one-way ANOVA in Section 7A, can help guard against making too many errors when multiple questions are asked. R. A. Fisher (1935) developed a multiple-comparison procedure sometimes referred to as the *protected least significant difference* (*LSD*) *test*. Its two stages are:

1. Conduct an overall (or "omnibus") F test at level α on the effects being compared, to determine whether there are significant differences among them.
2. Only if the test is significant, examine each pairwise difference of effects using a t test at level α.

Because of the extra protection afforded by the test, the LSD procedure helps to control the overall error rate under the null hypothesis of no differences in the set of effects. By "error rate," we mean the probability that one or more comparisons lead to an incorrect statistical conclusion of a significant difference.

For the percentage of Americans never married, Table 8-3 gives the ANOVA table corresponding to a full-effects analysis—that is, an analysis including all four main effects, all six two-factor interactions, and all four three-factor interactions. Although the values of F differ among the sources of variation, and although the numerator degrees of freedom also differ, every value of F is relatively large; all the values of P are smaller than .001, and only two are larger than .0001, so that each of the 14 sources of variation is highly significant (if we take the residuals as the definition of chance fluctuation).

For further illustration, we focus on the gender × year × race interaction term. With an F ratio of 9 and an associated P value of .0005, the analysis of variance gives convincing evidence of highly significant gender × year × race interaction effects. The test used is an *omnibus* test; it shows the presence of the interaction, but it does not give a more focused account of the nature or the details of the interaction. Once the related F test has found significance,

Table 8-3. Summary ANOVA Table for Full-Effects Analysis of the Percentage of Americans Who Have Never Married, by Gender, Year, Race, and Age Group

Source of Variation	d.f.	SS	MS	*F* Ratio	*P*
Gender	1	1,871	1,871	2,555	.0001
Year	4	2,868	717	979	.0001
Race	1	790	790	1,078	.0001
Age group	4	97,653	24,413	33,331	.0001
Gender × Year	4	113	28	38	.0001
Gender × Race	1	13	13	18	.0007
Gender × Age group	4	906	227	309	.0001
Year × Race	4	534	134	182	.0001
Year × Age group	16	1,759	110	150	.0001
Race × Age group	4	392	98	134	.0001
Gender × Year × Race	4	27	7	9	.0005
Gender × Year × Age	16	128	8	11	.0001
Gender × Race × Age	4	36	9	12	.0001
Year × Race × Age	16	138	9	12	.0001
Error (residual)	16	12	0.73		
			$R^2 = 99.989\%$		

Notes: R^2 is the percentage of the total squared-scale variation accounted for in the full-effects analysis. The analysis was performed in SAS, which gives .0001 whenever the actual P value is smaller than .0001. We have rounded the sums of squares, mean squares, and F ratios to the nearest integer.

we can be more comfortable in taking seriously at least some of the observed effects for the gender × year × race interaction (Table 8-2).

Bonferroni Method

Having noted that the preliminary F test protects the error rate in the null situation of no differences, we acknowledge that, if even a single comparison differs significantly from zero, that test does not protect us against finding chance differences among the other comparisons. The LSD approach is thus rather weak in protecting the null situation. A second multiple-comparison procedure, the *Bonferroni method* (Fisher 1935; Miller 1981, Chap. 1), is much more conservative in the way that it guards the null situation. (The penalty paid for this conservatism is its low power in detecting true departures from the null hypothesis; see Hochberg and Tamhane (1987, Chap. 1) for a detailed discussion.)

The Bonferroni inequality describes the probability of the simultaneous occurrence of a collection of K events, $A_1, A_2, \ldots, A_K$. It gives a lower bound on this probability.

Bonferroni Inequality

$$\Pr(A_1 \text{ and } A_2 \text{ and} \cdots \text{and } A_K) \geq 1 - \Pr(A'_1) - \Pr(A'_2) - \cdots - \Pr(A'_K),$$

where a prime (′) signifies the complement of an event.

To motivate this important inequality, we recall that the complement of the compound event (A_1 and A_2 and $\cdots$ and A_K) is (A'_1 or A'_2 or $\cdots$ or A'_K). In words, if events A_1 through A_K do not all occur, then at least one of their complements, A'_1 or A'_2 or $\cdots$ or A'_K, must occur. Then the probability that (A'_1 or A'_2 or $\cdots$ or A'_K) occurs is at most the sum of the probabilities that the A'_i occur. Altogether,

$$\begin{aligned}\Pr(A_1 &\text{ and } A_2 \text{ and} \cdots \text{and } A_K)\\ &= 1 - \Pr(A'_1 \text{ or } A'_2 \text{ or} \cdots \text{or } A'_K)\\ &\geq 1 - (\Pr(A'_1) + \Pr(A'_2) + \cdots + \Pr(A'_K)),\end{aligned}$$

and this result is the Bonferroni inequality.

When K tests are simultaneously carried out, the Bonferroni inequality gives a bound on the probability that no test will falsely reject the null hypothesis that all K null situations are true:

$$\begin{aligned}\Pr(\text{none of } K \text{ tests rejects}) \geq 1 - \Pr(\text{first rejects}) &- \Pr(\text{second rejects})\\ &- \cdots - \Pr(K\text{th rejects}).\end{aligned}$$

If, for example, all testing is done at level .05, and if there are $K = 10$ tests, then the Bonferroni inequality leads to

$$\Pr(\text{no false rejections}) \geq 1 - 10(.05) = .50.$$

The inequality suggests that the type I error rate could be as large as .50—substantially greater than the nominal .05—when we consider all tests simultaneously. The Bonferroni procedure compensates in a conservative way by performing each of the $K = 10$ individual tests at level α/K (here .005), so that the overall bound associated with K simultaneous tests is $K(\alpha/K) = \alpha$ (= .05).

As a small example, we return to the five differences between white women and white men aged 20–24 years in Figure 8-5b. Visually, when we ask about interaction, we compare the five vertical separations to see whether they appear to be roughly equal. In effect, we compare each separation with the others—a total of 10 pairwise comparisons. Application of Bonferroni's inequality would have us make each of these comparisons at the .005 (= .05/10) level. If one or more of the comparisons were significant at this level, we would suspect that something more than chance fluctuations is at work.

The vertical separations between the points for white men and white women in Figure 8-5b are, from left to right, 27.1, 25.2, 19.8, 18.3, and 18.9.

For example, from Table 8-1 the difference in 1960 is 52.6 − 27.4 = 25.2. Each of the five vertical separations is a difference, and so its variance is two times the variance of an individual observation. When we compare two of these numbers—say, 27.1 and 18.3—the variance of their difference is twice that of a simple difference. Altogether, the variance appropriate for comparing two vertical separations in Figure 8-5b is estimated by

$$4 \times (\text{residual mean square}) = 4 \times (.73) = 2.93,$$

and the corresponding standard deviation is 1.71 (with 16 degrees of freedom).

The difference between the largest (at 1950) and smallest (at 1980) separation—8.8—is (8.8/1.71 =) 5.1 standard deviation units. Because the associated P value (from Student's t on 16 d.f.) is less than .05/10 = .005, we conclude that the Bonferroni approach finds significant variation in the separation of these two traces. Our selection of the observed largest separation and smallest separation for this comparison makes a test at the usual .05 significance level invalid, and it points to the need for the Bonferroni adjustment (or a related approach). We conclude that the apparent interactions seen in Figure 8-5b are likely to be real, and not attributable to chance fluctuations. Further information about the standard errors of the percentages would let us modify and refine the analyses outlined here.

8F. EXPLORING RESIDUALS GRAPHICALLY

We examine the residuals from a fit:

1. To ask whether patterns remain in the residuals;
2. To compare the sizes of the residuals with the sizes of the various effects we have split out;
3. To look for unusual residuals;
4. To ask whether the residuals are symmetrically distributed, and how much they seem to depart from Gaussianity.

Sections 8A and 8B, as well as Chapter 5, use graphical methods to explore residuals:

Figure 5-2 shows a dotplot of the residuals by city;

Figure 5-3 gives a boxplot of residuals and identifies outliers;

Figure 5-6 plots residuals against fitted values for the smoking data to explore whether variance is homogeneous.

The graphical methods already introduced for exploring residuals are illustrated further in later examples. This section introduces additional graphical methods for exploring residuals.

Departures from Gaussian Shape

A plot of the quantiles of one distribution against those of another distribution helps us to see whether the two distributions are similar in shape, even when they may have widely different locations and spreads. These plots are called *quantile–quantile* (or Q-Q) *plots*. Often we use a Q-Q plot to compare a distribution of data—especially of residuals—to the (standard) Gaussian distribution. This plot is called a *normal probability plot*.

The median of the sampling distribution of the ith order statistic in a Gaussian sample of size n is very nearly given by the inverse cumulative distribution function evaluated at selected probability values:

$$\Phi^{-1}\left(\frac{i-\frac{1}{3}}{n+\frac{1}{3}}\right).$$

To illustrate the use of this expression, consider a Gaussian sample of size $n = 10$. For the first (smallest) order statistic, we obtain

$$\Phi^{-1}\left(\frac{2/3}{31/3}\right) = \Phi^{-1}(.0643) = -1.51.$$

Similarly, with 10 observations some other plotting positions are:

i	Φ^{-1}
2	−0.99
6	0.12
9	0.99
10	1.51

UREDA (Section 7B) and EDTTS (Section 10C) give more complete discussions. A normal probability plot of the ith ordered value in a Gaussian sample with n values against the above plotting position will resemble a straight line.

The normal probability plot displays departures from Gaussian shape and can indicate their nature. An S-shaped plot that is below the straight line at the extreme left, and above the straight line at the extreme right, indicates a heavy-tailed distribution. When the left tail is above the line and the right tail is below the line, the distribution is light-tailed. A curve that is concave up shows right-skewness, and one that is concave down shows left-skewness. More complex patterns are also common. Moore and McCabe (1989, Sec. 1.3) give a more complete discussion of the interpretation of normal probability plots; they use several data sets for illustrations. As we illustrate below, the plot is easily produced by any statistical software package that includes an inverse cumulative distribution function for the standard Gaussian distribution; we used Minitab to produce the plots shown.

Figure 8-8 gives, in two panels, normal probability plots: for the raw percentages of people who have never married, and for the residuals from a

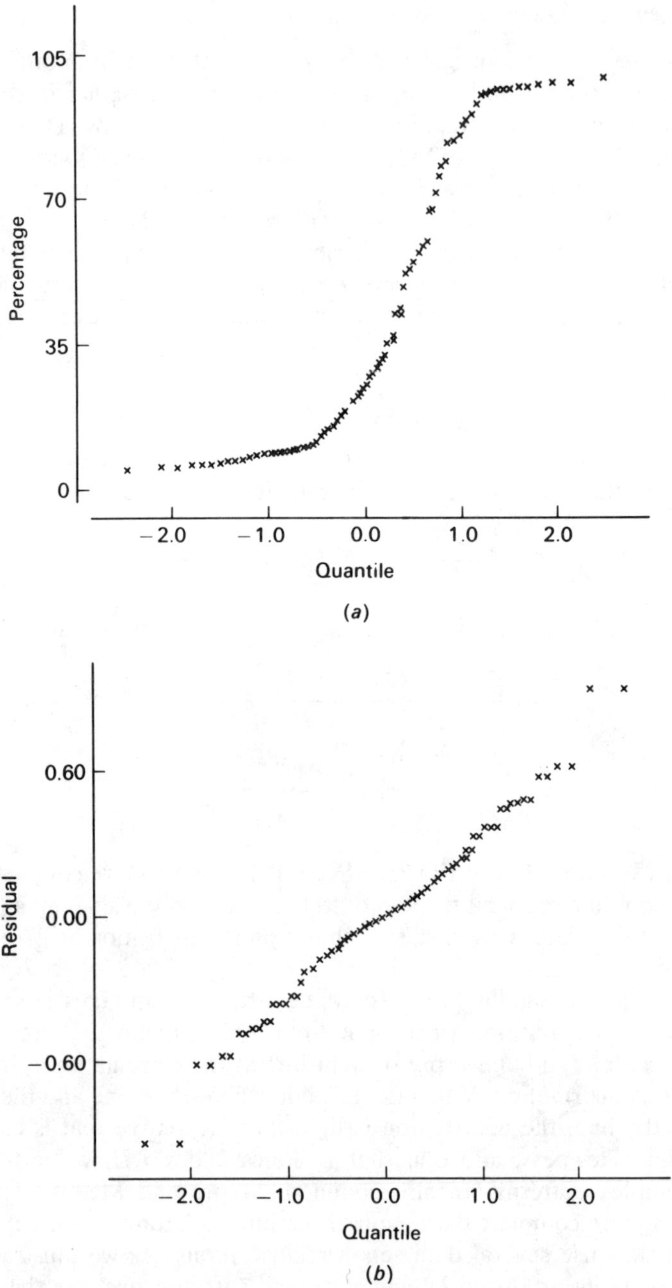

Figure 8-8. Normal probability plots of raw percentages and residuals in data on percentage never married. (*a*) Raw percentages. This plot illustrates severe departures from Gaussian shape. (*b*) Residuals from a full-effects analysis. The plot indicates near-Gaussian shape.

full-effects fit to these data. The linear pattern in the plot for the residuals suggests that the residuals are very nearly Gaussian in shape. (There is some evidence for heavy-tailedness at the very extreme edges of the residual distribution, but this may reflect the impact of a couple of cells that are not fitted well.)

Figure 8-8a, however, shows that the raw percentages are, as we might well expect, very far from Gaussian in shape. The behavior of the plot at the extremes suggests light-tailedness—due perhaps to a truncation effect for the percentage at both ends. The striking appearance of pronounced departures from linearity may reflect that the percentages comprise two or more distinct distributions. Our finding is consistent with the strong separation of the percentages for the different age groupings, seen in the display of the entire data set in Figure 8-5.

8G. SUMMARY

Graphical techniques, when suitably chosen, aid our understanding of systematic structure in data layouts. They can also help us effectively to convey and interpret that understanding.

Earlier chapters have made ample use of visual displays for raw data sets and of displays that can accompany a two-factor analysis of effects and residuals. After providing an overview of these techniques, this chapter develops a new side-by-side plot for use with mean squares and then moves to situations with three and more factors. We focus on the difficulties associated with asking questions simultaneously about many comparisons, and we briefly review two elementary techniques—Fisher's least significant difference and the Bonferroni procedure—that can help in starting to address multiplicity. Finally, we review graphical methods used in exploring residuals, and we introduce one additional display—the normal probability plot.

REFERENCES

Bianchi, S. M. and Spain, D. (1986). *American Women in Transition*. New York: Russell Sage Foundation.

Fisher, R. A. (1935). *The Design of Experiments*. Edinburgh and London: Oliver and Boyd.

Hochberg, Y. and Tamhane, A. C. (1987). *Multiple Comparison Procedures*. New York: Wiley.

Miller, R. G., Jr. (1981). *Simultaneous Statistical Inference*, 2nd ed. New York: Springer-Verlag.

Moore, D. S. and McCabe, G. P. (1989). *Introduction to the Practice of Statistics*. New York: Freeman.

Rosenthal, R. and Rosnow, R. L. (1985). *Contrast Analysis: Focused Comparison in the Analysis of Variance*. Cambridge University Press.

ADDITIONAL REFERENCES

Atkinson, A. C. (1985). *Plots, Transformations, and Regression*. Oxford: Clarendon Press.

Becketti, S. and Gould, W. (1987). "Rangefinder box plots," *The American Statistician*, 41, 149.

Chambers, J. M., Cleveland, W. S., Kleiner, B., and Tukey, P. A. (1983). *Graphical Methods for Data Analysis*. Belmont, CA: Wadsworth.

Cleveland, W. S. (1985). *The Elements of Graphing Data*. Monterey, CA: Wadsworth.

Gnanadesikan, R. (1980). "Graphical methods for interval comparisons in ANOVA and MANOVA." In P. R. Krishnaiah (Ed.), *Handbook of Statistics, Vol. 1 (Analysis of Variance)*. Amsterdam: North-Holland, pp. 133–177.

Johnson, E. and Tukey, J. W. (1987). "Graphical exploratory analysis of variance illustrated on a splitting of the Johnson and Tsao data." In C. L. Mallows (Ed.), *Design, Data, and Analysis by Some Friends of Cuthbert Daniel*. New York: Wiley.

Landwehr, J. M. and Watkins, A. E. (1986). *Exploring Data*. Palo Alto, CA: Dale Seymour Publications.

Moses, L. E. (1987). "Graphical methods in statistical analysis," *Annual Review of Public Health*, 8, 309–353.

Olson, C. L. (1987). *Statistics: Making Sense of Data*. Boston: Allyn & Bacon. [See especially Chapter 20, "Specialized Comparisons Among Means."]

Schmid, C. F. (1983). *Statistical Graphics: Design Principles and Practices*. New York: Wiley.

Tufte, E. R. (1983). *The Visual Display of Quantitative Information*. Cheshire, CT: Graphics Press.

Tukey, J. W. (1977). *Exploratory Data Analysis*. Reading, MA: Addison-Wesley.

Tukey, J. W. (1991). "The philosophy of multiple comparisons," *Statistical Science*, 6, 100–116.

Wainer, H. and Thissen, D. (1981). "Graphical data analysis," *Annual Review of Psychology*, 32, 191–241.

Wang, P. C. C. (Ed.) (1978). *Graphical Representation of Multivariate Data*. New York: Academic Press.

EXERCISES

1. Construct a parallel plot of the means, $\bar{x}_i$, with intervals $\bar{x}_i \pm s_i$ for the interest rates on new-car loans in six cities (Table 5-3). Compare your plot to Figure 5-1, and discuss the advantages of each plot.

2. Give an effective plot of the raw data, presented in Table 5-9, on percentage of smokers by age and income. Describe important features of these data.

Table 5-14 presents measurements of light absorbance for positive control samples in an ELISA test for HIV. The data come in 5 lots, with 5 runs in each lot and 3 replications within each run. Exercises 3 through 6 refer to these data.

3. Illustrate and compare the use of side-by-side boxplots for viewing differences across the five lots.

4. Develop a side-by-side display of effects and residuals that may be appropriate for the nested design used in this data set; use the effects and residuals presented in Table 5-15. Describe your findings from this display.

5. Use a suitable display to explore the variance of the residuals shown in Table 5-15. Do you think that your findings challenge an assumption of homogeneous variance? Why or why not?

6. Construct and interpret a normal probability plot for the residuals.

Table 6-1 presents data on actual and estimated cross-sectional areas of simulated solid tumors, as estimated by 13 oncologists. The simulated tumors were made of cork and rubber, and each came in three size-shape configurations. Exercises 7 through 9 refer to these data.

7. Give a visual display of the raw data, and summarize the features that seem most important.

8. Give a side-by-side boxplot display for the effects and residuals in Table 6-2; evaluate and compare it to Figure 6-1.

9. Construct a plot consisting of traces suitable for exploring material $\times$ form interaction effects. Discuss and interpret any interactions that you believe may be noteworthy; compare your findings to those presented near the end of Section 6A.

10. Figure 6-4 gives a side-by-side dotplot of the effects and the residuals for the initial full-effects analysis of the dental-gold data.

(a) Calculate the adjusted effects needed for the parallel plot attuned to mean squares.

(b) Construct the adjusted parallel plot.

(c) Describe whatever changes are brought by the adjustments; compare your plot to that in Figure 6-4.

11. Use graphical techniques to explore the residuals, presented in the southeast corner of Table 6-9, from the fit to the dental-gold data.

A 2^k factorial design has k factors, each at two levels. This design is the only layout for which the adjustments that attune the side-by-side plot to the mean squares do not make a difference. Exercises 12 through 16 examine this unusual situation for a 2^8 factorial design.

12. In a 2^8 design, suppose that the factors are A, B, C, D, E, F, G, and H and that the magnitude of the $ABCDEF$ interaction effects is Q. Describe the bordering table and the full interaction overlay for this interaction.

13. Show that the $ABCDEF$ interaction sum of squares is $256Q^2$ on 1 d.f., and thus that the mean square is also $256Q^2$.

14. Use the result of Exercise 12 to show that the adjusted side-by-side plot represents the $ABCDEF$ interaction effects as $\pm 16Q$.

15. Consider now a main effect for any factor. Describe the overlay. If the entries are $\pm U$, show that the adjusted main effects are $\pm 16U$.

16. Show that the common term and all interaction terms of all orders lead to adjusted effects of the form $16 \times$ effect. Conclude that the adjusted side-by-side plots are, in effect, no different from the unadjusted plots.

CHAPTER 9

Components of Variance

Constance Brown and Frederick Mosteller
Harvard University

To investigate the variability of observations, we may decompose their variance into parts labeled by factors we are studying: that is, we study the *components of variance*.

Section 9A introduces two structures leading to variance components in the one-way analysis of variance: fixed-effects models, where all the versions of interest are present, and random-effects models, where the data come from only a sample of the versions. In one-way analysis of variance, the variance components are the average of the within-group variances and the variance of the true group means. Section 9B gives an example using variance components to study blood pressure.

Because mean squares are so variable, the variance component estimates may be negative, and Section 9C offers some methods of avoiding such outcomes. Beyond point estimates of variance components, we provide ways of calculating confidence intervals for them in Section 9D.

When the groups are of unequal size—the unbalanced case—Section 9E provides a way to carry out the analysis.

Section 9F introduces and illustrates the analysis of two-way tables with one observation per cell. Section 9G studies the effect of national politics on local and state elections.

9A. STRUCTURES LEADING TO COMPONENTS IN ONE-WAY ANALYSIS OF VARIANCE

To illustrate the notion of variance components, we next discuss three similar examples of one-way ANOVA. They illustrate the ideas of fixed effects and random effects, of sampling from populations, and of several measurements on the same item.

Table 9-1. Mean and Standard Deviation of Forced Mid-expiratory Flow (FEF) for Five Groups Corresponding to Various Smoking Habits (200 People in Each Group)

Group	Mean	Standard Deviation
Nonsmokers	3.78	.79
Passive smokers	3.30	.77
Light smokers	3.23	.78
Medium smokers	2.73	.81
Heavy smokers	2.59	.82

Source: James R. White and Herman F. Froeb (1980). "Small-airways dysfunction in nonsmokers chronically exposed to tobacco smoke," *New England Journal of Medicine*, 302, 720–723 (data from Table 1, p. 721). In addition to the five groups of smokers listed above, the authors studied another group with a different sample size. We omitted that group so as to maintain a balanced example.

Fixed Effects

Example: smoking and lung function

White and Froeb (1980) used a measure of lung function (forced mid-expiratory flow or FEF) to investigate how smoking affected breathing. They chose five groups of people with various smoking habits and 200 people within each of these groups. Table 9-1 gives the mean and standard deviation in each group.

Glancing down the right-hand column, we see that the observed variation *within* the groups is nearly equal; the standard deviations are all approximately 0.8. If, contrary to fact, we knew the true mean and variance for each group and if we knew the shapes of the distributions of the measurements—if, for instance, they were approximately Gaussian—then we would have a complete description of each population. Also, we would be able to draw conclusions about relationships *between* the various populations; for example, we could estimate the percentage of time a randomly selected nonsmoker's FEF will be greater than a randomly selected heavy smoker's FEF (about 84%). (Actually, this particular result—because it involves moderate probabilities—is not very much affected by even substantial deviations from Gaussianity.)

The Balanced One-Way Model

In a model in the classical ANOVA style, we could consider writing the response y_{ij} for member j of group i as

$$y_{ij} = \mu_i + \varepsilon_{ij}, \qquad i = 1, 2, \ldots, I;\ j = 1, 2, \ldots, J,$$

where the uncorrelated ε_{ij} have some distributions with average zero and variance σ_i^2. Only the *lack of correlation* of the ε_{ij} is a serious assumption. We can always define μ_i to make the average of the ε_{ij} zero. Because the sample sizes are equal, the design is said to be *balanced*.

Sometimes we break up μ_i into two components so that

$$\mu_i = \mu + \alpha_i.$$

The model then parallels the overlays, with μ being the mean of the μ_i, the common value in the model overlays parallel to the grand mean of the original data. The Greek α alerts us that the study includes all the populations of interest. Later, when we sample populations, we use a_i (instead of α_i) to warn that the populations drawn into the study do not include all the populations in the superpopulation.

The smoking example has $I = 5$ and $J = 200$. Thus, for example, if the ε_{ij} were *independent* (and not just *uncorrelated*), the distribution of FEF for the nonsmokers would be $F_1(\mu_1, \sigma_1^2)$ and for the medium smokers $F_4(\mu_4, \sigma_4^2)$, for some F's. If the F's were Gaussian and if $\mu_1, \ldots, \mu_5$ and $\sigma_1^2, \ldots, \sigma_5^2$ were known, we could draw and compare pictures of these distributions, as in Figure 9-1 for the nonsmokers and the medium smokers, and we would have all the information required to understand each population of smokers individually, as well as the relationships *among* the various groups.

For much of analysis of variance, a Gaussian assumption has little role. For measurement data, distributions of averages are likely to be close to normal (Gaussian), even though that assumption fails for the individual measurements. We find the spot where the shoe sometimes pinches when we want to test the hypothesis that the population means are equal. Then the frequency interpretation of the theoretical F distribution conventionally used in the testing does depend, somewhat, on the normality assumption, though the extent of dependence, which is not easy to summarize, is only rarely large enough to bother us.

Similarly, independence matters only infrequently. When, as usual, we do not need the Gaussian assumption, we are working either with formulas

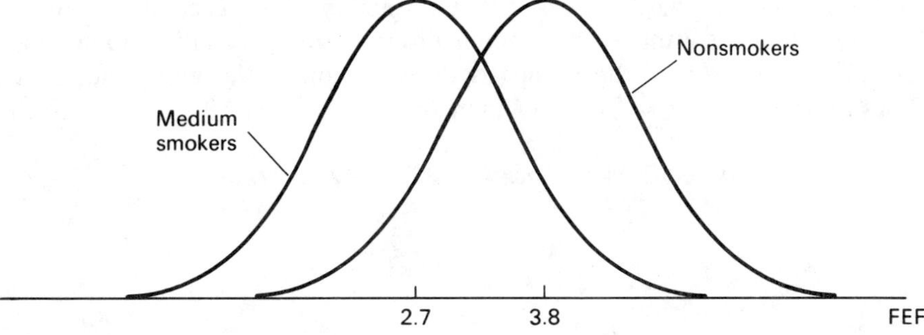

Figure 9-1. Comparing the distribution for the medium smokers with that for the nonsmokers.

obtained by equating averages, where lack of correlation surely suffices, or (in other chapters) with F ratios or Studentized comparisons of ranges, where lack of correlation usually seems to suffice.

For most of this chapter the assumptions of normality (Gaussianity) and independence do not need to be invoked. (Their consequences will, however, illuminate the discussion in Section 9D.)

In this smoking example, the groups such as "Nonsmokers" and "Passive smokers" have been selected because the investigators want to study people who behave in the way implied by the group titles. Although we assume that members *within* each group in the investigation form a random sample from their underlying population, we do *not* consider the populations themselves as sampled randomly from a collection of populations of smokers of various kinds. Because the five smoking groups have been chosen for their own sake, we say that the populations and the corresponding effects are *fixed*: that is, a given population from the five we are considering has a fixed μ_i associated with it, and the investigator cares about these μ_i. To put it another way, we want to think, calculate, and look at the five populations defined (perhaps implicitly) by the five samples that were studied. We do not plan to look beyond these to a larger superpopulation of populations.

If we have reason to believe that the variation *within* each group, σ_i^2, differs from group to group, then (to obtain full information about each group) we would also seek estimates of these σ_i^2 from each treatment group. Otherwise, if the variation *within* groups is nearly constant *across* the groups (as seems to happen in the example at hand), then we tend to pool information from all five populations to estimate the common σ^2. Because we deal with samples rather than populations, we cannot know exactly the means μ_i nor the σ_i^2. The analysis of variance does what it can to make up for our not knowing these population values. (Occasionally, when the σ_i^2 vary considerably, we may have reason to believe in relations other than equality among the σ_i^2, such as that $\sigma_1^2 = 2\sigma_2^2 = 3\sigma_3^2$, and then we can again combine information from the samples to improve estimates of σ_i^2.)

Variability Among Group Means

Until this point, we have discussed only variability within each group, but in addition we might want to know about group-to-group variation. One measure of the diversity of the groups is the variation of the group means μ_i about their average $\mu = \Sigma_{i=1}^{I}\mu_i/I$ namely,

FIXED EFFECTS: GROUP-TO-GROUP VARIANCE COMPONENT

$$\tau^2 = \sum_{i=1}^{I} (\mu_i - \mu)^2/(I - 1).$$

This quantity is essentially the variance of the μ_i when each μ_i is treated as a single measurement. For the pulmonary function example, $I = 5$. (It is rather arbitrary whether we define τ^2 by dividing by I or $I - 1$ or some other convenient number near I. Our formulas will be simpler, however, if we divide by $I - 1$.)

Thus the variability associated with the value of an FEF for a person in one of these groups has two components: (1) the component τ^2 for variation *between* the groups and (2) the component σ_i^2 $(i = 1, 2, \ldots, 5)$ for *within*-group variation. And, along with (or instead of) the μ_i, we may wish to estimate these "variance components." (Although "among" seems more grammatical with more than two groups, many people speak of "between.")

If we had more detailed data, we could break up the within-group variance component into parts, possibly one for day of test or for test within day, and one for long-run behavior of subject. If we had more groups and additional information about the groups, we could break up the between-group variance component into parts, possibly one for county of residence and one for group within county. No breakdown will be ultimately complete, so we must think about just what the name for each specific variance component means. We must not let the names mislead us into oversimplifying the sources of variation.

If the number of measurements in each group is small, then the estimates of τ^2 and of σ_i^2 may not be well determined; indeed, if the sample from each group's population is of size 1, we cannot sort out τ^2 from the σ_i^2, and we cannot estimate any σ_i^2 from the data alone. As the size of each group increases, the sample mean for the ith group, $y_{i\cdot}$, tends to estimate μ_i more closely, and we are able to get better estimates of τ^2 and of σ_i^2. If we were to estimate σ_i^2 separately for each group, we could use the customary unbiased estimate

VARIANCE ESTIMATE FOR POPULATION i

$$\hat{\sigma}_i^2 = \sum_{j=1}^{J} (y_{ij} - y_{i\cdot})^2/(J - 1), \qquad y_{i\cdot} = \sum_j y_{ij}/J.$$

(For printing convenience we place the ^ over σ rather than over the whole of σ^2, which some people would regard as a more descriptive—and more correct—notation.)

Random Effects

We now introduce a second example to illustrate a situation with a random sample of populations, each with a sample drawn from it.

EXAMPLE: PERFORMANCE SCORES

Consider the population of all comprehensive schools in England having at least 200 pupils who took their first nationwide public examinations. Ran-

Table 9-2. Observed Mean and Standard Deviation of ILEA Scores in Each of $I = 7$ English Comprehensive Schools, $J = 50$ Students

School	ILEA Score, $y_{i\cdot}$	Standard Deviation, s_i
1	28.3	13.4
2	21.1	14.2
3	14.3	11.8
4	16.3	14.9
5	26.5	13.4
6	20.3	12.1
7	16.8	13.9
	$y_{\cdot\cdot} = 20.5$	

Source: M. Aitkin and N. Longford (1986). "Statistical modelling issues in school effectiveness studies," *Journal of the Royal Statistical Society, Series A*, 149, 1–42 (with discussion) (data from Table 1, p. 4). The seven schools above are schools 3, 4, 7, 9, 10, 11, and 13 in the original paper. We have changed the data a little for the sake of clarity and simplicity. The seven schools used here actually had sample sizes ranging from 47 to 52, and we omitted data from 11 other schools. We did not choose the sample randomly, but chose schools with data from about 50 students.

domly select seven such schools. Then, within each of these schools, choose a random sample of 50 pupils and calculate a performance score on an examination (called the ILEA score) for each pupil. High numbers represent better performance. Table 9-2 gives the summary data from Aitkin and Longford (1986). As in the previous example, we can write y_{ij}, the ILEA score for pupil j from school i, as

$$y_{ij} = \mu_i + \varepsilon_{ij}, \qquad i = 1, 2, \ldots, I;\ j = 1, 2, \ldots, J,$$

where ε_{ij} has a distribution with mean zero and variance σ_i^2, and μ_i represents the mean ILEA score, and σ_i^2 the true variance, for pupils from school i. Because ε_{2j} and ε_{3k}, for instance, occur within different schools and the schools are separately sampled, assuming absence of correlation between them leads to no difficulty. In the example $I = 7$, $J = 50$.

One important difference distinguishes our two examples: in the study of lung function, the investigators cared about the five groups representing various smoking habits for their own sake; but for the comprehensive schools, attention focuses not solely on the seven schools chosen in the sample, but on the mean ILEA score over all the schools in the described population and on the variation of the individual schools' mean ILEA scores about the population mean score.

Here we want our view of the sample of seven schools to inform us about the whole population of schools from which they were drawn. Formally, we do this by proceeding as if the individual school mean scores μ_i were drawn randomly from a population, having mean scores with overall mean μ and variance σ_A^2, made up of the N eligible schools in England. We sometimes (although this is not always necessary) assume that the μ_i are normally distributed, $\mu_i \sim \text{Gau}(\mu, \sigma_A^2)$. In some problems this assumption is fairly accurate; in others, it is just a convenient way to obtain rough approximations. For example, for many distributions the interval $\mu \pm 2\sigma$ contains approximately 95% of the distribution, whether or not the shape of the distribution is normal. The σ_i^2 represent the "within" variance for school i, and σ_A^2 represents the "between" variance for a population of true means. The individual schools might have a common "within" variance σ^2, or we might treat them as if they did, or we might recognize that their variances presumably differ.

We could also write the model as

$$y_{ij} = \mu + a_i + \varepsilon_{ij},$$

where

$$\mu_i = \mu + a_i$$

and the use of the letter a (as opposed to the Greek α) reminds us that the populations (schools) drawn into the sample do not exhaust the populations in the superpopulation (of English schools), and that the population means a_i are themselves drawn from a population with mean 0 and variance σ_A^2.

Just as the previous example was a "fixed-effects one-way ANOVA" because the groups (the people with various smoking habits) being considered were fixed and important in themselves, this example is a "random-effects one-way ANOVA" because the groups being considered are randomly selected from a "population of groups" and we want to know about *this population* of schools rather than the specific seven schools at hand. In the first example, we may wish to estimate the mean for each group, $\mu_1, \ldots, \mu_5$, and the components of variance, which are the within-group variances $\sigma_1^2, \ldots, \sigma_5^2$ and the between-group variance τ^2. In this "random-effects" example we may wish to estimate overall characteristics, like the *overall mean* μ (the mean of the population of N true school means) rather than estimating the true means, μ_1 to μ_7, corresponding to the seven schools in the study. Similarly, we may want to estimate the variance component between schools (more generally, between populations),

RANDOM EFFECTS: VARIANCE COMPONENT BETWEEN POPULATIONS

$$\sigma_A^2 = \frac{\Sigma(\mu_i - \mu)^2}{N - 1},$$

where i runs over the N populations in a superpopulation. In our example i runs over all N schools, not just over the seven schools in the sample, and μ

is the average of the N true mean ILEA scores for the schools. This definition of σ_A^2 is identical with that for τ^2, and so the only difference is a reminder that we may be sampling some but not all of the populations. If the variances within schools have the same value σ^2, we may wish to estimate that common variability. If the variances for schools differ, matters become more complicated. (Also, we have not here made explicit allowance for the fact that a student might get a different ILEA score each time he or she takes the examination; that is, we have not explicitly considered within-pupil variability.)

For the following discussion we define the notion of "precision" as the reciprocal of the variance.

Assessing Precision with a Random-Effects Model

Let us consider another example where the random-effects model may be appropriate.

EXAMPLE: MASS-PRODUCED ITEMS

A metal-working machine produces items, ideally all alike. We wish to investigate the distribution of the weights of the items produced. To this end, eight are randomly drawn for special attention from the population of 800 produced during a day. We have no special interest in these particular eight items; rather we would like to view them as representative of all the items produced by the machine in a day. This is why the random-effects (as opposed to the fixed-effects) model is appropriate. Each of the eight items is very precisely weighed twice. The two measurements on the same item are likely to differ just a little, especially when they are very precise and taken independently. In principle, the measurement process has a variance of σ^2, and that variance may well be the same for each measurement on each piece. The pieces themselves may differ slightly in their true weights, μ_i. We can use the average $y_{i\cdot}$ of the two measurements for an item to estimate μ_i. We may want to know how much the μ_i vary over the lot of 800, and so we may want to estimate σ_A^2, the between-means component of variation. We use our 16 measurements on the eight items to make estimates for the finite population of 800 items that was produced.

We might have taken more measurements on each piece, but no matter how many we take, we still get only an observed mean $y_{i\cdot}$ for piece i, not the true μ_i. Thus the information about the mean weight μ_A of all 800 items that we get from the eight pieces, even if repeatedly weighed, must be less than we would get from knowing exactly eight randomly drawn μ_i, if we had a way to know true values of μ_i. In other words, the within-item variance σ^2 (through the variance, σ^2/J, of the mean $y_{i\cdot}$ of J measurements on an item) enlarges the variability of our estimate of μ_i, and hence enlarges the variability of the overall sample mean $y_{\cdot\cdot}$.

It is convenient and instructive to think of the hypothetical population of all the items that might have been made during the day and to think of the 800 produced as a random sample from that infinite population. Then the random sample of eight from the 800 would also be a random sample from the hypothetical infinite population. More precisely, the infinite-population variance of $y_{..}$, the average of the 16 independent measurements, is

$$\sigma_{y_{..}}^2 = \frac{\sigma_A^2}{8} + \frac{\sigma^2}{16}$$

so that the degradation mentioned above is $\sigma^2/16$. Note that the variance of $y_{..}$ has two components: $\sigma^2/16$ representing the "within" variation (from the measurement process) and $\sigma_A^2/8$ representing the "between-item" variance.

We provide short asides on finite sampling and correlated measurements. Some readers may prefer to skip to the subsection on Estimation of Variance Components by the Expected-Mean-Square Method (page 203).

With the usual finite-population correction, the variance of our estimate of the mean of all 800 items is

$$\left(\frac{1}{8} - \frac{1}{800}\right)\sigma_A^2 + \frac{1}{16}\sigma^2.$$

More generally, if we draw I items randomly from an infinite population with between-means variance σ_A^2 and measure each item J times using a measurement process with error variance σ^2 and no correlation between measurement errors, then the sample mean of the IJ measurements has infinite-population variance

Variance of Grand Mean of I Items, Each Measured J Times

$$\sigma_{y_{..}}^2 = \frac{\sigma_A^2}{I} + \frac{\sigma^2}{IJ}$$

and finite-population variance

$$\left(\frac{1}{I} - \frac{1}{N}\right)\sigma_A^2 + \frac{1}{IJ}\sigma^2.$$

If we are realistic and admit a correlation $\rho > 0$ for measurement errors whenever a single item is measured and remeasured, the variance of measurement for each item will be

VARIANCE OF MEAN FOR CORRELATED MEASUREMENTS ON AN ITEM

$$\left(\frac{1}{J} + \left(1 - \frac{1}{J}\right)\rho\right)\sigma^2 = \frac{1}{J}(1 + (J - 1)\rho)\sigma^2 > \rho\sigma^2.$$

Even a quite small $\rho > 0$, say $\rho = .05$, makes it difficult to reduce the measurement error term to a modest fraction of itself—requiring $J = 19$ instead of $J = 10$ to reach $\sigma^2/10$—and impossible to make it as small as $\rho\sigma^2$ (here $\sigma^2/20$).

Thus the (infinite-population) variance of the sample mean will be

VARIANCE OF GRAND MEAN FOR CORRELATED MEASUREMENTS

$$\sigma_{\bar{y}_{..}}^2 = \frac{\sigma_A^2}{I} + (1 + (J - 1)\rho)\frac{\sigma^2}{IJ} > \frac{\sigma_A^2}{I} + \rho\frac{\sigma^2}{I}.$$

We could think about our precision for the sample mean in comparison with what we would have obtained if we had been able to measure the μ_i exactly for the I items in the sample. We could call the ratio of variances the (idealized) measurement efficiency. Thus our (idealized) efficiency is

MEASUREMENT EFFICIENCY FOR UNCORRELATED MEASUREMENTS

$$\frac{\sigma_A^2}{I} \bigg/ \left(\frac{\sigma_A^2}{I} + \frac{\sigma^2}{IJ}\right) = \frac{\sigma_A^2}{\sigma_A^2 + \sigma^2/J}.$$

The real measurement efficiency can be less because the real measurements will be correlated. To bring this ratio near 1, J must be large enough and/or σ^2 small enough to make σ^2/J much smaller than σ_A^2. This result shows why it is difficult for highly fallible measurements to give us reliable results. If $\sigma_A^2 = (.001)^2$, $\sigma^2 = (.001)^2$, and $J = 5$, then the efficiency of measuring the I items is $5/6$ or about .83. The average idealized efficiency if each item is measured once is much less, of course.

Finite Populations

By considering the infinite case we set aside the complications of sampling from finite populations. For example, if we draw eight items from 800 without replacement, we would get a little credit for not having drawn with replacement if the population whose mean we want to estimate actually is the 800 items. The formula for the variance of the mean of a sample of size J drawn from an infinite population with variance σ^2 is σ^2/J, whereas that for the mean of a sample of size J drawn from a population of size N is $(1/J - 1/N)\sigma^2$. The extra factor is $1 - J/N$. The simplicity of this result is one of the benefits of our definition of variance in the population—that is, of dividing by $N - 1$. If we write the sampling fraction $f = J/N$, this finite-population correction is $1 - f$. Thus in drawing eight items from 800, $1 - f = 0.99$, and so the between-item part of the variance of the mean is reduced by only 1%.

Before leaving this issue, we note that it is sometimes useful to view the fixed-effects model as a set of N populations, from which we draw into the sample all N populations without replacement. Then we draw samples from each of the populations.

Estimation of Variance Components by the Expected-Mean-Square Method

We have defined and discussed the components of variance in one-way ANOVA for both fixed-effects and random-effects models. How might we estimate these quantities? We begin with the first example, which gives lung function of various smoking groups (see Table 9-1). Recall that we modeled the lung function of member j in group i as

$$y_{ij} = \mu_i + \varepsilon_{ij}, \qquad i = 1, 2, \ldots, I;\ j = 1, 2, \ldots, J,$$

where $I = 5$, $J = 200$. Here the ε_{ij} have mean 0, and we assume that their variances all equal σ^2. Strictly speaking, the model has only one component of variance, σ^2. However, we may also wish to consider the variation in the group means μ_i as measured by $\tau^2 = \Sigma(\mu_i - \mu)^2/4$. Section 5D introduced the mean squares (and other entries) in an analysis-of-variance table for a one-way layout. In this chapter we focus on the Between-Group Mean Square, BMS, and the Within-Group Mean Square, WMS. In this example (with five groups, each consisting of 200 observations) recall that

$$\text{BMS} = J \sum_i (y_{i\cdot} - y_{\cdot\cdot})^2/(I - 1),$$

$$\text{WMS} = \sum_{i=1}^{I} \sum_{j=1}^{J} (y_{ij} - y_{i\cdot})^2/I(J - 1),$$

$$y_{i\cdot} = \sum_j y_{ij}/J, \qquad y_{\cdot\cdot} = \sum_i y_{i\cdot}/I.$$

A common method for estimating variance components equates the average of the mean squares (or equivalently, the expected values of the mean squares) to the observed values of the mean squares and then solves the resulting equations (in our example, two equations) for the desired estimates. This is the "equating-average-values method." The technical literature in statistics generally uses the term "method of moments." In this chapter we prefer the more descriptive "expected-mean-square method."

Fixed Effects: Estimation of Variance Components

For our example, we have the expected values

$$E(\text{BMS}) = \sigma^2 + J\tau^2, \qquad \tau^2 = \sum_i (\mu_i - \mu)^2/(I-1),$$

$$E(\text{WMS}) = \sigma^2.$$

Thus we see immediately that we can equate average values and solve to get estimates (indicated in this display by a "hat" ($\hat{}$) over the population value):

$$\hat{\sigma}^2 = \text{WMS},$$

$$\hat{\tau}^2 = \frac{\text{BMS} - \text{WMS}}{J}.$$

From the mean and standard deviation of FEF for the five smoking habits (Table 9-1), we find

$$\begin{aligned} \text{WMS} &= \left[(.79)^2 + (.77)^2 + (.78)^2 + (.81)^2 + (.82)^2\right]/5 \\ &= 3.1539/5 = 0.631, \\ y_{..} &= [3.78 + 3.30 + 3.23 + 2.73 + 2.59]/5 = 3.126, \\ \text{BMS} &= \frac{200}{4}\left[(3.78 - 3.126)^2 + (3.30 - 3.126)^2 + \cdots + (2.59 - 3.126)^2\right] \\ &= \frac{200}{4}(.91292) = 45.646, \end{aligned}$$

so that in summary

$$\hat{\sigma}^2 = 0.631$$

and

$$\hat{\tau}^2 = (45.646 - 0.631)/200 = 0.225.$$

It can be shown that $\hat{\sigma}^2$ and $\hat{\tau}^2$ are unbiased estimates of σ^2 and τ^2, respectively. This tells us that the variance *within* each of the groups is about 0.63 (or the standard deviation is 0.79) and that the variance of the observed group means about the overall mean is about 0.23 (a standard deviation of about 0.48).

We are making two "approximations" in using 0.63 as the within-group variance. First, we make the *assumption* that the variance within each group is constant across groups, and second, we can only *estimate* this common variance σ^2. We might treat each group separately and estimate five within-group variances; but if the assumption of equal variances is indeed reasonable, we naturally obtain a more precise estimate by pooling information across all the groups to calculate $\hat{\sigma}^2$.

In many circumstances, even if the σ_i^2 differ, we are content to estimate the average of the σ_i^2:

$$\sigma_.^2 = \frac{1}{I} \sum \sigma_i^2.$$

Pooling the sums of squares for the groups, as we did above to get the WMS, will always estimate this. If the σ_i^2 are not equal, we lose some precision, but not very much. If half the σ_i^2 equal $\frac{2}{3}\sigma_.^2$ and the other half equal $\frac{4}{3}\sigma_.^2$ (the worst case when all the σ_i^2 differ by at most a factor of two), the inequality of the σ_i^2 will yield an efficiency of 90%: we still get 90% of the number of formal degrees of freedom. Thus we can accommodate moderate ratios among the σ_i^2 without losing much, especially because we tend to have more —often many more—degrees of freedom for the within mean square than for any other mean square in a particular problem.

Paralleling the random-effects situation, we can look on the μ_i as coming from a distribution with mean μ and variance τ^2. To summarize, with random effects we assume the μ_i to be a random sample from a population of group means, but in dealing with fixed effects we view the set of μ_i as a 100% sample, drawn without replacement.

To tie the idea of variance components in with the value-splitting approach of previous chapters, the value splitting for the detailed data would go as in Table 9-3 with $I = 5$ groups and $J = 200$ replications. We see that the variance component τ^2 is related to the overlay of group effects; it estimates the variability of the group population means that underlie these effects—these observed group means (as deviations). More simply, the variance component σ^2 measures the variability in the residual overlay. Of course, no variance component is associated with the overlay repeating only the overall mean, because every entry is identical. To get something from the first overlay, we need μ, which we do not have, so as to compute $y_{..} - \mu$; or else we need two or more replications of the whole data pattern, including new versions of everything that might contribute to $y_{..}$. (Replications within cells would *not* suffice.)

Value splitting provides a way of drawing out all the components in the model: the overall mean, the group effects, and the residuals. Similarly, equating averages allows us to estimate the *variability* associated, under the lack-of-correlation assumption, with each term represented by an overlay in the value splitting (except for the overall mean). Both parts of the estimation process are important in themselves; each yields information about the population in a different form.

Table 9-3. Schematic Description of Reduced Overlays for the Data on Smokers from Table 9-1

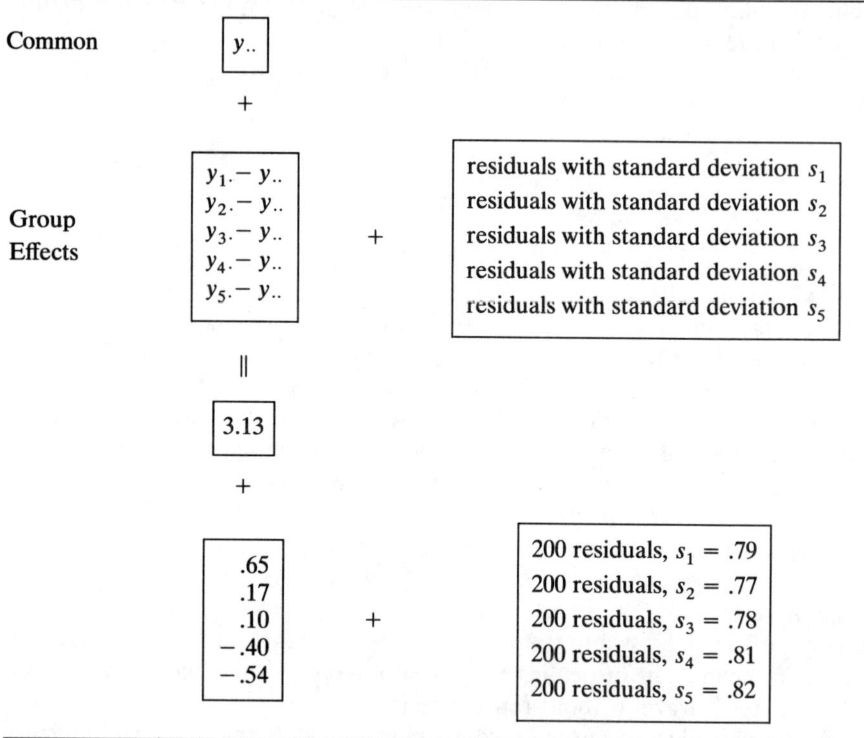

Random Effects: Estimation of Variance Components

The estimation of variance components in the random-effects model follows closely what we have described above, with one difference. Let us return to the second example, concerning performance scores in seven comprehensive schools. The model is the same as the one we used for the lung-function example, except for the extra specification that the μ_i are drawn from a distribution with mean μ and variance τ^2. We write

$$y_{ij} = \mu_i + \varepsilon_{ij}, \qquad i = 1, 2, \ldots, 7;\ j = 1, 2, \ldots, 50,$$

where, as before, the ε_{ij} have mean 0 and variance σ^2, and we assume that the within-school variance is constant across schools. (We need not actually do this. For a balanced design, the formulas below work all right if we replace σ^2 by $\sigma_{\cdot}^2$, the mean of the σ_i^2.) The μ_i and the ε_{ij} are all assumed to be uncorrelated.

In the random-effects model not all the possible populations are drawn into the sample. We often label the mean and variance of the population under discussion by the abbreviation for the factor associated with the population, commonly A, B, and so on. Here we use A for the factor associated with schools, and then we can write $\mu = \mu_A$, $\tau^2 = \sigma_A^2$, and

$$\mu_i = \mu + a_i, \qquad i = 1, 2, \ldots, I,$$

with the understanding that not all possible levels of factor A, here schools, are included in our data. We employ averaging (the expected-mean-square method) as before to estimate the variance components σ^2 and σ_A^2, noting that, although $E(\text{WMS}) = \sigma^2$ still, the expectation of the Between-Groups Mean Square has now become

$$E(\text{BMS}) = \sigma^2 + 50\sigma_A^2.$$

Thus

$$\hat{\sigma}^2 = \text{WMS}$$

and

$$\hat{\sigma}_A^2 = (\text{BMS} - \text{WMS})/50.$$

Using the data in Table 9-2 in these formulas, we have

$$y_{..} = (28.3 + 21.1 + \cdots + 16.8)/7 = 20.5,$$

$$\hat{\sigma}^2 = \text{WMS} = \frac{\sum_{i=1}^{7}\sum_{j=1}^{50}(y_{ij} - y_{i.})^2}{7 \times 49} = \frac{\sum_{i=1}^{7}(\text{std. dev. in group } i)^2}{7}$$

$$= \left[(13.4)^2 + (14.2)^2 + (11.8)^2 + (14.9)^2 + (13.4)^2 + (12.1)^2 + (13.9)^2\right]/7$$

$$= 1261.63/7 = 180.23,$$

and because

$$\text{BMS} = 50\sum_{i=1}^{7}(y_{i.} - y_{..})^2/6 = 50\left[(28.3 - 20.5)^2 + \cdots + (16.8 - 20.5)^2\right]/6$$

$$= 1391.74,$$

it follows that

$$\hat{\sigma}_A^2 = \frac{1391.74 - 180.23}{50} = 24.23.$$

We see considerable variability in the performance scores. The amount of variation within a given school (as measured by $\hat{\sigma}^2$) is about seven times the variation that occurs between the schools (as measured by $\hat{\sigma}_A^2$). We might have anticipated these large values by casting our eye down the column of rather uneven mean performance scores (for σ_A^2) or the column of substantial school standard deviations (for σ^2).

A check. When the sample size is less than about 16, we can get a fair approximation to the standard deviation of a set of measurements from a simple formula. This formula is that

RANGE ESTIMATE OF σ

$$\text{estimated } \sigma = \frac{\text{range}}{\sqrt{\text{number of measurements}}}.$$

(In this instance the measurements happen to be means.) Since the standard error of the mean (SEM) is $1/\sqrt{n}$ times the estimated standard deviation, we have

RANGE ESTIMATE OF $\sigma/\sqrt{n}$

$$\text{SEM} \approx \frac{\text{range}}{\text{number of measurements}} \quad (\text{sample size} \leq 16).$$

Looking back at Table 9-2, we might make a rough assessment of the standard deviation of the schools' performance scores. The range is the largest minus the smallest measurement (here means),

$$\text{range} = 28.3 - 14.3 = 14, \qquad J = 7,$$

and so the rough estimate of the standard deviation is $14/\sqrt{7} = 5.3$, which is not far from our earlier estimate $\sqrt{24.23} = 4.9$. The median of the seven standard deviations in Table 9-2 is 13.4, which is almost too close to $\sqrt{180.23} = 13.4$.

The three pieces of information $\hat{\mu}_A = y_{..} = 20.5$, $\hat{\sigma}_A^2 = 24.2$, and $\hat{\sigma}^2 = 180$ give us a neat summary of the situation. They tell us that the mean performance score in a given English comprehensive school comes from a distribution with mean of about 20.5 and variance of about $24.2 + 180/50 = 27.8$ (where 50 is the sample size per school), giving a standard deviation of about 5, and that the individual scores *within* a school fall around the mean of that particular school with a standard deviation of $\sqrt{180} = 13.4$. The standard error of a school mean is $13.4/\sqrt{\text{sample size of school}}$, and the standard error of the grand mean is

$$\sqrt{4.9^2/(\text{number of schools}) + 13.4^2/(\text{total number of children})}\,.$$

Value splitting would proceed exactly as in the fixed-effects example and, once again, $\sigma_A^2 + \sigma^2/(\text{sample size per school})$ corresponds to the overlay of (observed) group effects and σ^2 to the overlay of residuals. Although τ^2 is not always important in the fixed-effects model, σ_A^2 often has great importance in the random-effects model.

To generalize this procedure for estimating variance components in the one-way analysis of variance, suppose we consider the random-effects model

$$y_{ij} = \mu_i + \varepsilon_{ij}, \qquad i = 1, 2, \ldots, I;\ j = 1, 2, \ldots, J.$$

As usual, the ε_{ij} have mean 0 and variance σ^2, and the μ_i come from a distribution with mean μ_A and variance σ_A^2. It can be shown that the following expectations hold:

$$E(\text{WMS}) = E\left[\frac{\sum_{i=1}^{I}\sum_{j=1}^{J}(y_{ij} - y_{i\cdot})^2}{I(J-1)}\right] = \sigma^2,$$

$$E(\text{BMS}) = E\left[\frac{J\sum_{i=1}^{I}(y_{i\cdot} - y_{\cdot\cdot})^2}{I-1}\right] = \sigma^2 + J\sigma_A^2,$$

so that equating average values (the expected-mean-square method) estimates the variance components as

ESTIMATE OF σ^2

$$\hat{\sigma}^2 = \text{WMS}$$

and

ESTIMATE OF σ_A^2

$$\hat{\sigma}_A^2 = \frac{\text{BMS} - \text{WMS}}{J}$$

just as before.

When the groups have different variances σ_i^2, with $\Sigma\sigma_i^2/I = \sigma_{\cdot}^2$, the formulas immediately above become

UNEQUAL σ^2: EXPECTED WMS AND BMS

$$E(\text{WMS}) = \sigma_{\cdot}^2,$$

$$E(\text{BMS}) = \sigma_{\cdot}^2 + J\sigma_A^2.$$

Thus we find when the σ_i^2 are unequal

$$\hat{\sigma}_{\cdot}^2 = \text{WMS},$$

$$\hat{\sigma}_A^2 = (\text{BMS} - \text{WMS})/J.$$

Negative Estimates of Variance

A feature of this method for estimating the variance of the population means is that the estimate often turns out to be negative, which might seem a bit embarrassing for a parameter that must be positive. Indeed, when the true variance of the population means is zero (τ^2 or σ_A^2), the estimate will be negative about half the time, and this means that whenever that true variance is not very large, the estimate will still often be negative. The simplest remedy is to replace such a negative estimate by zero, and that is what we shall do here. This move gives up unbiasedness and suggests that some more drastic action might be appropriate. In Section 9C we discuss some more complicated options that have attractive properties, while still necessarily giving up unbiasedness.

Similarly, if we are dealing with *fixed* effects, the estimates are

$$\hat{\sigma}^2 = \text{WMS}$$

and

$$\hat{\tau}^2 = (\text{BMS} - \text{WMS})/J,$$

where

$$\tau^2 = \frac{\sum_{i=1}^{I} (\mu_i - \mu)^2}{I - 1}$$

unless BMS < WMS, and then we set $\hat{\tau}^2 = 0$. When we do this, we again lose unbiasedness.

9B. EXAMPLE: VARIANCE COMPONENTS FOR BLOOD PRESSURE

To illustrate how variance-component analysis arises in a more complicated practical situation, this section gives a separate example.

High blood pressure creates hazards that medicine now has ways to prevent, and early identification of persons with this condition improves their prognosis. To provide better background information for doctors who interpret children's blood pressure, a sizable study looked into the reproducibility of these measurements from week to week. Large numbers of children had their blood pressures, systolic and diastolic, taken at weekly visits for four successive weeks, three measurements per visit. After stratifying the data according to sex and age group (8–12 years, 13–18 years), the researchers (Rosner et al. 1987) used variance-component analysis to estimate the variability of systolic and diastolic blood pressure between persons and within person. They further allotted variability within person to variance components between visits and within visit. In addition, they compared these results for children with results they had obtained earlier for adults. Table 9-4 displays many of their results (rounded to whole numbers for easier comparisons).

Focus first on the line for Systolic BP, Male, Between-person. Note that older males vary more from one another than do younger males. A similar result holds for females. For diastolic BP, although the variance component is smaller, the increase with age is similar for both sexes.

Focus next on the lines for Within-person variability, systolic. They combine both Between- and Within-visit variation. For males and females, the Within-person variability is fairly similar except for adults.

For diastolic Within-person, the variation decreases with age for both sexes. Furthermore, for the children the sizes of the diastolic Within-person variation are larger than those for systolic. This is surprising because the mean diastolic pressure is considerably smaller than the mean for systolic pressure. As mentioned elsewhere, we usually find that groups of larger quantities are more variable than groups of smaller quantities of the same kind. Thus when we look at ages 13–18, we see that the Within-person variation is about the same for each sex and the two types of blood pressure —variance components for males 53 versus 56 and for females 47 versus 48. The authors of the original article conclude that, for young children, estimates of blood pressure may better come from systolic measures because of smaller Within-person variability, or possibly from a combination of systolic and diastolic, though they do not recommend a particular combination.

The limited value of repeat testing during a visit is shown by the ratio of Between-visit to Within-visit variance components, which is roughly between 2.5 and 4. One visit, one measurement gives 75–80% as much information as one visit, many measurements. (In other words, the Within-visit correlation of measurement error is about 0.7.)

Table 9-4. Comparisons of Variance Components for Standard Blood Pressure (BP) Measurements for Two Age Groups in Adults and in Children.[a]

Variance Component	Children 8–12	Children 13–18	Adults 30–49	Adults 50–69
	Systolic BP			
Male				
[Mean[b]]	[102]	[112]	[123]	[133]
Between-person	**70** 61%	**118** 69%	**163** 77%	**327** 83%
Within-person	**45** 39%	**53** 31%	**49** 23%	**65** 17%
Between-visit	33	41	35	50
Within-visit	12	12	14	16
(Sample size)	(349)	(365)	(262)	(179)
Female				
[Mean]	[101]	[105]	[113]	[125]
Between-person	**88** 67%	**87** 65%	**204** 79%	**370** 83%
Within-person	**42** 33%	**47** 35%	**56** 21%	**78** 17%
Between-visit	31	35	43	65
Within-visit	11	12	13	14
(Sample size)	(397)	(419)	(172)	(196)
	Diastolic BP[c]			
Male				
[Mean]	[67]	[68]	[81]	[82]
Between-person	**19** 19%	**95** 63%	**87** 71%	**105** 77%
Within-person	**79** 81%	**56** 37%	**36** 29%	**31** 23%
Between-visit	58	41	28	24
Within-visit	21	16	8	7
(Sample size)	(226)	(365)	(262)	(179)
Female				
[Mean]	[66]	[68]	[75]	[78]
Between-person	**25** 24%	**80** 63%	**90** 75%	**101** 77%
Within-person	**79** 76%	**48** 37%	**30** 25%	**30** 23%
Between-visit	57	36	23	24
Within-visit	23	11	7	6
(Sample size)	(255)	(419)	(172)	(196)

[a] Each of the four systolic–sex and diastolic–sex subtables has Between-person and Within-person variance components for each age group.
[b] Based on average BP across all available visits for specific individuals.
[c] K4 was used for ages 8–12; K5 was used for ages 13–18 and for adults.

Source: B. Rosner, N. R. Cook, D. A. Evans, M. E. Keough, J. O. Taylor, B. F. Polk, and C. H. Hennekens (1987). "Reproducibility and predictive values of routine blood pressure measurements in children: comparison with adult values and implications for screening children for elevated blood pressure," *American Journal of Epidemiology*, 126, 1115–1125 (data for means and sample sizes for children from Table 2, p. 1119; data for Between-person, Within-person, Between-visit, and Within-visit from Table 5, p. 1121). Means and sample sizes for Adults were kindly provided by a personal communication from Bernard Rosner.

Table 9-5. Ratios for Between-person VC to $\frac{1}{2}$(Within-person VC)[a]

	Children		Adults	
	8–12	13–18	30–49	50–69
		Systolic		
Male	3.1	4.5	6.7	10.1
Female	4.2	3.7	7.3	9.5
		Diastolic		
Male	(.5)	3.4	4.8	6.8
Female	(.6)	3.3	6.0	6.7

[a] Large values correspond to more easily measured Between-person behavior; variance components apply to one measurement at each of two visits. Computed from Table 9-4.

Note: Within-person VC taken as Within-visit VC + Between-visit VC. VC denotes variance component.

When we want to distinguish among blood pressures of different people or different groups, the between-person variance component has special interest. If we can afford two visits, one measurement each, routinely, the quantities we need to look at further are (where VC denotes variance component):

the Between-person VC (what we have to measure)

and

$\frac{1}{2}$(Between-visit VC + Within-visit VC) (our variance of measurement).

Table 9-5 shows the ratios of these quantities. Differences between persons seem more easily measured for systolic than for diastolic pressures, even in adults. In all cases, accumulation of information over several visits improves the measurement situation.

Rosner et al. proceed to offer methods for using these data to decide which children have true blood pressures at or above the 90% point of the distribution. We do not pursue that aspect here.

9C. ALTERNATIVE METHODS FOR ESTIMATING VARIANCE COMPONENTS

In Section 9A we derived the following estimates of variance components via the expected-mean-square approach:

$$\hat{\sigma}^2 = \text{WMS},$$
$$\hat{\sigma}_A^2 = (\text{BMS} - \text{WMS})/J.$$

We noted that one drawback of this estimation method was the possibility of negative estimates of σ_A^2. We suggested that whenever $\hat{\sigma}_A^2$ was negative, it should be adjusted to zero because we know that σ_A^2 must be nonnegative. It turns out that incorporating this simple adjustment in the estimation procedure greatly improves the performance of the estimator in terms of its mean squared error (MSE). When comparing different estimators of a given quantity, the MSE is one of the criteria commonly employed to distinguish a good estimator from a poor one. By adding together the variance of the estimator and the square of the bias, it allows for both variance and bias. Thus, for example, when comparing two unbiased estimators, the one with smaller variance has smaller MSE and therefore may be the preferred estimator. Adjusting $\hat{\sigma}_A^2$ to zero whenever it is negative produces an estimator with a much smaller MSE than the expected-mean-square approach gives. Although the adjustment to zero causes the estimator to become biased, thus increasing the squared-bias component of the MSE, the reduction in variance more than offsets that increase. So, if we are concerned about performance as measured by MSE, we would redefine our estimator of σ_A^2 as

$$\hat{\sigma}_A^2 = \begin{cases} (\text{BMS} - \text{WMS})/J, & \text{BMS} > \text{WMS} \\ 0, & \text{BMS} \leq \text{WMS}. \end{cases}$$

But we can do better yet. Klotz, Milton, and Zacks (1969) (hereafter KMZ) compare the performance, in terms of mean squared error, of several methods of estimating variance components. It turns out that further fine-tuning of $\hat{\sigma}_A^2$ yields a further payoff—in mean squared error. This result is analogous to what happens if we focus on mean squared error of the estimated variance for a simple sample. KMZ present a variety of estimators derived by using Bayesian techniques (which we do not describe here) and show, via plots of MSE, that one in particular performs consistently well. The form of this estimator is

$$\hat{\sigma}_{A\text{-KMZ}}^2 = \begin{cases} \dfrac{1}{J}\left(\dfrac{\text{BSS}}{I+2} - \dfrac{\text{WSS}}{I(J-1)+2}\right), & \dfrac{\text{BSS}}{I+2} > \dfrac{\text{WSS}}{I(J-1)+2} \\ 0, & \text{otherwise}. \end{cases}$$

Recalling that BMS = BSS/$(I-1)$ and WMS = WSS/$I(J-1)$, we see that the only difference between the KMZ estimator and the expected-mean-square estimator (adjusted to zero) is that $I-1$ is replaced by $I+2$ and $I(J-1)$ is replaced by $I(J-1)+2$, and the corresponding change in making a zero estimate.

We can write $\hat{\sigma}_{A\text{-KMZ}}^2$ in terms of $\hat{\sigma}_A^2$ and $\hat{\sigma}^2$ as follows:

$$\hat{\sigma}_{A\text{-KMZ}}^2 = \max\left[\frac{I-1}{I+2}\hat{\sigma}_A^2 - \left(\frac{3}{I+2} - \frac{2}{I(J-1)+2}\right)\frac{\hat{\sigma}^2}{J}, 0\right]$$

which will be slightly less than $(I-1)/(I+2)$ times $\hat{\sigma}_A^2$.

In the example of the comprehensive schools, we get

$$\hat{\sigma}^2_{A\text{-KMZ}} = \max\left[\frac{6}{9}(24.23) - \left(\frac{3}{9} - \frac{2}{345}\right)\frac{180.23}{50}, 0\right]$$
$$= \max[16.15 - 1.18, 0] = 14.97,$$

which leads us to make several remarks:

1. If $\hat{\sigma}^2_A = 24.23$ and $\hat{\sigma}^2_{A\text{-KMZ}} = 14.97$ are both reasonable estimates, then we have considerable uncertainty about the size of σ^2_A, as we might expect from our discussion in Section 7D.
2. If $\hat{\sigma}^2 = 0$, we recover the minimum-MSE estimate of σ^2_A for a simple sample, namely, $((I - 1)/(I + 2))\hat{\sigma}^2_A$.
3. Because many analyses of variance will have small values of I, the ratio of $I - 1$ to $I + 2$ need not be close to 1.
4. If $\hat{\sigma}^2 = 0$ and $I = 2$—or if we have a simple sample of two observations—$(I - 1)/(I + 2) = \frac{1}{4}$ and $\hat{\sigma}^2_{A\text{-KMZ}} = \hat{\sigma}^2_A/4$, so that the average value of $\hat{\sigma}^2_{A\text{-KMZ}}$ is $\sigma^2_A/4$, not σ^2_A. In this situation, where both estimates are distributed like multiples of χ^2 on one degree of freedom, the probability of $\hat{\sigma}^2_A \leq \sigma^2_A$ is 68.3%, and the probability of $\hat{\sigma}^2_{A\text{-KMZ}} \leq \sigma^2_A$ is 84.3%.
5. When we feel we should use such a shrunken estimate, it will ordinarily be sufficient to use the approximation

$$\hat{\sigma}^2_{A\text{-KMZ}} \approx \max\left[\hat{\sigma}^2_A - \frac{3}{I + 2}\left(\hat{\sigma}^2_A + \frac{1}{J}\hat{\sigma}^2\right), 0\right]$$
$$= \max\left[\hat{\sigma}^2_A - \frac{3}{I + 2}\frac{\text{BMS}}{J}, 0\right],$$

which gives $14.95 = 24.23 - 9.28$ instead of 14.97 in the example.

Clearly, we need to think carefully about using $\hat{\sigma}^2_{A\text{-KMZ}}$ for small I.

There is very little to be gained by changing $\hat{\sigma}^2$, so we do not consider alternatives to taking $\hat{\sigma}^2 = \text{WMS}$. Derivation of the KMZ estimates is by no means ad hoc (the reader familiar with the Bayesian philosophy may wish to consult the KMZ article), but it mainly involves the factor $(I - 1)/(I + 2)$ that arises for a simple sample when we minimize the MSE.

9D. CONFIDENCE INTERVALS FOR VARIANCE COMPONENTS

Consider the balanced one-way ANOVA model with random effects:

$$y_{ij} = \mu + a_i + \varepsilon_{ij}, \qquad i = 1, 2, \ldots, I;\ j = 1, 2, \ldots, J,$$

where

$$\varepsilon_{ij} \sim \text{Gau}(0, \sigma^2)$$

or follows some distribution other than normal, still with mean 0 and variance σ^2, and

$$a_i \sim \text{Gau}(0, \sigma_A^2),$$

or follows some other distribution with mean 0 and variance σ_A^2. (Recall that the symbol $\sim$ means "is distributed according to.") We showed in Section 9A that the expected-mean-square estimation procedure yields the following estimates of the variance components:

$$\hat{\sigma}^2 = \text{WMS},$$
$$\hat{\sigma}_A^2 = (\text{BMS} - \text{WMS})/J.$$

When we applied this method in the comprehensive schools example (see Section 9A), we found that

$$\hat{\sigma}^2 = 180.23 \quad \text{and} \quad \hat{\sigma}_A^2 = 24.23.$$

Next, we would like to learn something about the *sampling variability* of these estimates, so that we can confidently make statements like "The between-group variance component varies between 22.23 and 26.23," or "between 14.23 and 34.23," whichever is appropriate. A point estimate has more interest when accompanied by its standard error. From the distributional assumptions we made in setting up the model, it would follow that

$$\frac{(I-1)\text{BMS}}{J\sigma_A^2 + \sigma^2} \sim \chi_{I-1}^2 \quad \text{and} \quad \frac{I(J-1)\text{WMS}}{\sigma^2} \sim \chi_{I(J-1)}^2, \tag{1}$$

where χ_d^2 denotes the chi-squared distribution with d degrees of freedom. We use these distributions to construct *confidence intervals* for both σ^2 and σ_A^2 so as to obtain some idea about the variability of our estimates. We begin with the within-groups variance component σ^2.

Confidence Intervals for σ^2

At first glance, knowing that $\hat{\sigma}^2 = \text{WMS}$, it appears that the construction of a confidence interval would be straightforward. Using relation (1) above, we have, as we discussed in Chapter 7,

$$1 - \alpha = P\left(\chi^2_{I(J-1),L} \leq \frac{\hat{\sigma}^2}{\sigma^2} I(J-1) \leq \chi^2_{I(J-1),U}\right)$$
$$= P\left(\frac{\hat{\sigma}^2 I(J-1)}{\chi^2_{I(J-1),U}} \leq \sigma^2 \leq \frac{\hat{\sigma}^2 I(J-1)}{\chi^2_{I(J-1),L}}\right),$$

where $\chi^2_{d,L}$ and $\chi^2_{d,U}$ are, respectively, the $\alpha/2$ and the $1 - \alpha/2$ critical points for a chi-squared distribution with d degrees freedom, so that $P(\chi^2_{d,L} \leq \chi^2_d \leq \chi^2_{d,U}) = 1 - \alpha$. Thus a plausible $100(1-\alpha)\%$ confidence interval for σ^2 would be

First Approximate Confidence Limits for σ^2

$$\mathrm{CI}_1(\sigma^2, 1-\alpha) = \left(\frac{\hat{\sigma}^2 I(J-1)}{\chi^2_{I(J-1),U}}, \frac{\hat{\sigma}^2 I(J-1)}{\chi^2_{I(J-1),L}}\right)$$
$$= ((1 - k_L)\hat{\sigma}^2, (1 + k_U)\hat{\sigma}^2)$$

with

$$1 - k_L = \frac{I(J-1)}{\chi^2_{I(J-1),U}} \quad \text{and} \quad 1 + k_U = \frac{I(J-1)}{\chi^2_{I(J-1),L}}.$$

However, as we said in Chapter 7, the validity of the interval rests on the assumption of normality in the model, and it turns out that this interval is not very robust to departures from that assumption. If the data *do* come from a normal distribution, the interval is satisfactory. Otherwise, it matters whether the tails of the distribution are substantially lighter or substantially heavier than those of a normal distribution. In the present setting we judge such departures by the distribution's kurtosis, γ_2, defined in terms of the fourth moment: $\gamma_2 = E(X - \mu)^4/\sigma^4 - 3$. The normal distribution has $\gamma_2 = 0$, lighter tails give $\gamma_2 < 0$, and heavier tails give $\gamma_2 > 0$. As examples, uniform distributions have lighter tails, and Student t distributions have heavier ones.

Suppose γ_2 is the *actual* kurtosis of the distribution of the ε_{ij}. Then it can be shown that

$$\mathrm{var}(\hat{\sigma}^2) = \sigma^4\left(\frac{2}{I(J-1)} + \frac{\gamma_2}{IJ}\right)$$

which reduces to $2\sigma^4/I(J-1)$ if the distribution is normal. Thus the confidence interval based on the chi-squared distribution could potentially be too wide (if $\gamma_2 < 0$) or too narrow (if $\gamma_2 > 0$) (Miller 1986, p. 264). An intuitively appealing way to deal with this is to investigate the sample kurtosis of the ε_{ij}:

$$\hat{\gamma}_2 = \frac{IJ\sum_{i=1}^{I}\sum_{j=1}^{J}(\varepsilon_{ij} - \varepsilon_{i\cdot})^4}{\left(\sum_{i=1}^{I}\sum_{j=1}^{J}(\varepsilon_{ij} - \varepsilon_{i\cdot})^2\right)^2} - 3.$$

(For calculations we replace $\varepsilon_{ij} - \varepsilon_{i\cdot}$ by $y_{ij} - y_{i\cdot}$.) Then, noting the approximation

$$\operatorname{var}(\hat{\sigma}^2) \approx \frac{2\sigma^4}{I(J-1)}\left(1 + \tfrac{1}{2}\gamma_2\right)$$

and recalling that the variance of a chi-squared random variable is twice its degrees of freedom, we may either adjust our first confidence limits by a factor of $(1 + \hat{\gamma}_2/2)^{1/2}$ or choose to match the distribution of $\hat{\sigma}^2$ by a multiple of a chi-squared distribution such that the first two moments agree. In the latter case, we write

$$\hat{\sigma}^2 \sim \frac{\sigma^2}{dI(J-1)}\chi^2_{dI(J-1)},$$

where

$$d = \left(1 + \tfrac{1}{2}\gamma_2\right)^{-1}.$$

Then

$$\begin{aligned}\operatorname{var}(\hat{\sigma}^2) &= \frac{\sigma^4}{d^2(I(J-1))^2}2dI(J-1)\\ &= \frac{2\sigma^4}{dI(J-1)}\\ &= \frac{2\sigma^4}{I(J-1)}\left(1 + \tfrac{1}{2}\gamma_2\right),\end{aligned}$$

as desired. Finally, taking $\hat{d} = (1 + \hat{\gamma}_2/2)^{-1}$, our second (approximate) $100(1-\alpha)\%$ confidence interval for σ^2 is

SECOND APPROXIMATE CONFIDENCE LIMITS FOR σ^2

$$\mathrm{CI}_2(\sigma^2, 1-\alpha) = \left(\frac{\hat{\sigma}^2 \hat{d}I(J-1)}{\chi^2_{\hat{d}I(J-1),U}}, \frac{\hat{\sigma}^2 \hat{d}I(J-1)}{\chi^2_{\hat{d}I(J-1),L}} \right).$$

If we merely spread the confidence interval by $(1 + \hat{\gamma}_2/2)^{1/2}$, we get

SPREADING APPROXIMATION

$$\mathrm{CI}_{2*}(\sigma^2, 1-\alpha) = \left(\left(1 - (1 + \hat{\gamma}_2/2)^{1/2} k_L\right)\hat{\sigma}^2, \left(1 + (1 + \hat{\gamma}_2/2)^{1/2} k_U\right)\hat{\sigma}^2 \right).$$

Before adjusting the degrees of freedom or spreading the confidence interval, we suggest examining a Q-Q plot of the residuals. If the plot is more or less linear, it may not be worthwhile (or appropriate) to use $\mathrm{CI}_2(\sigma^2, 1-\alpha)$ or $\mathrm{CI}_{2*}(\sigma^2, 1-\alpha)$ instead of $\mathrm{CI}_1(\sigma^2, 1-\alpha)$. And if the plot resembles either Figure 9-2 or Figure 9-3, then use of one of the latter may be required if we are to be close to the nominal confidence level.

Let us assume that we *do* suspect nonnormality of the errors. If we knew the true kurtosis, γ_2, of the errors, then we could rest assured that "fine-tuning" the degrees of freedom would improve the coverage of the chi-squared confidence interval. However, we do not know γ_2 in general and must

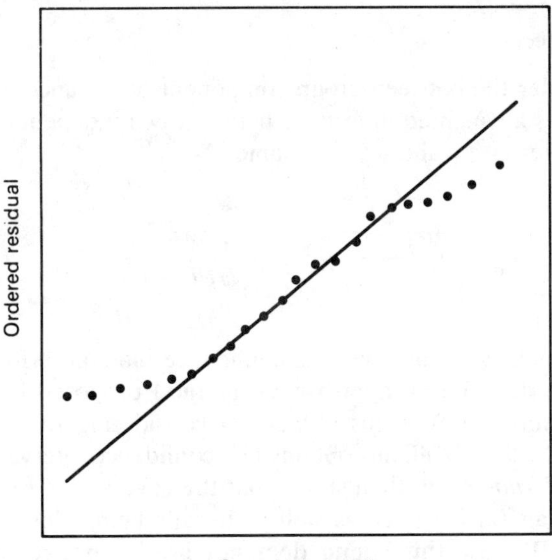

Figure 9-2. Q-Q plot of residuals suggesting $\gamma_2 < 0$.

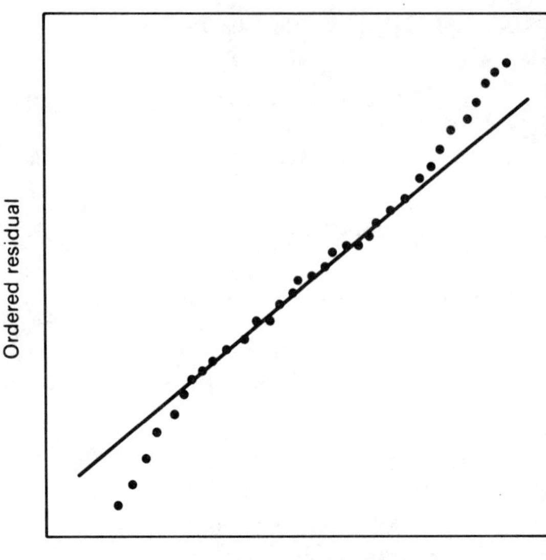

Figure 9-3. Q-Q plot of residuals suggesting $\gamma_2 > 0$.

estimate it by $\hat{\gamma}_2$. Thus, the effectiveness of using the second set of confidence limits instead of the first rests a great deal on how good an estimate of γ_2 we have.

Confidence Intervals for σ_A^2

We now consider the between-group component of variance σ_A^2. Because $\hat{\sigma}_A^2$ is distributed as a weighted difference between two independent chi-squared random variables, χ_{I-1}^2 and $\chi_{I(J-1)}^2$, namely,

$$\hat{\sigma}_A^2 \sim \frac{J\sigma_A^2 + \sigma^2}{J(I-1)}\chi_{I-1}^2 - \frac{\sigma^2}{IJ(I-1)}\chi_{I(J-1)}^2,$$

it is not obvious how to construct the confidence interval. Without going into the details, we describe an approximate method proposed by Tukey (1951) and independently by Williams (1962). By considering the distributions of $\hat{\sigma}_A^2/\hat{\sigma}^2$ and $\hat{\sigma}^2 + J\hat{\sigma}_A^2$, Williams obtains two confidence intervals for σ_A^2 when σ^2 is *fixed and known*. He then shows that the coverage of the common part of these intervals (at level α), as well as usually being close to $1 - \alpha$, is at least $1 - 2\alpha$. Because this bound does not involve σ^2, it holds across all values of the within component of variance. And it happens very conveniently that the intersection of these two intervals does *not* involve σ^2 either: the intersection yields the following interval for σ_A^2 based on the F distribution, with superscripts L and U standing for lower and upper critical values:

WILLIAMS–TUKEY CONFIDENCE INTERVAL FOR σ_A^2

$$\mathrm{CI}_1(\sigma_A^2, 1-\alpha) = \left(\frac{(I-1)}{J\chi^2_{I-1,U}}\left(\mathrm{BMS} - \mathrm{WMS}\, F^U_{I-1,I(J-1)}\right), \frac{(I-1)}{J\chi^2_{I-1,L}}\left(\mathrm{BMS} - \mathrm{WMS}\, F^L_{I-1,I(J-1)}\right) \right),$$

where

$$P\left(F^L_{k,j} \le F_{k,j} \le F^U_{k,j}\right) = 1 - \alpha \quad \text{and} \quad F^L_{k,j} = 1/F^U_{j,k}$$

is the notation used. If the lower limit of CI_1 is negative, it is set to zero; and if both limits of CI_1 are negative, σ_A^2 is taken to be zero. This interval, with a nominal $100(1-\alpha)\%$ coverage, has *at least* a $100(1-2\alpha)\%$ coverage rate.

Several possibilities for confidence limits on σ_A^2 have been suggested in the literature. We have chosen to describe the Williams–Tukey method because of results presented by Boardman (1974) from a Monte Carlo study designed to compare the various methods. Boardman considers the one-way balanced random-effects model and nine different approaches to interval construction for σ_A^2. Only two of the methods perform consistently well—the Williams–Tukey method and the Moriguti–Bulmer method—and these two methods are practically identical. With $\alpha = .05$ the coverage rates for these two methods were always very close to 95%, irrespective of I and J or the size of the ratio σ_A^2/σ^2. Some of the other methods were quite accurate when the ratio σ_A^2/σ^2 was "large," but for smaller ratios, their coverage could fall as low as 35%. Figures 9-4 and 9-5 reproduce the information on four methods from two of the six plots given in Boardman's article. Both plot the percent coverage of the 95% confidence interval for σ_A^2 (for each method) versus the ratio σ_A^2/σ^2. The first is for $I = 4$, $J = 5$, and the second is for $I = 6$, $J = 10$. The Williams–Tukey method is number 6. We see in both graphs that it maintains a virtually constant coverage of 95% over the entire range of σ_A^2/σ^2. Method 5 also performs well, and Method 1 has a coverage of 95% in Figure 9-5. However, in Figure 9-4 the coverage of Method 1 falls as low as 80% for small values of σ_A^2/σ^2, so its performance is not consistent.

When we discussed the choice of a confidence interval for σ^2, we were concerned about the effects of nonnormality. This concern exists also when considering σ_A^2, but there is no analog to the "kurtosis adjustment" we described for σ^2. A confidence interval for σ_A^2 that is robust to departures from normality can be obtained by using the *jackknife* technique; Arvesen and Schmitz (1970) and Miller (1986, Sec. 3.6.3) give more details.

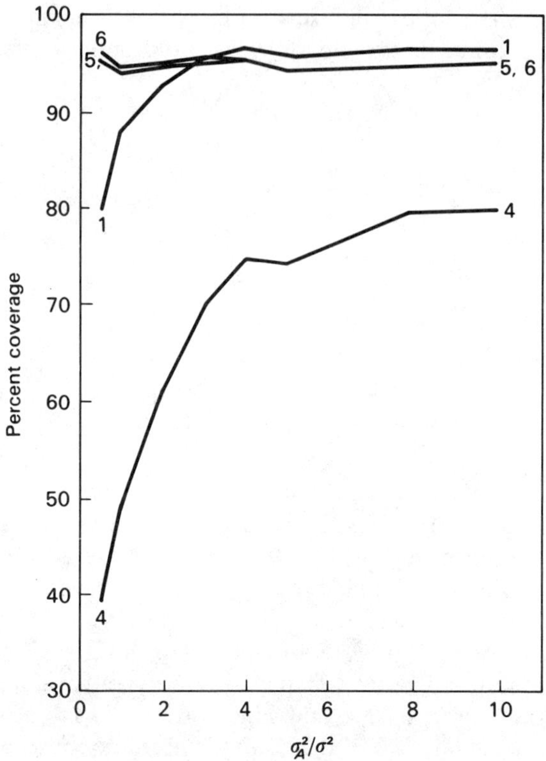

Figure 9-4. Percent coverage of a 95% confidence interval for σ_A^2 from four possible methods, for $I = 4$ and $J = 5$. The methods are: 1, Satterthwaite (1941); 4, Welch (1956), unbiased; 5, Moriguti (1954), Bulmer (1957); and 6, Tukey (1951), Williams (1962). Reproduced from: Thomas J. Boardman, "Confidence Intervals for Variance Components—a Comparative Monte Carlo Study." *Biometrics* 30: 251–262, 1974 [from Figure 1, p. 257]. With permission from The Biometric Society.

Example: performance scores

For an example we construct the two intervals $\text{CI}_1(\sigma^2, 1 - \alpha)$ and $\text{CI}_1(\sigma_A^2, 1 - \alpha)$ for the data on performance scores in comprehensive schools (Section 9A). We do not have access to the detailed data, so we cannot check the normality assumption. Recall that $\hat{\sigma}^2 = 180.23$, $\hat{\sigma}_A^2 = 24.23$, $\text{BMS} = 1391.74$, $I = 7$, and $J = 50$. Thus the 90% confidence intervals are

$$\begin{aligned}\text{CI}_1(\sigma^2, .9) &= \left(\frac{\hat{\sigma}^2 I(J-1)}{\chi^2_{I(J-1),U}}, \frac{\hat{\sigma}^2 I(J-1)}{\chi^2_{I(J-1),L}} \right) \\ &= \left(\frac{180.23 \times 7 \times 49}{386.9}, \frac{180.23 \times 7 \times 49}{300.8} \right) \\ &= (159.8, 205.5),\end{aligned}$$

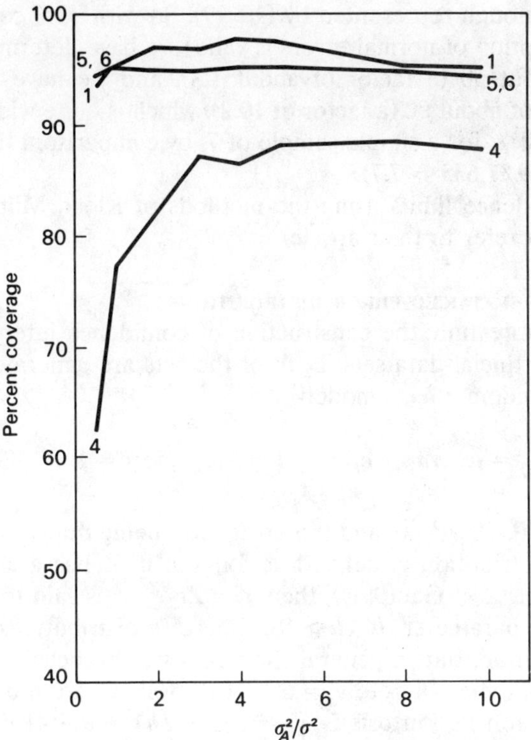

Figure 9-5. Percent coverage of a 95% confidence interval for σ_A^2 from four possible methods, for $I = 6$ and $J = 10$. The methods are: 1, Satterthwaite (1941); 4, Welch (1956), unbiased; 5, Moriguti (1954), Bulmer (1957); and 6, Tukey (1951), Williams (1962). Reproduced from: Thomas J. Boardman, "Confidence Intervals for Variance Components—a Comparative Monte Carlo Study." *Biometrics* 30: 251–262, 1974 [from Figure 4, p. 259]. With permission from The Biometric Society.

which is well enough represented by (160, 206), and

$$\begin{aligned}
\text{CI}_1(\sigma_A^2, .9) &= \left(\frac{(I-1)}{J\chi^2_{I-1,U}} \left(\text{BMS} - \text{WMS } F^U_{I-1,I(J-1)}\right), \right. \\
&\qquad \left. \frac{(I-1)}{J\chi^2_{I-1,L}} \left(\text{BMS} - \text{WMS } F^L_{I-1,I(J-1)}\right) \right) \\
&= \left(\frac{6}{50 \times 12.59} (1391.74 - 180.23 \times 2.10), \right. \\
&\qquad \left. \frac{6}{50 \times 1.635} (1391.74 - 180.23 \times 0.272) \right) \\
&= (9.66, 98.55),
\end{aligned}$$

which is well enough represented by (10, 99). So with 90% confidence (given that the assumption of normal errors is valid) we have determined σ^2 within a range of about 50 (a factor of about 1.3), and we have determined σ_A^2 within a range of about 90 (a factor of 10.2), which is quite wide (as we had to expect for small I; for a simple sample of 7, by comparison, the factor would have been $12.59/1.635 = 7.7$).

To set confidence limits using the methods of Klotz, Milton, and Zacks, the reader may refer to their article.

Example: heavier-tailed error distributions

To illustrate more fully the construction of confidence intervals for σ^2, we consider two artificial data sets. Both of the sets are generated according to the one-way random-effects model

$$y_{ij} = \mu + a_i + \varepsilon_{ij}, \qquad i = 1, \ldots, 5;\ j = 1, \ldots, 30,$$

with $a_i \sim \text{Gau}(0, 4)$, $\mu = 0$, and the error term being drawn from an *h-distribution*. This distribution is defined as follows: if Z has a standard normal distribution (i.e., $Z \sim \text{Gau}(0,1)$), then $Y = Ze^{hZ^2/2}$ is said to have an h-distribution with parameter h ($h \geq 0$). These h-distributions resemble the normal distribution, but in general their tails will be heavier, depending on the value of h (when $h = 0$, $Y = Z$). If Y follows an h-distribution with parameter h, then its kurtosis is $\gamma_2 = 3(1-2h)^3/(1-4h)^{5/2} - 3$ for $0 \leq h < \frac{1}{4}$, and the variance is $\sigma^2 = 1/(1-2h)^{3/2}$ (Hoaglin 1985). The two data

Table 9-6. Estimates of σ^2 and γ_2 Obtained for Five Simulations of the One-Way Model with $I = 5$ Groups and $J = 30$ Observations per Group by Sampling from an h-Distribution with $h = 0.1$ ($\sigma^2 = 1.40, \gamma_2 = 2.51$)

Simulation	$\hat{\sigma}^2$	$\hat{\gamma}_2$	Expansion Factor[a] $(1 + \hat{\gamma}_2/2)^{1/2}$
1	1.47	1.85	1.39
2	1.35	3.17	1.61
3	1.39	2.79	1.55
4	1.85	5.53	1.94
5	1.42	1.23	1.27

[a] On the basis of the population value of γ_2 we would like to have the expansion factor equal to $(1 + 2.51/2)^{1/2} = 1.50$. Here

$$\hat{\sigma}^2 = \sum_i \sum_j (y_{ij} - y_{i\cdot})^2/145 \quad \text{and} \quad \hat{\gamma}_2 = \frac{150 \sum_i \sum_j (y_{ij} - y_{i\cdot})^4}{\left(\sum_i \sum_j (y_{ij} - y_{i\cdot})^2\right)^2} - 3.$$

sets correspond to two different error distributions: one with parameter $h = 0.1$ and the other with $h = 0.2$.

When $h = 0.1$, the true kurtosis of the errors is $\gamma_2 = 3(1 - 0.2)^3/(1 - 0.4)^{5/2} - 3 = 2.51$, and the true error variance is $\sigma^2 = 1/(1 - 0.2)^{3/2} = 1.40$. We simulated the y_{ij} from the one-way model with $I = 5$ groups and $J = 30$ observations per group. We repeated the simulation several times to get an idea of the variability of the estimates $\hat{\sigma}^2$ and $\hat{\gamma}_2$. The values of $\hat{\sigma}^2$ and $\hat{\gamma}_2$ for five simulated one-way layouts are given in Table 9-6. Although $\hat{\sigma}^2$ is fairly stable, we can see that $\hat{\gamma}_2$ ranges from 1.23 to 5.53 and the estimated expansion factor (required for CI_{2*}) varies from 1.3 to 1.9. Using an estimated expansion factor improves on doing nothing, though it has not totally solved the problems that arise because $\gamma_2 \neq 0$. A typical Q-Q plot of the residuals is shown in Figure 9-6. We have plotted only 30 of the 150 possible points for the sake of clarity. The plot exhibits a shape similar to that in Figure 9-3, as we expect because $\gamma_2 > 0$.

Let us calculate the intervals $\mathrm{CI}_1(\sigma^2, .95)$ and $\mathrm{CI}_2(\sigma^2, .95)$ for Sample 2 ($\hat{\sigma}^2 = 1.35, \hat{\gamma}_2 = 3.17$).

$$\begin{aligned}\mathrm{CI}_1(\sigma^2, .95) &= \left(\frac{1.35 \times 5 \times 29}{\chi^2_{145,U}}, \frac{1.35 \times 5 \times 29}{\chi^2_{145,L}}\right)\\ &= \left(\frac{195.75}{180.23}, \frac{195.75}{113.56}\right)\\ &= (1.09, 1.72) = ((1 - .195)1.35, (1 + .277)1.35)\end{aligned}$$

which covers the true value of 1.40.

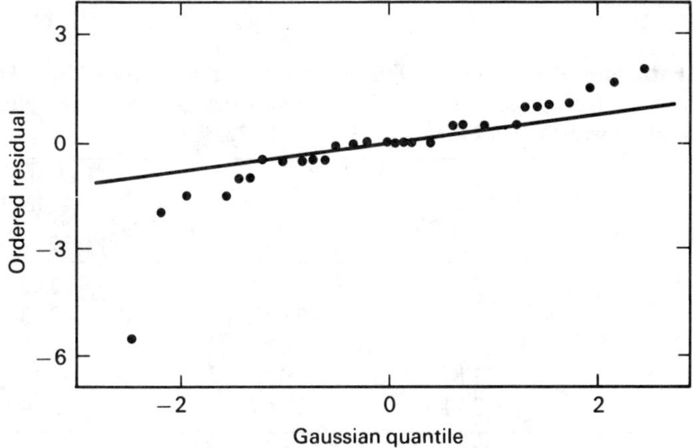

Figure 9-6. Q-Q plot of the residuals from Simulation 2 of the one-way model when $h = 0.1$.

Next we consider the second form of interval, which adjusts for $\hat{\gamma}_2$. First, $\hat{d} = (1 + \hat{\gamma}_2/2)^{-1} = (1 + 3.17/2)^{-1} = 0.387$, so the adjusted degrees of freedom are $145(.387) \approx 56$, and

$$\begin{aligned}
\mathrm{CI}_2(\sigma^2, .95) &= \left(\frac{1.35 \times 0.387 \times 5 \times 29}{\chi^2_{56,U}}, \frac{1.35 \times 0.387 \times 5 \times 29}{\chi^2_{56,L}} \right) \\
&= \left(\frac{75.75}{78.57}, \frac{75.75}{37.21} \right) \\
&= (0.964, 2.04).
\end{aligned}$$

This interval also includes the true value of $\sigma^2 = 1.4$ but is about 1.67 times as wide as the first interval because it adjusts for an error distribution whose tails are heavier than the Gaussian.

The second example differs from the first only in having $h = 0.2$ instead of 0.1. This means that $\gamma_2 = 3(1 - 0.4)^3/(1 - 0.8)^{5/2} - 3 = 33.22$ and $\sigma^2 = 1/(1 - 0.4)^{3/2} = 2.15$. The sample values of these two quantities, $\hat{\sigma}^2$ and $\hat{\gamma}_2$, are given in Table 9-7 for five samples. Once again we witness large variation in the values of $\hat{\gamma}_2$. We choose the fifth sample, with $\hat{\gamma}_2 = 29.07$ and $\hat{\sigma}^2 = 3.71$, for illustrating the construction of the intervals. The first interval gives

$$\begin{aligned}
\mathrm{CI}_1(\sigma^2, .95) &= \left(\frac{3.71 \times 5 \times 29}{\chi^2_{145,U}}, \frac{3.71 \times 5 \times 29}{\chi^2_{145,L}} \right) \\
&= \left(\frac{537.95}{180.23}, \frac{537.95}{113.56} \right) \\
&= (2.98, 4.74) = ((1 - .195)3.71, (1 + .277)3.71).
\end{aligned}$$

Table 9-7. Estimates of σ^2 and γ_2 Obtained for Five Simulations of the One-Way Model with $I = 5$ Groups and $J = 30$ Observations per Group by Sampling from an h-Distribution with $h = 0.2$ ($\sigma^2 = 2.15, \gamma_2 = 33.22$)

Simulation	$\hat{\sigma}^2$	$\hat{\gamma}_2$	Expansion Factor[a] $(1 + \hat{\gamma}_2/2)^{1/2}$
1	2.26	6.24	2.03
2	2.20	11.82	2.63
3	1.64	0.69	1.16
4	1.87	9.52	2.40
5	3.71	29.07	3.94

[a] On the basis of the population value of γ_2 we would like to have the expansion factor equal to $(1 + 33.22/2)^{1/2} = 4.20$.

This interval does not include $\sigma^2 = 2.15$. For the second interval, $\hat{d} = (1 + 29.07/2)^{-1} = 0.064$, so that the adjusted degrees of freedom are $145(.064) \approx 9$ and

$$\begin{aligned} \mathrm{CI}_2(\sigma^2, .95) &= \left(\frac{3.71 \times 0.064 \times 5 \times 29}{\chi^2_{9,U}}, \frac{3.71 \times 0.064 \times 5 \times 29}{\chi^2_{9,L}} \right) \\ &= \left(\frac{34.43}{19.02}, \frac{34.43}{2.70} \right) \\ &= (1.81, 12.75). \end{aligned}$$

The second interval does include 2.15, whereas the first did not. We might conclude from these examples that taking γ_2 into account is a wise move when constructing intervals for σ^2; but the effectiveness of the method relies on how well $\hat{\gamma}_2$ estimates γ_2, and the value of $\hat{\gamma}_2$ involves a great deal of variability.

9E. UNBALANCED CASES (EXPECTED-MEAN-SQUARE METHOD)

When the sample sizes are unequal, the estimates obtained by the expected-mean-square method are more complicated than in the balanced case. Moreover, they do not accommodate different σ_i^2 nearly as well.

Fixed Effects

Let the sample sizes be J_i, $i = 1, 2, \ldots, I$, and let the populations have a common variance σ^2. As before,

$$y_{ij} = \mu + \alpha_i + \varepsilon_{ij}$$

subject to the constraint making the mean effect zero:

$$\sum J_i \alpha_i = 0.$$

(Sometimes the constraint $\Sigma \alpha_i = 0$ is used; then the calculations get more complicated.) Let the total sample size be $J_+ = \Sigma_i J_i$. The estimate of σ^2 by the expected-mean-square method is

$$\hat{\sigma}^2 = \mathrm{WMS} = \frac{1}{J_+ - I} \sum_{i=1}^{I} \sum_{j=1}^{J_i} (y_{ij} - y_{i\cdot})^2.$$

The mean square between populations, BMS, is

$$\mathrm{BMS} = \frac{1}{I-1} \sum_{i=1}^{I} J_i (y_{i\cdot} - \bar{y}_{\cdot\cdot})^2, \qquad \text{with } \bar{y}_{\cdot\cdot} = \sum J_i y_{i\cdot} / J_+.$$

(The bar in $\bar{y}_{..}$ serves to distinguish the grand mean in an unbalanced data layout from the grand mean $y_{..}$ in a balanced one.) The expected values of BMS and WMS are

$$E(\text{BMS}) = \sigma^2 + \frac{\sum_{i=1}^{I} J_i\alpha_i^2}{I-1},$$

$$E(\text{WMS}) = \sigma^2.$$

If we think of the J_i as frequencies associated with the α_i, then we can use the weights $w_i = J_i/J_+$ on the α_i^2 to give the equivalent of a variance for the set of means, weighted by their observed occurrences (because the constraint implies that the weighted sum of the α_i has zero mean). When the J_i differ because the numbers of occurrences in the samples reflect proportions in the real world, this seems a desirable definition. But when the J_i reflect the investigator's needs or desires—as when the σ_i^2 are expected to differ and the J_i have been made large for i's with plausibly large σ_i^2—we are less comfortable. We are also less comfortable with weighting when the differences in the J_i reflect the accidents that occur in everyday experimentation.

We want to define τ^2 in such a way that it matches our previous definition for the balanced case and fits in with the expected mean squares. We want something like

$$\tau^2 = \frac{\Sigma J_i\alpha_i^2}{J_+ - J^*},$$

where J^* is some sort of average of the J_i. Although we could choose $J^* = J_+/I$, the definition will fit in better with later work if we choose

$$J^* = \frac{\Sigma J_i^2}{J_+},$$

which is close to J_+/I whenever the J_i are similar. Note that J^* weights the J_i by their relative frequency in the samples. The choice of J^* comes from an analysis of the expected values of BMS and WMS given above. Then we might estimate this between-group variance as

Estimate of τ^2

$$\hat{\tau}^2 = \frac{(I-1)(\text{BMS} - \text{WMS})}{J_+ - J^*} = \frac{\Sigma J_i\hat{\alpha}_i^2}{J_+ - J^*}$$

to go with

$$\hat{\sigma}^2 = \text{WMS}.$$

Random Effects

In the random-effects model, we have

$$y_{ij} = \mu + a_i + \varepsilon_{ij},$$

where the a_i and the ε_{ij} are all independent, the a_i come from a distribution with mean 0 and variance σ_A^2, and the ε_{ij} come from a distribution with mean 0 and variance σ^2. For this model, the expected values of the mean squares are

$$E(\text{BMS}) = \sigma^2 + \frac{\left(J_+^2 - \sum_{i=1}^{I} J_i^2\right)\sigma_A^2}{J_+(I-1)}$$

$$E(\text{WMS}) = \sigma^2.$$

Note that when the J_i all equal J, $E(\text{BMS}) = \sigma^2 + J\sigma_A^2$, in agreement with the result in Section 9A.

To estimate the common variance σ^2 by the expected-mean-square method, we again use

$$\hat{\sigma}^2 = \text{WMS} = \frac{\sum_{i=1}^{I}\sum_{j=1}^{J_i}(y_{ij} - y_{i\cdot})^2}{J_+ - I},$$

and to estimate σ_A^2 we use

$$\hat{\sigma}_A^2 = \frac{(I-1)(\text{BMS} - \text{WMS})}{J_+ - J^*}$$

or zero, whichever is greater.

EXAMPLE: PERFORMANCE SCORES

Let use consider the British school example, with the data on 16 schools as published originally (Table 9-8). Applying the random-effects formulas gives

$$\hat{\sigma}_A^2 = 15.56,$$

$$\hat{\sigma}^2 = 196.85,$$

both well within the confidence intervals based on seven schools (10 to 99 and 160 to 206, respectively).

Thus, compared with the information in Table 9-2 from seven schools with about 50 students each, the 16 schools have a within-school variance estimate

Table 9-8. Summary Information for the ILEA Scores from 16 Schools

School	Number of Pupils, J_i	Mean, $y_{i\cdot}$	Standard Deviation, s_i	$s_i^2/J_i =$ $(\text{SEM})^2$
1	65	26.9	16.0	3.94
2	79	18.7	13.7	2.38
3	48	28.3	13.4	3.74
4	47	21.1	14.2	4.29
5	66	17.5	14.9	3.36
6	41	11.8	12.1	3.57
7	52	14.3	11.8	2.68
8	67	21.2	13.3	2.64
9	49	16.3	14.9	4.53
10	47	26.5	13.4	3.82
11	50	20.3	12.1	2.93
12	41	22.6	19.3	9.09
13	49	16.8	13.9	3.94
14	29	22.6	15.7	8.50
15	72	18.0	12.9	2.31
16	62	19.2	12.9	2.68

$$J_+ = \sum J_i = 864 \qquad \text{BSS} = \sum J_i y_{i\cdot}^2 - C = 15{,}508$$

$$\sum J_i y_{i\cdot} = 17{,}329.5 \qquad \text{WSS} = \sum (J_i - 1)s_i^2 = 166{,}933$$

$$\sum J_i y_{i\cdot}^2 = 363{,}091 \qquad \text{BMS} = \frac{15{,}508}{16 - 1} = 1033.87$$

$$\bar{y}_{\cdot\cdot} = \left(\sum J_i y_{i\cdot}\right)/J_+ = 20.0573 \qquad \text{WMS} = \frac{166{,}933}{864 - 16} = 196.85 = \hat{\sigma}^2$$

$$C = \frac{\left(\sum J_i y_{i\cdot}\right)^2}{J_+} = 347{,}583 \qquad \text{BMS} - \text{WMS} = 837.02$$

$$\hat{\sigma}_A^2 = \frac{J_+(16 - 1)(\text{BMS} - \text{WMS})}{J_+^2 - \sum J_i^2} = \frac{864(15)(837.02)}{(864)^2 - 49{,}270} = 15.558$$

Source: M. Aitkin and N. Longford (1986). "Statistical modelling issues in school effectiveness studies," *Journal of the Royal Statistical Society, Series A*, 149, 1–42 (with discussion) (data from Table 1, p. 4).

about 9% larger and a between-school component of variance about 36% smaller. A further comparison would look at confidence intervals calculated from the more extensive data.

Using these results as our estimates, we regard the standard deviation of the population of school means as perhaps $\hat{\sigma}_A = 4$, and the corresponding standard deviation of the distribution of scores of students within schools as very tightly measured at $\hat{\sigma} = 14$. Thus there appears to be more variability within schools than between schools.

9F. TWO-WAY TABLES

Thus far we have emphasized the one-way analysis of variance with several observations in each group. Sometimes we have additional stratification. For example, in the two-way analysis of variance, we may have rows and columns crossed to handle two descriptive or experimental variables. Either or both of these variables may have a fixed set of effects or a random set. As an example, we might have 10 laboratories, each of which has been asked independently to measure the potassium content of 10 distinct specimens, the same 10 for each laboratory. Each laboratory gets all 10 specimens, and the distributing center identifies the results by both laboratory and specimen. If each laboratory gives us a value for each sample, then we have a 10×10 table of measurements as shown in Table 9-9, with one measurement per cell. (We have split the 10 specimens into first five and last five, in successive time periods.)

The 10 specimens may have arbitrarily been chosen by the distributing laboratory, and then their values could be regarded as fixed. Had their means been chosen randomly from a population of means, the specimens would be regarded as random, and we could be interested in the variance of that population.

Similarly, the laboratories might be a total population, such as a set belonging to a single firm or all hospitals associated with a single medical school, or they might be a (more or less) random sample from a set of laboratories. In the former situations the laboratories would be regarded as producing fixed effects, and in the latter, random effects. In practice, the laboratories are likely to be a "scattering" rather than a random sample. In such cases it seems better (1) to begin by treating them as if they were a random sample, and then (2) to add "a grain of salt," increasing the estimated component of variance to allow for nonrandom selection.

All told then, if we neglect interaction, we have three kinds of models in the two-way ANOVA: fixed row effects and fixed column effects, fixed row effects and random column effects, and random row effects and random column effects. We may refer to these more briefly as the fixed model, the mixed model, and the random model, respectively.

Fixed Row Effects and Fixed Column Effects

The model is

$$y_{ij} = \mu + \alpha_i + \beta_j + \varepsilon_{ij}$$

in which $i = 1, 2, \ldots, I$, $j = 1, 2, \ldots, J$, $\Sigma\alpha_i = \Sigma\beta_j = 0$, and the ε_{ij} are uncorrelated and drawn from a distribution with mean 0 and variance σ^2. We can assign a finite-population variance to the row effects and the column

Table 9-9. Potassium Measurements for 10 Laboratories Using Analytic Method 5 in Two Sets of Five Specimens, and Mean Square Calculations for the First Set

(a) Data

	First Set of Five						Second Set of Five					
Laboratory	1	2	3	4	5	Total	1	2	3	4	5	Total
1	5.4	4.5	4.8	4.0	3.5	22.2	5.6	5.2	4.8	7.5	6.2	29.3
2	5.4	4.5	4.9	4.1	3.6	22.5	5.7	5.1	4.6	7.7	6.1	29.2
4	5.3	4.5	4.9	4.2	3.5	22.4	5.7	5.2	4.6	7.7	6.1	29.3
7	5.4	4.4	4.8	4.1	3.5	22.2	5.6	5.1	4.6	7.5	6.1	28.9
15	5.2	4.4	4.8	4.0	3.4	21.8	5.6	5.0	4.3	7.5	5.9	28.3
16	5.4	4.6	5.0	4.1	3.6	22.7	5.7	5.1	4.6	7.6	6.0	29.0
18	5.3	4.3	4.8	4.0	3.4	21.8	5.5	5.1	4.6	7.1	6.1	28.4
19	5.1	4.4	5.0	4.2	3.6	22.3	5.7	5.1	4.6	7.7	6.2	29.3
21	5.3	4.5	4.8	4.0	3.5	22.1	5.6	5.0	4.7	7.5	6.0	28.8
25	5.3	4.3	5.0	4.1	3.7	22.4	5.8	5.1	4.6	7.6	6.1	29.2
Total	53.1	44.4	48.8	40.8	35.3	222.4	56.5	51.0	46.0	75.4	60.8	289.7

(b) Calculations for First Set

$\sum(\text{Row total})^2 = 4946.92$	$\sum(\text{Row total})^2/5 = 989.3840$	$\sum(\text{Column total})^2 = 10083.14$
	$C = 989.2352$	$\sum(\text{Column total})^2/10 = 1008.314$
	BRSS = .1488	$C = 989.2352$
$C = \dfrac{(222.4)^2}{50} = 989.2352$	BRMS = .01653	BCSS = 19.0788
		BCMS = 4.7697

Source: M. J. R. Healy (1985). "24. Quality control data in clinical chemistry." In D. F. Andrews and A. M. Herzberg, *Data: A Collection of Problems from Many Fields for the Student and Research Worker*. New York: Springer-Verlag, pp. 151–155 (data from Table 24.2, pp. 154–155). His source: National Quality Control Scheme, Queen Elizabeth Hospital Birmingham.

effects as we have done before:

Components of Variance for Rows and Columns

$$\tau^2 = \frac{\sum_i \alpha_i^2}{I-1}, \qquad \lambda^2 = \frac{\sum_j \beta_j^2}{J-1}$$

and then we have quantities that correspond to components of variance for rows and for columns, respectively. If we use the expected-mean-square method to estimate τ^2, λ^2, and σ^2, we are motivated by the expected mean squares

$$\begin{aligned}
&\text{Rows:} & E(\text{BRMS}) &= \sigma^2 + J\tau^2,\\
&\text{Columns:} & E(\text{BCMS}) &= \sigma^2 + I\lambda^2,\\
&\text{Residual:} & E(\text{Residual MS}) &= \sigma^2,
\end{aligned}$$

to use the following estimates because they are unbiased:

$$\begin{aligned}
\hat{\tau}^2 &= (\text{BRMS} - \text{Residual MS})/J,\\
\hat{\lambda}^2 &= (\text{BCMS} - \text{Residual MS})/I,\\
\hat{\sigma}^2 &= \text{Residual MS}.
\end{aligned}$$

Fixed Row Effects and Random Column Effects

For fixed row effects and random column effects we use the model

$$y_{ij} = \mu + \alpha_i + b_j + \varepsilon_{ij}$$

with $i = 1, 2, \ldots, I$, $j = 1, 2, \ldots, J$, $\tau^2 = \Sigma_i \alpha_i^2/(I - 1)$, $\Sigma\alpha_i = 0$, $E(b_j) = E(\varepsilon_{ij}) = 0$, $\sigma_B^2 = E(b_j^2)$, $\sigma^2 = E(\varepsilon_{ij}^2)$. The motivating expected mean squares are now

$$\begin{aligned}
&\text{Rows:} & E(\text{BRMS}) &= \sigma^2 + J\tau^2,\\
&\text{Columns:} & E(\text{BCMS}) &= \sigma^2 + I\sigma_B^2,\\
&\text{Residuals:} & E(\text{Residual MS}) &= \sigma^2.
\end{aligned}$$

Thus our method-of-moments estimates are

$$\begin{aligned}
\hat{\tau}^2 &= (\text{BRMS} - \text{Residual MS})/J,\\
\hat{\sigma}_B^2 &= (\text{BCMS} - \text{Residual MS})/I,\\
\hat{\sigma}^2 &= \text{Residual MS}.
\end{aligned}$$

We have the same numbers, but one of them, $\hat{\sigma}_B^2$, is now regarded as an estimate of a new parameter.

If the random effects are in the rows and the fixed effects are in the columns, we simply interchange I for J, $\hat{\sigma}_A^2$ for $\hat{\sigma}_B^2$, and BRMS for BCMS in the above formulas.

Random Row Effects and Random Column Effects

The model may be written

$$y_{ij} = \mu + a_i + b_j + \varepsilon_{ij}$$

with $i = 1,2,\ldots,I$, $j = 1,2,\ldots,J$, $E(a_i) = E(b_j) = E(\varepsilon_{ij}) = 0$, $\sigma_A^2 = E(a_i^2)$, $\sigma_B^2 = E(b_j^2)$, $\sigma^2 = E(\varepsilon_{ij}^2)$. Again the a_i, the b_j, and the ε_{ij} are independently drawn. For estimating the components of variance σ_A^2, σ_B^2, and σ^2, the motivating mean squares are

$$\begin{aligned} &\text{Rows:} && E(\text{BRMS}) = \sigma^2 + J\sigma_A^2, \\ &\text{Columns:} && E(\text{BCMS}) = \sigma^2 + I\sigma_B^2, \\ &\text{Residual:} && E(\text{Residual MS}) = \sigma^2, \end{aligned}$$

leading to the estimates

$$\begin{aligned} \hat{\sigma}_A^2 &= (\text{BRMS} - \text{Residual MS})/J, \\ \hat{\sigma}_B^2 &= (\text{BCMS} - \text{Residual MS})/I, \\ \hat{\sigma}^2 &= \text{Residual MS}. \end{aligned}$$

Again we use the same numbers, but differently named parameters. Note that, in the above sets of equations, the pairs σ_A^2 and τ^2, and σ_B^2 and λ^2 play identical roles as far as calculations are concerned.

In actually making the calculations we need only one set of sums of squares and one set of mean squares for all the models.

Calculations

For hand calculations some convenient formulas for TSS (total sum of squares), BRSS, BCSS, Residual SS, and the correction term C are as follows:

Total: $$\text{TSS} = \sum_i \sum_j y_{ij}^2 - C,$$

where

$$C = \left(\sum_i \sum_j y_{ij}\right)^2 / IJ;$$

Rows: $$\text{BRSS} = \frac{\left(\sum_j y_{1j}\right)^2}{J} + \cdots + \frac{\left(\sum_j y_{Ij}\right)^2}{J} - \text{C};$$

Columns: $$\text{BCSS} = \frac{\left(\sum_i y_{i1}\right)^2}{I} + \cdots + \frac{\left(\sum_i y_{iJ}\right)^2}{I} - \text{C};$$

Residual: Residual SS = TSS − BRSS − BCSS.

It may be helpful to note that, in taking sums and squaring them in these

Table 9-10. Analysis-of-Variance Table and Estimates of Components of Variance for the Potassium Measurements on the First Set of Five Specimens

Source	d.f.	Sum of Squares	Mean Square	
Total	49	19.4648		
Rows (labs)	9	.1488	.01653	$\hat{\sigma}^2_{\text{Labs}} = .001989$
Columns (specimens)	4	19.0788	4.7697	$\hat{\tau}^2_{\text{Specimens}} = .4763$
Residual	36	.2372	.00659	$\hat{\sigma}^2 = .00659$

Estimates of standard deviations

$\hat{\sigma}_{\text{Labs}} = .045$ $\hat{\tau}_{\text{Specimens}} = .69$ $\hat{\sigma} = .081$

formulas, we always, for C as well as for other sums, divide by the number of original y_{ij} that are summed.

An Interlaboratory Example

M. J. R. Healy contributed (Andrews and Herzberg 1985, p. 154) several sets of data for 100 laboratories that make chemical analyses of a specimen from a homogeneous pool of serum every two weeks, a different pool each time. Although the laboratories use different methods of analysis, we have chosen the first 10 laboratories that used Method 5 to measure potassium and reported all 10 successive specimens.

We analyzed the data as two sets of five specimens, for these 10 laboratories. Thus we have two parallel analyses for the same 10 laboratories during two successive time periods. Table 9-9 gives the reported measurements of potassium. The laboratory numbers are labels that Healy assigned. Tables 9-10 and 9-11 give the analysis-of-variance tables for the two sets of specimens, including estimates of the variance components and standard deviations.

Table 9-11. Analysis-of-Variance Table and Estimates of Components of Variance for the Potassium Measurements on the Second Set of Five Specimens

Source	d.f.	Sum of Squares	Mean Square	
Total	49	51.1882		
Rows (labs)	9	.2482	.02758	$\hat{\sigma}^2_{\text{Labs}} = .00353$
Columns (specimens)	4	50.5832	12.6458	$\hat{\tau}^2_{\text{Specimens}} = 1.2636$
Residual	36	.3568	.00991	$\hat{\sigma}^2 = .00991$

Estimates of standard deviations $\hat{\sigma}_{\text{Labs}} = .059$ $\hat{\tau}_{\text{Specimens}} = 1.12$ $\hat{\sigma} = .100$

Visually Examining the Data

Attending first to Table 9-9, we observe very little variation within a column. Ordinarily the range is 0.3 or less. This close agreement suggests strongly that the laboratories are measuring much the same thing in the same way. Looking across the columns, the numbers are considerably more variable, and so we think the specimens have different amounts of potassium (or at least different amounts of whatever is being measured). Because these are arbitrarily chosen, we have no special interest in their values. (If we wished to relate their sizes to the different methods of analysis, our interest in Healy's larger table would perk up.)

The row totals vary little. Laboratories 15 and 18 have the lowest sums in both sets. Because the rows totals vary so little, we might want first to look for outliers without computing residuals for each cell. We do this mainly by looking down columns. Subject to more careful analysis, second set column 3, laboratory 15, and second set column 4, laboratory 18 seem low. There may be merit in a re-analysis that would take these outliers into account.

Ignoring the possibility of outliers for now, we turn to Table 9-10. It shows, as we expect on the basis of our visual impressions, very small variation among the laboratories, and substantial variation among the specimens. The estimated standard deviation among the laboratories is .045. The standard deviation among specimens is estimated as about .7. (This fits well with our rough estimate: range/ $\sqrt{J} = (5.31 - 3.53)/\sqrt{5} = .80$.)

Because the data are recorded only to the nearest tenth, at least in our source, it is worth thinking about the contribution of rounding error. If we consider a rounding error as uniformly distributed over an interval of length L, then from statistical theory its variance is $L^2/12$. In our problem $L = .1$, and so the variance due to rounding error is $.01/12 = .000833$. This means the rounding error accounts for about 13% (= 100(.000833)/.00659) of the residual variance. Keeping another decimal place, if it is feasible, would capture almost all this information. However, the residual standard deviation would decrease only from .081 to .076, and so perhaps it is not worth the paper required.

Table 9-11 presents the information parallel to Table 9-10 for the second set of five specimens. As before, the variance among laboratories is small and for specimens large. The estimated laboratory component of variance has nearly doubled—something that often happens when dealing with such small samples. The estimated specimens component of variance is up by about a factor of 2.5. The estimated residual variance is up by a factor of 1.5. Part of these increases may be attributable to the two outliers that we noted. And, of course, some of the variation between the two sets of specimens may come from plain old sampling variation. Indeed, the two outliers may be consequences of conventional variation, just as they may, alternatively, spring from unusual circumstances.

We might consider the possibility that, because the measurements in the second set are generally larger than the measurements in the first set, they may produce larger variance. Detailed examination of the distributions in the columns leaves this possibility confounded (confused) with the possibility that the outliers go far to explain the changes. For the moment, we do not invoke the empirically well-established principle that larger numbers are often accompanied by larger variability.

Detailed Examination of Residuals

Table 9-12 gives to two decimals the residuals obtained by fitting the grand mean and row and column means to the data in Table 9-9. And then, to show how these residuals are distributed, we supply stem-and-leaf displays for each set of 50 residuals. Earlier, we had commented that laboratory 18, second set column 4, seemed very low, and it stands out again as an outlier, the most extreme in two panels. Laboratory 15, second set column 3, is second most extreme in the right-hand panel; but it nestles up to the rest of the data in that set and so would not seem surely to be an outlier.

In the first panel two residuals stand out as isolated, −.22 and −.17. Depending on the outlier rule used, one or both would be regarded as

Table 9-12. Residuals from a Two-Way Additive Fit to the Two Sets of Potassium Measurements, Together with Stem-and-Leaf Displays for the Residuals

	First Set					Second Set				
Lab	1	2	3	4	5	1	2	3	4	5
1	.10	.07	−.07	−.07	−.02	−.12	.03	.13	−.11	.05
2	.04	.01	−.03	−.03	.02	.00	−.05	−.05	.11	−.03
4	−.04	.03	−.01	.09	−.06	−.02	.03	−.07	.09	−.05
7	.10	−.03	−.07	.03	−.02	−.04	.01	.01	−.03	.03
15	−.02	.05	.01	.01	−.04	.08	.03	−.17	.09	−.05
16	−.00	.07	.03	−.07	−.02	.04	−.01	−.01	.05	−.09
18	.08	−.05	.01	.01	−.04	−.04	.11	.11	−.33	.13
19	−.22	−.05	.11	.11	.06	−.02	−.07	−.07	.09	.05
21	.02	.09	−.05	−.05	−.00	−.02	−.07	.13	−.01	−.05
25	−.04	−.17	.09	−.01	.14	.10	−.05	−.05	.01	−.03

Stem	First Set leaves	Stem	Second Set leaves
−3*		−3*	3
−2·		−2·	
−2*	2	−2*	
−1·	7	−1·	7
−1*		−1*	21
−0·	777765555	−0·	977775555555
−0*	444433322221100	−0*	44333222111
0*	11111223334	0*	011133334
0·	56778999	0·	5558999
1*	00114	1*	0111333

Note: The original values were given to only one decimal. Thus attention to the second should not be overdone, though it is true in some tables of mathematical functions that a careful empirical analysis can retrieve much of an additional decimal.

borderline outliers. Had this picture been available when the data were fresh, we certainly would have investigated these two observations further.

Failure of additivity: *Interaction*. We have already suggested that the residual variance σ^2 has as part of its makeup a rounding error whose magnitude we assessed. Beyond this, no matter which model we used, we could have had an interaction term, whose notation for random laboratories as rows and random specimens as columns might be $(ab)_{ij}$, so the full model would be

$$y_{ij} = \mu + a_i + b_j + (ab)_{ij} + \varepsilon_{ij}.$$

Unless outside information or assumptions come to our aid, as they did in looking at the rounding error, we have no way of isolating either the individual $(ab)_{ij}$ or an estimate of some property of them such as $\Sigma_{ij}(ab)^2_{ij}/(I - 1)(J - 1)$. The expression "lack of identifiability" is sometimes used in this situation. As long as the structure of the investigation has only one observation per cell, no matter how many rows and columns we have, we cannot get estimates relevant to only the interaction terms $(ab)_{ij}$ (or only the errors ε_{ij}) from the experimental data alone; those terms cannot be separated—unless we impose some mathematical structure. If the structure were different, so that we had more than one independent observation per cell, then we could get estimates of the size of the interaction. Our potassium example has only one observation per cell.

Although the last paragraph is, word-by-word, correct as written, it still may mislead. What we can get are functions of the combination

$$\text{interaction} + \text{error} = (ab)_{ij} + \varepsilon_{ij}.$$

We cannot avoid having some of both. If one dominates the other sufficiently, and we can be sure this is so, we may be able to judge fairly well how large the dominant one is.

Comparing Methods 5 and 6. We took the first set of 10 laboratories that used Method 6 on the first set of five measurements, thus parallel to the first set in Table 9-9, which used Method 5. Table 9-13 gives the potassium values.

Table 9-14 gives the analysis of variance for Method 6 and compares the estimates of components of variance for Method 6 with the components of variance from the Method 5 results. The estimated component of variance for laboratories is nearly six times as large for Method 6 as for Method 5. The estimate of variance between specimens is nearly identical for the two methods, and the error variance is about 1.7 times as big for Method 5 as Method 6.

From Table 9-9, the row totals of the first set run from 21.8 to 22.7, with range 0.9, whereas those of Table 9-13 for Method 6 run from 21.2 to 23.4, with range 2.2. The ratio of squared ranges $2.2^2/0.9^2 = 6.0$ is even larger than 5.9, the ratio of the two s^2 terms.

Table 9-13. Healy's Data on Potassium for the Set of First Five Specimens from Laboratories using Method 6

	First Set					
Lab	1	2	3	4	5	Total
11	5.2	4.5	4.8	4.0	3.4	21.9
27	5.3	4.5	4.8	4.1	3.5	22.2
33	5.2	4.6	4.8	4.1	3.5	22.2
41	5.4	4.6	4.9	4.1	3.5	22.5
44	5.6	4.8	5.0	4.2	3.8	23.4
49	5.3	4.5	4.8	4.0	3.4	22.0
53	5.3	4.5	4.6	4.0	3.5	21.9
57	5.3	4.6	4.8	4.1	3.5	22.3
66	5.2	4.4	4.8	4.0	3.6	22.0
83	5.2	4.3	4.6	3.8	3.3	21.2
Total	53.0	45.3	47.9	40.4	35.0	221.6

Source: M. J. R. Healy (1985). "24. Quality control data in clinical chemistry." In D. F. Andrews and A. M. Herzberg, *Data: A Collection of Problems from Many Fields for the Student and Research Worker*. New York: Springer-Verlag, pp. 151–155 (data from Table 24.2, pp. 154–155). His source: National Quality Control Scheme, Queen Elizabeth Hospital Birmingham.

We note that laboratory 44 has especially high values. Even setting it aside, the next highest, laboratory 41, sums to 22.5, leaving a range of 1.3, about 1.4 times that of Method 5. To get a comparable range from Method 6, we would need to set aside not only laboratory 44's high value but also laboratory 83's low one, and then the range of the remaining eight laboratories would be 0.6. Thus, in this simple instance, we can see the higher apparent variability among laboratories directly in the row totals. We say "apparent" on general principles. If the underlying values behave like samples from a Gaussian distribution, so that an unmodified F distribution applies, a ratio of 5.9 (on 9 and 9 degrees of freedom) would go beyond the (upper) 1% point.

Turning now to column totals, the range for the first set in Table 9-9 is 17.8, whereas that for Table 9-13 is 18.0, and so we are not expecting a large

Table 9-14. Analysis of Variance for Data from Method 6 Given in Table 9-13, Together with Comparisons of Components of Variance for Method 6 and Method 5 (from Table 9-10)

Source	d.f.	SS	MS	Method 6 Component	Method 5 Component
Rows (labs)	9	.5568	.06187	$\hat{\sigma}^2_{\text{Labs}} = .01165$	.001989
Columns (specimens)	4	19.1348	4.7837	$\hat{\tau}^2_{\text{Specimens}} = .4780$	.4763
Within	36	.1372	.00381	$\hat{\sigma}^2_{\text{Error}} = .00381$	.00659

difference in the estimated component of variance for specimens, nor did we get one. Our uncertainty about the variability between samples is mainly confined to how well the five specimens represent any population of specimens whose variance we wish to consider or assess.

9G. EXAMPLE: NATIONALIZATION OF ELECTORAL FORCES

A study of voting patterns furnishes a separate example of a more complicated situation involving both fixed and random effects.

Political scientists have historical as well as current interests in voting patterns in various countries. One question that has concerned political analysts is the comparative impacts on elections of local districts within states, of states within the United States, and of the nation as a whole. To what extent does national politics contribute to the outcomes in states and local districts and to what extent the reverse? We want to appreciate what is going on as a process changing through time.

To explore the effect of national, state, and local constituencies on turnout and on variability of party vote, Donald E. Stokes (1967) used a components-of-variance model. The model was

$$y_{ijk} = \alpha + \beta_i + \gamma_{ij} + a_k + b_{ik} + c_{ijk},$$

where

$$\begin{aligned}
\alpha &= \text{fixed effect of national politics,}\\
\beta_i &= \text{fixed effect of politics in state } i,\\
\gamma_{ij} &= \text{fixed effect of politics in district } j \text{ of state } i,\\
a_k &= \text{random effect of national politics in year } k,\\
b_{ik} &= \text{random effect of politics in state } i \text{ in year } k,\\
c_{ijk} &= \text{random effect of politics in district } j \text{ of state } i \text{ in year } k.
\end{aligned}$$

In some analyses

$$y_{ijk} = \text{proportion turning out in congressional district } j \text{ of state } i \text{ in year } k,$$

and in another

$$y_{ijk} = \text{proportion Republican of the two-party vote in congressional district } j \text{ of state } i \text{ in year } k.$$

Some analyses took each decade separately, five congressional elections per decade, so in a decade $k = 1, 2, 3, 4, 5$. Stokes computed for each decade the

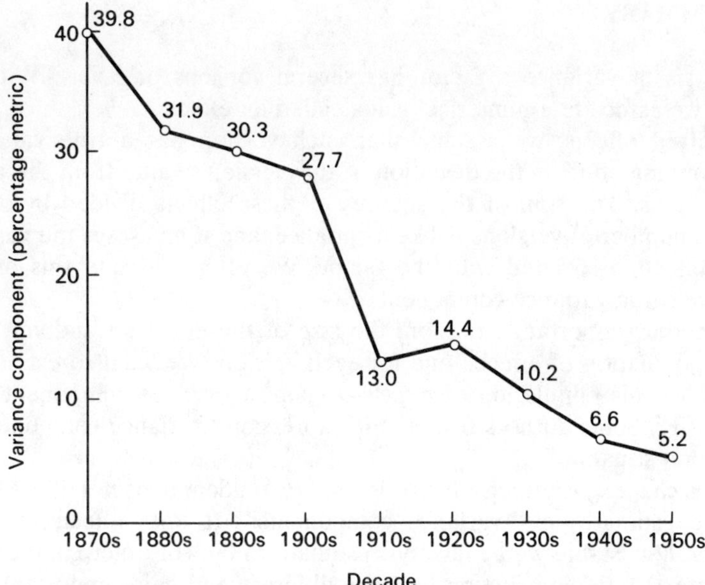

Figure 9-7. Changes in constituency influence on congressional-election turnout by decade. *Note:* Only congressional districts with Democratic–Republican contests in all five elections of the decade are included. From "Parties and the Nationalization of Electoral Forces," by Donald E. Stokes from *The American Party Systems*, First Edition, edited by William N. Chambers and Walter Burnham. Copyright © 1967, 1965 by Oxford University Press, Inc. Reprinted by permission.

percentage of the variance represented by the component associated with congressional districts for the election of congressional representatives.

Figure 9-7 shows how the percentage of variance in voter turnout owing to the component for districts decreased drastically during the decades from the 1870s through the 1950s. Essentially, the variance of turnout has moved primarily to the national level. Indeed, in the decade 1952–1960, 86% of the variance was associated with the national or a_k level, and 8% with the state or b_{ik} level, leaving 6% for the district or c_{ijk} level.

For the same decade (1950s), the corresponding analysis of the variance in the proportion Republican (or Democrat) of the two-party vote gave 32% to the national level, 19% for the state, and 49% for the constituency districts.

To compare these results for the U.S. Congress with the results of a similar analysis for the British parliament, the corresponding analysis of the party vote for the years 1950–1966 gave 47% to the national level, 13% to the regional, and 40% to the constituency. Thus for the party vote, the districts appeared somewhat more important in the United States than they were in Britain.

The mathematical treatment appears in Stokes (1969).

9H. SUMMARY

In analysis of variance a factor has several versions or levels. With each version we associate a numerical value called an effect.

For fixed effects, we assume that each version has a true value. The corresponding effect is the deviation of this version's value from the average over versions. The sum of the squares of these effects divided by one less than the number of versions is like a variance, and it measures the variability of the effects associated with the factor. We often estimate this quantity, which we call a variance component.

In another important situation, the size of the effect is randomly drawn from a population of effects, one for each version. We want to estimate the variance of this population of effects—again, a variance component, which we label σ_A^2. Some authors reserve the expression "variance component" for this second situation.

In this chapter, whether a factor is fixed or random does not affect how we calculate estimates of "variance components." It does affect, often very seriously, how stable we believe our estimate to be. (In other chapters, such as Chapter 10, this distinction has very different and quite important consequences.)

In both situations, we are concerned with variability due to errors of measurement or variability of individuals in a population. We almost always need to estimate the error variance σ^2, which is often treated as if it were the same for each version of the factor. We often want to estimate within-version variance or, if it varies, its average across versions.

The expected values for the between mean square, BMS, and the within mean square, WMS, guide us to equating-averages ("expected-mean-square method") estimates for both the component of variance for versions and that for error variance. We know from Chapter 7 that the BMS will ordinarily be based on few degrees of freedom and therefore that we cannot expect very reliable estimates of the component for levels or versions.

The unbiased estimate of the component of variance for versions often generates negative estimates of that component, especially when the true component for versions is small compared with σ^2 for error. By giving up the unbiasedness, and disallowing negative estimates by replacing them by zero, we can considerably reducc the mean squared error of the estimate of the component for versions and increase its usefulness.

When the error variances within the several versions differ, the WMS estimates the average of those error variances.

Confidence limits for variance components, in view of the above considerations, can be expected often to lead to wide intervals. We provide formulas that reflect these.

If the residuals come from distributions with fourth moments far from those of the Gaussian distribution, special formulas provided may give more appropriate (still wider) confidence limits for the within variance component.

These may be needed even when within-version distributions are Gaussian but the σ_i^2 vary widely.

For the unbalanced one-way analysis of variance, a slight modification of the formulas for estimating variance components gives reasonable estimates for both fixed and random effects, though variation of σ_i^2 may produce more concern.

When the layout is a two-way table with one observation per cell, we may have both factors fixed, both random, or a mixed case, one factor random and one fixed. Each case requires some special treatment. All start with the same mean squares and the same variance-component-like numbers. With only one observation per cell, we cannot separate the cell error variance from the interaction effects, and so these both contribute to the residual mean square.

We provide two special illustrations of applications of analyses of variance to practical problems. A study of blood pressures focuses on variation within and between occasions for young and adult human males and females. A special study of voting behavior shows how congressional politics in the United States became more and more nationalized over an 80-year period.

REFERENCES

Aitkin, M. and Longford, N. (1986). "Statistical modelling issues in school effectiveness studies," *Journal of the Royal Statistical Society*, *Series A*, 149, 1–42 (with discussion).

Andrews, D. F. and Herzberg, A. M. (Eds.) (1985). *Data*: *A Collection of Problems from Many Fields for the Student and Research Worker*. New York: Springer-Verlag.

Arvesen, J. N. and Schmitz, T. H. (1970). "Robust procedures for variance component problems using the jackknife," *Biometrics*, 26, 677–686.

Boardman, T. J. (1974). "Confidence intervals for variance components—a comparative Monte Carlo study," *Biometrics*, 30, 251–262.

Bulmer, M. G. (1957). "Approximate confidence limits for components of variance," *Biometrika*, 44, 159–167.

Healy, M. J. R. (1985). "24. Quality control data in clinical chemistry." In D. F. Andrews and A. M. Herzberg (Eds.), *Data*: *A Collection of Problems from Many Fields for the Student and Research Worker*. New York: Springer-Verlag, pp. 151–155.

Hoaglin, D. C. (1985). "Summarizing shape numerically: The *g*-and-*h* distributions." In D. C. Hoaglin, F. Mosteller, and J. W. Tukey (Eds.), *Exploring Data Tables*, *Trends*, *and Shapes*. New York: Wiley, pp. 461–513.

Klotz, J. H., Milton, R. C., and Zacks, S. (1969). "Mean square efficiency of estimators of variance components," *Journal of the American Statistical Association*, 64, 1383–1402.

Miller, R. G., Jr. (1986). *Beyond ANOVA*, *Basics of Applied Statistics*. New York: Wiley.

Moriguti, S. (1954). "Confidence limits for a variance component," *Reports of Statistical Application Research, Japanese Union of Scientists and Engineers*, 3, 7–19.

Rosner, B., Cook, N. R., Evans, D. A., Keough, M. E., Taylor, J. O., Polk, B. F., and Hennekens, C. H. (1987). "Reproducibility and predictive values of routine blood pressure measurements in children," *American Journal of Epidemiology*, 126, 1115–1125.

Stokes, D. E. (1967). "Parties and the nationalization of electoral forces." In W. Chambers and W. Burnham (Eds.), *The American Party Systems*. London: Oxford University Press, pp. 182–202.

Stokes, D. E. (1969). "A variance components model of political effects." In J. M. Claunch (Ed.), *Mathematical Applications in Political Science I*. Dallas: Southern Methodist University Press, pp. 61–85.

Tukey, J. W. (1951). "Components in regression," *Biometrics*, 7, 33–69.

White, J. R. and Froeb, H. F. (1980). "Small-airways dysfunction in nonsmokers chronically exposed to tobacco smoke," *New England Journal of Medicine*, 302, 720–723.

Williams, J. S. (1962). "A confidence interval for variance components," *Biometrika*, 49, 278–281.

EXERCISES

1. In discussing Table 9-1, the text says that a randomly selected nonsmoker's FEF will be greater than a randomly selected heavy smoker's FEF in about 84% of the paired draws. Assuming the FEF for each group is normally distributed with the true standard deviation given in the table, verify this percentage.

2. What is meant by a balanced one-way table?

3. If $I = 3$ and $\mu_1 = 3$, $\mu_2 = 4$, and $\mu_3 = 8$, find μ and α_i, $i = 1, 2, 3$, for a one-way table.

4. Use Table 9-1 and Figure 9-1 to find what percentage of nonsmokers have FEFs larger than those of medium smokers.

5. Explain to a journalist the distinction between inference for a fixed set of populations and inference from a set of populations drawn at random from a superpopulation.

6. Suppose, as suggested in the text, that three populations are known to have $\sigma_1^2 = 2\sigma_2^2 = 3\sigma_3^2$. If we have independent samples of size n from each group, how would you use the sample variances s_i^2 to estimate the σ_i^2?

7. In one-way ANOVA we speak of two variance components. What are they?

8. Explain the meanings of y_{ij}, $y_{i\cdot}$, and $y_{\cdot\cdot}$. Do these notations have the same meaning for fixed-effects and random-effects models?

9. Three groups of 10 soldiers each (privates, corporals, and sergeants) are observed for their average amounts of sleep. What would the two variance components in this set of data describe?

10. Since σ_A^2 and τ^2 describe the same thing—variance of the population means in the collection of populations—why is it useful to give them different names?

11. What is implied by the notation $\mu + a_i$ that differs from what is implied by $\mu + \alpha_i$?

12. For three groups of independent measurements taken on different days, the variance component between days is $\sigma_A^2 = 10$, and the common within-group variance is 2. If each group has 20 measurements, find the variance of the grand mean of the 60 measurements. Comment on the comparative effect of σ_A^2 and σ^2 in determining the final value.

13. In fixed-effects inference, we estimate the variability of all the population means, and in random-effects inference we do the same. But the basis for the inference differs. How?

14. When we take a random measurement to estimate a group mean μ_i instead of evaluating μ_i itself, we lose efficiency. If σ^2 and σ_A^2 are constant, what can we do to bring the efficiency near 1? If $J = 1$ and $\sigma_A^2 = \sigma^2$, what is the efficiency?

15. In discussing sampling from finite populations, the text compares the variance for sampling with replacement with that for sampling without replacement and finds a factor of $(1 - J/N)$. Derive this factor using the information given in the text.

16. The mean is the first moment of the distribution, and the variance is the second moment (about the mean). The mean is the expected value of the variable; the variance is the expected value of $(X - \mu)^2$. What are the third and fourth moments about the mean?

17. In the balanced case, given that the expected value of the sample variance for all groups is σ^2 (the population variance), prove that $E(\text{WMS}) = \sigma^2$. What is $E(\text{WMS})$ when they are not equal?

18. How many overlays does a one-way table have? Identify them.

19. In a sample of size 9, about how big will the range be in σ units?

20. If the range is 18 in a sample of size 9, about how big is the standard error of the mean?

21. Four "pours" of metals have had two core samples each analyzed for the amount of a trace element. The results are

	Pours			
Cores	***A***	***B***	***C***	***D***
1	1.0	1.4	1.3	.8
2	1.2	1.2	1.7	1.0

Regard the pours as a random four drawn from a large number of possible pours.

(a) Estimate the common variance σ^2 from within pours.

(b) Estimate the between-pour variance σ_A^2.

(c) Compare $\hat{\sigma}_A^2$ and $\hat{\sigma}^2$ and interpret the result in the light of the investigator's intent that the pours be identical.

22. In a sample of four groups of five measurements each, the WMS is 10, and the BMS is 20. Estimate the common variance within groups and the between-means variance σ_A^2.

23. (Continuation) In Exercise 22 let WMS be 20 and BMS be 10. Now determine the estimates of the variance components. Discuss.

24. (Continuation) For Exercise 22 find the KMZ estimate of σ_A^2.

25. (Continuation) For Exercise 23 find the KMZ estimate of σ_A^2.

26. When $I = 5$, $J = 9$, WMS $= 10$, and $\hat{\gamma}_2 = 4$:

(a) Compute "the first" 95% confidence interval for σ^2, $\text{CI}_1(\sigma^2, .95)$.

(b) Use the tabular information in part (a) to compute k_L and k_U for later use.

(c) Compute the second confidence interval for σ^2, $\text{CI}_2(\sigma^2, .95)$.

(d) Use the constants from part (b) to compute the spreading approximate confidence interval for σ^2, $\text{CI}_{2*}(\sigma^2, .95)$.

27. Refer to Exercise 22. Set 90% confidence limits on σ_A^2.

28. In the confidence limits for σ_A^2, explain how you know that the upper limit cannot be negative when the lower limit is positive.

29. Suppose in the unbalanced case that J_+ (the total number of measurements) is 12 and $I = 4$. We want to compare $J^* = \Sigma J_i^2/J_+$ with $\bar{J} = J_+/I$. Examine the values of J^* and $\bar{J}$ when each group has three measurements, when they are 2, 2, 4, 4, and when they are 1, 1, 5, 5. How about 2, 3, 3, 4 and 1, 1, 1, 9?

30. The following set of observations was drawn from normal distributions with mean 0, 2, and 4, so that the true $\sigma_A^2 = 4$ had the design been balanced, and $\sigma^2 = 1$.

Group 1	Group 2	Group 3
−.6	3.5	3.6
−.6	2.9	3.4
.7		4.2
2.6		5.4

(a) What is the true τ^2 for this situation? (Warning: Usually you won't know the μ_i, so this question won't arise.)
(b) Estimate σ^2 from the data and compare it with the true value.
(c) Estimate σ_A^2 from the data and compare with your true value from part (a).
(d) Compare J^* with $\bar{J}$.

31. Four groups of data are drawn from uniform distributions of range 1 with means 0, .5, 1.0, and 1.5. The kurtosis index γ_2 is actually −1.2.

Group 1	Group 2	Group 3	Group 4
.02	.92	1.40	1.67
.41	.71	.90	1.00
.47	.30	1.21	1.34
−.08	.12	.65	1.82
.39	.21	.51	1.20

Use the text formula to estimate γ_2.

32. (Continuation) Use the data of Exercise 31 to estimate the common σ^2 and compare it with the true value of $\frac{1}{12}$.

33. (Continuation) Estimate σ_A^2 for the data of Exercise 31 and compare it with the true value.

34. (Continuation) Use the information from Exercises 31, 32, and 33 to set 90% confidence limits on σ^2.

35. (Continuation) Use the information from the previous exercises to set 90% confidence limits on σ_A^2.

36. Use the right-hand data of Table 9-9 to verify the calculations in Table 9-11.

37. In a two-way table with one observation per cell, what devices have we for separating the measurement error variance from the interaction effects?

38. In estimating the variance components for row effects in the two-way analysis, we divide the difference in mean squares by the number of _____. (Fill in the blank.)

39. Random digits are uniformly distributed, so that the digits $0, 1, 2, \ldots, 9$ are equally likely. Compute for the distribution of random digits the mean μ, variance σ^2, the fourth moment $\Sigma_{x=0}^{9}(x - \mu)^4/10$, and γ_2.

40. Consider again the data on smoking and lung function in Table 9-1.

(a) Suppose the mean for the Heavy Smokers were changed from 2.59 to 1.5 but the rest of the data remained unchanged. Would you expect this change to have any effect on the expected-mean-square-method estimate of σ^2? Or τ^2? Why? (You need not carry out any calculations.)

(b) Suppose that the mean for the Heavy Smokers is set back to 2.59, but the standard deviation is increased to 0.90. Again without carrying out any calculations, what kind of change would you expect to see in $\hat{\sigma}^2$ or $\hat{\tau}^2$? Why?

(c) Carry out the appropriate calculations to check your answers to parts (a) and (b).

41. Consider again the smoking data in Table 9-1. The design is a balanced one-way ANOVA with fixed effects. The sample size in each group is $J = 200$. Suppose that the design were actually unbalanced, with sample sizes (for Nonsmokers through Heavy Smokers) $J_1 = 150$, $J_2 = 150$, $J_3 = 200$, $J_4 = 180$, and $J_5 = 180$. Using the formulas on page 228, recalculate estimates of τ^2 and σ^2 for this unbalanced design.

42. Consider the data related to AIDS testing in Table 5-14. Ignoring lots B through E, take the data corresponding to Lot A as a one-way layout with five runs and three samples per run. Treat run as a random effect.

(a) Calculate the expected-mean-square-method estimates of σ^2 and σ_A^2 for these data. Interpret them in terms of within-run and between-run variation.

(b) With $\alpha = .05$, construct the confidence intervals $\mathrm{CI}_1(\sigma^2, 1-\alpha)$ and $\mathrm{CI}_1(\sigma_A^2, 1-\alpha)$ for σ^2 and σ_A^2, respectively.

(c) We know that the interval for σ^2, $\mathrm{CI}_1(\sigma^2, 1-\alpha)$, depends on the assumption that the data are at least approximately Gaussian. Check this assumption with the data from Lot A by constructing a Q-Q plot of the y_{ij}. Is it more or less linear or does it suggest nonnormality? Estimate the kurtosis by $\hat{\gamma}_2$ and construct the alternative confidence interval $\mathrm{CI}_2(\sigma^2, 1-\alpha)$. Compare it with $\mathrm{CI}_1(\sigma^2, 1-\alpha)$ in light of your Q-Q plot.

43. Use the same data as in Exercise 42. Calculate the Klotz–Milton–Zacks estimate of σ_A^2, $\hat{\sigma}^2_{A\text{-KMZ}}$. How does it compare to $\hat{\sigma}_A^2$?

44. Once again consider the AIDS testing data as in Exercise 42. Suppose a mistake was made at the lab and the second reading in the first run, 1.708, was lost, so that group 1 has only $J_1 = 2$ readings and the other four groups ($i = 2, 3, 4, 5$) have $J_i = 3$ observations each. Calculate estimates of σ^2 and σ_A^2 for this unbalanced case using the formulas on page 229. Note that 1.708 is the largest observation in Lot A, with most observations falling around 1.0. Without doing any calculations, say whether you would expect the omission of a less extreme observation like 0.977 to have as much impact on the estimated values of σ^2 and σ_A^2, respectively.

45. Suppose an experiment was carried out to test the effectiveness of three types of baby formula on infant weight gain. Thirty babies of roughly the same age, weight, and health were randomized into three groups of size 10. The babies in group A were fed Formula A, the babies in group B got Formula B, and those in group C, Formula C. After a suitable period, the babies were carefully reweighed, and an analysis of variance was carried out on the gains, treating formula type as a fixed effect. The expected-mean-square-method estimates were $\hat{\sigma}^2 = 10$ and $\hat{\tau}^2 = 4$. Interpret these values and say what they mean for drawing conclusions from the experiment about which baby formula is the best.

46. Consider a one-way design with $I = 3$ groups and $J = 3$ observations per group. Suppose that each observation can be either a or b, where a and

b are numbers such that b is much larger than a. Examine the three outcomes to the experiment listed below and rank them from largest to smallest, first in terms of their between-group variance estimate $\hat{\sigma}_A^2$ and then in terms of the within variance estimate $\hat{\sigma}^2$.

Outcome I				**Outcome II**				**Outcome III**			
Group	1	2	3	Group	1	2	3	Group	1	2	3
	a	a	a		b	b	b		b	b	a
	a	a	b		a	a	a		b	a	a
	a	a	b		a	a	a		a	a	a

47. Consider the 3×3 multiplication table:

	1	2	3
1	1	2	3
2	2	4	6
3	3	6	9

Suppose we think of it as a two-way layout with the rows as factor A and the columns as factor B. Both factors have three levels: 1, 2, and 3. Suppose we wish to consider the additive model

$$y_{ij} = \mu + \alpha_i + \beta_j + \varepsilon_{ij}$$

in which $i, j = 1, 2, 3$ and α_i and β_j both represent fixed effects.

(a) Estimate the components of variance for both rows and columns. Are they equal and, if so, why?

(b) Write down the values you expect to see as residuals from the additive model.

(c) We know that a more appropriate model is

$$y_{ij} = \alpha'_i \beta'_j$$

because the table is a multiplication table. What would be an appropriate model if we were to consider $\log(y_{ij})$ as the response rather than y_{ij}?

48. Consider the following 3×3 block of random digits

9	8	2
3	9	8
2	1	2

and suppose we wish to treat the data exactly as we did in the previous exercise. What might we expect to see if we compute the components of variance? What interpretation can be given to the row and column variance components?

49. The following is a 5 × 5 block of random digits extracted from *A Million Random Digits with 100,000 Normal Deviates* by the RAND Corporation (New York: The Free Press, 1955):

Group A	B	C	D	E
6	2	8	9	7
4	4	8	9	7
6	6	7	7	1
9	7	9	1	4
1	8	5	3	3

Let us consider it as a one-way layout with the columns representing groups and the rows representing replication within groups.

(a) Treating the columns as random effects, compute the expected-mean-square-method estimates of between-column and within-column variance.

(b) Compute the confidence interval $CI_1(\sigma^2, .95)$. On what assumption about the data does the validity of this interval depend? Is the assumption reasonable in this context?

(c) Compute $\hat{\gamma}_2$ for these data. Does it suggest nonnormality? Compute the confidence interval $CI_2(\sigma^2, .95)$ and compare it with the interval from part (b).

50. The following data give blood glucose concentration (after fasting) for five pregnant (Group 1) and five nonpregnant (Group 2) women. These data are adapted from Data Set 35 in Andrews and Herzberg (1985).

Pregnant	60	56	80	55	62
Nonpregnant	45	59	71	80	76

(a) Treating the groups as fixed, estimate the within-group and between-group variance components. By simply scanning the data, would you predict more variance between than within the groups? Is this reflected in your estimates of the variance components?

(b) Suppose that one data value is missing—namely, the third observation in Group 1 (80). Re-estimate the variance components. Have they changed from part (a)? Why? Would you get similar results if the first observation in Group 1 (60), rather than the third, were missing?

CHAPTER 10

Which Denominator?

Thomas Blackwell, Constance Brown, and Frederick Mosteller
Harvard University

In appreciating possible structures for the analysis of variance (Section 10A) we review the classical approach, starting with the general pigeonhole model in Section 10B, and show how various two-factor models lead to somewhat different expected mean squares, depending on whether factors are fixed or random. In the usual layout some lines are regarded as "above" others (Section 10C), leading to tests for effects of various kinds. In Section 10D, we handle the expected mean squares for the four main three-factor cases. We meet some difficulty in finding appropriate error terms for assessing some effects, and so Section 10E shows how to build an approximate error term. We show how to estimate variance components in the two-factor situation in Section 10F and build an artificial example to illustrate the concept of interaction when one factor is continuous. Then in Section 10G we offer an alternative construction for interaction that is more appropriate when the factors are all discrete. Some oncologists' measurements on a set of simulated tumors give us an opportunity in Sections 10H and 10I to see what factors affect the measurements and to estimate some variance components. Although we mainly deal with balanced designs, we give some suggestions for dealing with mildly unbalanced situations in Section 10J.

10A. ANALYZING THE STRUCTURE

In exploring the structures that a set of data may have, we often look for major simplifications or approximations that apply when some factor—or some interaction—matters little or not at all. Classical tests of significance may help with such exploration, because we need to allow satisfactorily both for the natural variability of the individual observations and for additional sources of variability. Analysis-of-variance methods have especially been designed to aid this search for simplification or, alternatively, to establish firmly the presence of complexity and evaluate it.

In presenting these ideas, largely from a classical point of view, we first generalize on the kinds of models presented in previous chapters, so as to gain a better appreciation of the unity of the work. Then for most practical applications, we draw back to a simpler set of models in wide everyday use. For these we present the mean squares and the average or expected mean squares needed to guide analysis. In most problems these average mean squares create a hierarchical order that helps with the assessment of sources of variation. When the question is "Might this variance component be very small, close to zero?", F ratios help make the assessment. When matters become complex, we may have to be careful—or crafty—in putting together an appropriate error term to serve as a denominator for a particular F ratio. These average-mean-square formulas give forceful guidance for setting up such tests. They show how much the sources of variation recognized in the design contribute, on the average, to each mean square.

10B. THE SAMPLING OR PIGEONHOLE MODEL

Repeatedly we have used fixed and random models to describe data sets. A more general model exposited by Cornfield and Tukey (1956) includes both ideas as special cases. To capture the main concepts, we need a conveniently simple but sufficiently general prototype situation. Two crossed factors serve well for this.

As examples of possibilities, a factory might have many machines of a given type (factor A) and many operators of the machines (factor B), and we could study a measure of production for machine–operator pairs. Or we might have a collection of farms (factor A) and of varieties of wheat (factor B) to be grown on the farms, and study the yield for farm–variety pairs. Again we might have a population of secondary schools (factor A) and several different methods for teaching foreign languages (factor B), and we could investigate school–method proficiency. More generally, suppose that factor A has N_A possible versions or levels and factor B has N_B possible versions, where N_A or N_B or both may be infinite. As usual, "infinite" will sometimes mean "finite, but so many that it no longer matters just how many." Infinitely many possible versions readily arise when a factor has continuously measured values, as, for example, the amount of water or fertilizer delivered to seedlings in raising plants.

In an investigation leading to inferences about the effects associated with factor A and factor B, we might draw a random sample from the possible versions of factor A and another random sample from the possible versions of factor B and then study all the pairs generated by these two samplings. (We would have to assign probabilities or a distribution to the versions. Often equal probabilities will do.) Such pairings or crossings in a two-way design are the simplest example of *factorials*. Because the sampling is random, we are dealing with random-effects models, and some special cases lead us back to the familiar fixed and random ideas from earlier chapters. One extreme occurs when the study includes all the possible levels of a factor. Then the factor corresponds to what we have called "fixed." Another extreme occurs when the number of possible versions of a factor is so large that the proportion in the chosen sample of versions is zero or nearly so—when we are ready to say that the number of possible versions is "infinite." Then the factor corresponds to what we have been calling "random." When the number of versions in the sample is large enough to be a substantial fraction of the total, but not all or practically all (as when one draws three schools from the eight-member Ivy League), we could need to attend more carefully to the idea of sampling from a finite population.

Let us return to the situation where, in the universe, we have N_A levels of factor A and N_B levels of factor B. Let the general name for a version in the A population be I and that for one in B be J ($I = 1, 2, \ldots, N_A$; $J = 1, 2, \ldots, N_B$). (Even if the possible versions are continuous, we act as if a very large discrete list is an adequate representation.) Each of the $N_A N_B$ pairs of versions has an associated population with its own number of items, N_{IJ} (often infinite). For example, we could have Ivy League schools as factor A and faculty departments as factor B, and students in the departments as the populations in each school–department cell. If these students produce measurements (such as SAT scores), we could represent them as Y_{IJK}, where $K = 1, 2, \ldots, N_{IJ}$. Such measurements have a distribution and moments, which may naturally be finite in any event—and must be finite if $N_{IJ} < \infty$. Each cell I, J is a *pigeonhole*, and this whole model of the two-way universe of populations is called a *pigeonhole model*. The pigeon in each hole is the population associated with the pigeonhole or cell. In this discussion it is convenient to think of the versions as finite in number and equally weighted.

For the universe we adopt the following notation:

$$Y_{IJ-} = \text{mean of the population of measurements in cell } I, J,$$

$$Y_{I--} = \text{mean of the means of factor } B \text{ for level } I \text{ of factor } A,$$

$$Y_{-J-} = \text{mean of the means of factor } A \text{ for level } J \text{ of factor } B,$$

$$Y_{---} = \text{grand mean of all the means.}$$

The dash in place of a dot distinguishes means over one or more populations from means over samples and reemphasizes that we are dealing with the universe. (A dot in place of a subscript indicates that we take the *sample mean* over the values in the sample of that subscript, as in $y_{i\cdot\cdot}$, where the lowercase y and i indicate a sample value.)

If we had all the values for all the populations, we could write out a set of overlays, containing, respectively,

Y_{---}	(the same everywhere),
$Y_{I--} - Y_{---}$	(the same throughout all the pigeonholes of any one row),
$Y_{-J-} - Y_{---}$	(the same throughout all pigeonholes of any one column),
$Y_{IJ-} - Y_{I--} - Y_{-J-} + Y_{---}$	(the same throughout any one pigeonhole),
$Y_{IJK} - Y_{IJ-}$	(individual deviations in each pigeonhole),

which would add up to reproduce each value in the universe. Although we cannot do this in practice, thinking about what would happen if we did guides us in defining variance components for the universe.

Given overlays, we define mean squares in the usual way. The unusual feature of the situation comes from the fact that we are working with whole populations, whole rows or columns of populations, or even the whole universe. The interaction overlay, for instance, is based on cell averages Y_{IJ-} that, because they average over the whole of the population of each cell, should be thought of as unaffected by the variation of the individual values in that cell. This means that all the variation among (universe) interaction values is to be thought of as interaction variance, so the (universe) interaction variance component is the same as the (universe) interaction mean square.

Similar arguments apply to row, column, and common, and we define the universe variance components by the usual formulas for mean squares, namely:

VARIANCE COMPONENT FOR FACTOR A

$$\sigma_A^2 = \sum_I (Y_{I--} - Y_{---})^2/(N_A - 1);$$

VARIANCE COMPONENT FOR FACTOR B

$$\sigma_B^2 = \sum_J (Y_{-J-} - Y_{---})^2/(N_B - 1);$$

INTERACTION VARIANCE COMPONENT

$$\sigma_{AB}^2 = \sum_{I,J} (Y_{IJ-} - Y_{I--} - Y_{-J-} + Y_{---})^2/(N_A - 1)(N_B - 1);$$

WITHIN-CELL VARIANCE COMPONENT

$$\sigma_{\text{Error}}^2 = \sum_{I,J} \sigma_{IJ}^2/N_A N_B,$$

where

$$\sigma_{IJ}^2 = \sum_K (Y_{IJK} - Y_{IJ-})^2/(N_{IJ} - 1)$$

is the variance within the I, J pigeonhole.

Samples

Corresponding to the universe analysis, once we have data, we also have an analysis for a sample. We let y_{ijk} denote measurements in the sampled set. It will be convenient, once we have taken the sample of rows and the sample of columns, to renumber them and index them by i and $j, i = 1, 2, \ldots, n_A, j = 1, 2, \ldots, n_B$, where n_A and n_B are, respectively, the number of rows and columns in the sample (note the change from N_A to n_A and N_B to n_B as we move from universe to sample).

Change of Notation

We now simplify our notation slightly. It is convenient to think of a rectangular array with R rows and C columns. Let us regard the levels of one factor, say A, as associated with rows, and let the number of rows in the population be R (formerly N_A). Similarly, let the levels of factor B be associated with columns, and let the number of columns be C (formerly N_B). Let I be the general name for a row and J that for a column. We shall suppose the pigeonholes all have populations containing N elements (formerly N_{IJ}, possibly unequal).

It will be a convenience also to let R and C in a subscript be abbreviations for "row" and "column," so that σ_A^2 will be replaced by σ_R^2, σ_B^2 by σ_C^2, and σ_{AB}^2 by σ_{RC}^2. Recall that σ_{IJ}^2 is the population variance in the I, J cell or pigeonhole. We shall seldom refer specifically to the variance of the population in the last row and last column, and so the possible confusion between the interaction definition of σ_{RC}^2 and the variance of that particular population in this slightly cheating notation should not bother us appreciably. Figure 10-1 gives a schematic two-way layout for the populations in the pigeonhole model.

When we draw randomly r from R rows and c from C columns and n replicates from the N elements in the pigeonholes, the usual analysis-of-variance table for the resulting data has lines that give degrees of freedom, sum of squares, and mean square for rows, columns, interaction, and error. We want to know the average (or expected) mean square across repetitions of the process in terms of various population values that we next describe.

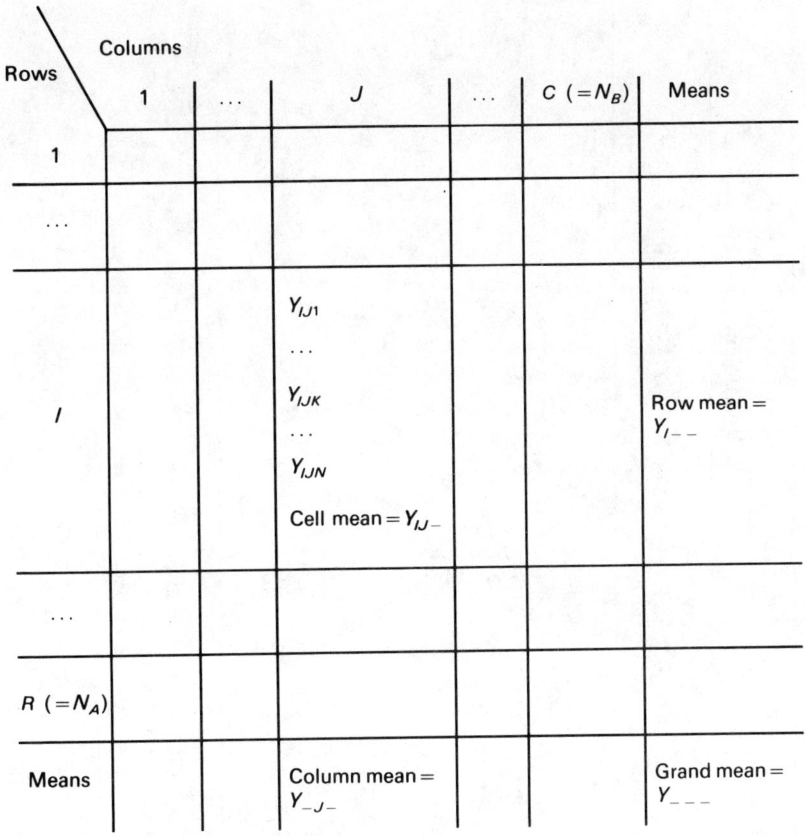

Rows \ Columns	1	...	J	...	$C\ (=N_B)$	Means
1						
...						
I			Y_{IJ1} ... Y_{IJK} ... Y_{IJN} Cell mean $= Y_{IJ-}$			Row mean $=$ Y_{I--}
...						
$R\ (=N_A)$						
Means			Column mean $=$ Y_{-J-}			Grand mean $=$ Y_{---}

Figure 10-1. Schematic drawing of the two-way layout of populations in the pigeonhole model.

The populations in the pigeonholes may have different variances, σ_{IJ}^2. The overall within-cell variance is defined as before as

$$\sigma_E^2 = \frac{1}{RC} \sum_I \sum_J \sigma_{IJ}^2.$$

Table 10-1 summarizes the average (or expected) mean squares for the sampled tables, both for the general case and for three cases of special interest. Note first that the quantities

$$1 - \frac{r}{R}, \quad 1 - \frac{c}{C}, \quad 1 - \frac{n}{N}$$

Table 10-1. General Results and Special Cases[a]

Line in the Analysis of Variance	Average Value of Mean Squares: General Case	Special Cases: (Fixed) $c = C$, $r = R$, N Infinite	Special Cases: (Random) c, r Finite, N, C, R, Infinite[b]	Special Cases: (Mixed[c]) $c = C$, r Finite, N, R Infinite[d]
Rows	$\left(1 - \frac{n}{N}\right)\sigma_E^2 + \left(1 - \frac{c}{C}\right)n\sigma_{RC}^2 + nc\sigma_R^2$	$\sigma_E^2 + nc\sigma_R^2$	$\sigma_E^2 + n\sigma_{RC}^2 + nc\sigma_R^2$	$\sigma_E^2 + nc\sigma_R^2$
Columns	$\left(1 - \frac{n}{N}\right)\sigma_E^2 + \left(1 - \frac{r}{R}\right)n\sigma_{RC}^2 + nr\sigma_C^2$	$\sigma_E^2 + nr\sigma_C^2$	$\sigma_E^2 + n\sigma_{RC}^2 + nr\sigma_C^2$	$\sigma_E^2 + n\sigma_{RC}^2 + nr\sigma_C^2$
Interaction	$\left(1 - \frac{n}{N}\right)\sigma_E^2 + n\sigma_{RC}^2$	$\sigma_E^2 + n\sigma_{RC}^2$	$\sigma_E^2 + n\sigma_{RC}^2$	$\sigma_E^2 + n\sigma_{RC}^2$
Error	σ_E^2	σ_E^2	σ_E^2	σ_E^2

[a] N is the number of elements in a pigeonhole, n the number sampled from it; C is the number of columns available, c the number drawn; R is the number of rows available, r the number drawn.

[b] In practice, both r/R and c/C are close to zero.

[c] Interchange r and c, as well as R and C, to get one other mixed case; see Table 10-11.

[d] In practice, r/R closes to zero and $c/C = 1$.

Source: Corrected from Jerome Cornfield and John W. Tukey (1956). "Average values of mean squares in factorials," *Annals of Mathematical Statistics*, 27, 907–949 (Table 1, p. 926). Reproduced by permission of the Institute of Mathematical Statistics.

play key roles in these formulas. For example, when $r/R = 1$, all rows have been included, and we have the "fixed" case for rows. Then $1 - r/R = 0$, and certain population variance components do not appear in the expected values. Similarly for columns and for cells (pigeonholes).

When $r/R = 0$ (i.e., when we draw a finite sample from an infinite set of rows), terms multiplied by $(1 - r/R)$ have the full coefficient of 1 and are treated as if rows were "random." When rows are fixed and columns random, or vice versa, we have the mixed case.

Random Sampling Versus Real Life

Our general case covers drawing finite samples without replacement from finite populations with $1 \le r \le R$ and $1 \le c \le C, 1 \le n \le N$. Although situations can arise in practice where r/R or c/C or n/N is not near either 0 or 1, the analysis is rarely treated according to the general case. Whenever the general case is needed, however, the formulas for it will be helpful.

As a practical matter, we usually find ourselves analyzing situations where we regard each variable as being either fixed or random, and so the general case, even when we extend the analysis to more sources of variability than rows and columns, mainly (1) helps us understand the overall process and (2) generates the results for the various special cases in ordinary use.

The average mean square results are a valuable guide in deciding what mean squares in the analysis-of-variance table are to be compared. We now explain. For the two-way analysis, let us compare in Table 10-1 the average mean squares in the three special cases: (1) both factors fixed, (2) both random, and (3) mixed.

1. *Within-Cell.* The average mean square for error is always the same, σ_E^2, even in the general case.

2. *Interaction.* The average mean square for interaction is the same in each of the three special cases, $\sigma_E^2 + n\sigma_{RC}^2$, where σ_{RC}^2 is the population interaction component of variance. If the populations in each cell are infinite, finite values for R or C will not disturb this fact. Thus in testing for the presence of interaction we compare the observed interaction mean square with the observed error mean square (unless we need to allow for $N < \infty$). The idea is that if $\sigma_{RC}^2 = 0$ and $N = \infty$, then the ratio of observed interaction mean square to observed error mean square should vary around 1, because both are estimating σ_E^2, though often with different numbers of degrees of freedom. More specifically, if the observations are independent and normally distributed with common variance and $\sigma_{RC}^2 = 0$, then the ratio of these two observed mean squares would be distributed as F with appropriate degrees of freedom.

3. *Columns.* The average mean square for columns for the three special cases given includes $\sigma_E^2 + nr\sigma_C^2$, but neither the fixed case nor the mixed case

with $c = C$ includes $n\sigma_{RC}^2$ as the others do. Thus for the fixed case, or for the mixed case with $r = R$, in testing for column effects we compare the observed column mean square with the observed within-cell mean square. In the random and mixed (with $c = C$) cases, we examine the average column mean square and average interaction mean square and note that both include the within-cell mean square and the same multiple of the interaction variance component. If in the random or the mixed case with $c = C$ we want to test whether $\sigma_C^2 = 0$, we compute F, the ratio of the observed column mean square to the observed interaction mean square, because they both estimate the same quantity when columns are alike in effect. What matters when looking at columns is whether the "other factor"—here rows—is fixed or random.

4. *Rows.* We need only interchange "rows" and "columns," concluding that what matters when looking at rows is whether the "other factor"—here columns—is random or fixed.

Three Factors

When we add a third factor (called slice, rather than layer, in Table 10-2 and abbreviated S), we get the average mean squares shown in Table 10-2. In the expression for the general mean, σ_G^2 is the variance component for the general mean arising from the various samplings, M is the level from which the measurements are considered to depart, and ave$\{y\}$ is the population average of the general mean, so that ave$\{(y - M)^2\} = (\text{ave}\{y\} - M)^2 + \sigma_G^2$. Table 10-2 gives the general case; we discuss special cases in Section 10D.

10C. THE NOTION OF "ABOVE"

In making comparisons of mean squares, in many routine situations, as in the two-way analyses (with $N = \infty$) described earlier, the terms in the average mean square of one line of the ANOVA table are totally included in another line. When this occurs, the *including* line is said to be "above" the other. For example, in the mixed case of Table 10-1 the line for rows is "above" the line for error, and the line for columns is "above" the line for interaction.

When one line is "above" another, the ratio of their two mean squares can help us to judge whether the extra components of the "above" line exceed zero. Continuing the example, the ratio of the mean square for rows to the mean square for error tells about whether σ_R^2 is zero. Similarly, the ratio of the mean square for columns to the mean square for interaction indicates whether σ_C^2 is zero. We do not regard the line for columns as directly "above" the line for error, however, because we are not interested in testing the combination $n\sigma_{RC}^2 + nr\sigma_C^2$.

Randomization theory (see Chapter 7) shows that the distribution of the ratio of mean squares will resemble the tabulated F distribution quite

Table 10-2. Average Values of Mean Squares in the General (Replicated) Three-Way Classification

Item	d.f.	Average Value of Mean Square
General mean	1	$\left(1-\frac{n}{N}\right)\sigma_E^2+n\left(1-\frac{r}{R}\right)\left(1-\frac{c}{C}\right)\left(1-\frac{s}{S}\right)\sigma_{RCS}^2$ $+nr\left(1-\frac{c}{C}\right)\left(1-\frac{s}{S}\right)\sigma_{CS}^2+nc\left(1-\frac{r}{R}\right)$ $\cdot\left(1-\frac{s}{S}\right)\sigma_{RS}^2+ns\left(1-\frac{r}{R}\right)\left(1-\frac{c}{C}\right)\sigma_{RC}^2$ $+ncr\left(1-\frac{s}{S}\right)\sigma_S^2+ncs\left(1-\frac{r}{R}\right)\sigma_R^2$ $+nrs\left(1-\frac{c}{C}\right)\sigma_C^2+ncrs\sigma_G^2$ $+ncrs(\text{ave}\{y\}-M)^2$
Rows (R)	$r-1$	$\left(1-\frac{n}{N}\right)\sigma_E^2+n\left(1-\frac{c}{C}\right)\left(1-\frac{s}{S}\right)\sigma_{RCS}^2$ $+nc\left(1-\frac{s}{S}\right)\sigma_{RS}^2+ns\left(1-\frac{c}{C}\right)\sigma_{RC}^2+ncs\sigma_R^2$
Columns (C)	$c-1$	$\left(1-\frac{n}{N}\right)\sigma_E^2+n\left(1-\frac{r}{R}\right)\left(1-\frac{s}{S}\right)\sigma_{RCS}^2$ $+nr\left(1-\frac{s}{S}\right)\sigma_{CS}^2+ns\left(1-\frac{r}{R}\right)\sigma_{RC}^2+nrs\sigma_C^2$
Slices (S)	$s-1$	$\left(1-\frac{n}{N}\right)\sigma_E^2+n\left(1-\frac{r}{R}\right)\left(1-\frac{c}{C}\right)\sigma_{RCS}^2$ $+nr\left(1-\frac{c}{C}\right)\sigma_{CS}^2+nc\left(1-\frac{r}{R}\right)\sigma_{RS}^2+ncr\sigma_S^2$
RC	$(r-1)(c-1)$	$\left(1-\frac{n}{N}\right)\sigma_E^2+n\left(1-\frac{s}{S}\right)\sigma_{RCS}^2+ns\sigma_{RC}^2$
RS	$(r-1)(s-1)$	$\left(1-\frac{n}{N}\right)\sigma_E^2+n\left(1-\frac{c}{C}\right)\sigma_{RCS}^2+nc\sigma_{RS}^2$
CS	$(c-1)(s-1)$	$\left(1-\frac{n}{N}\right)\sigma_E^2+n\left(1-\frac{r}{R}\right)\sigma_{RCS}^2+nr\sigma_{CS}^2$
RCS	$(r-1)(c-1)(s-1)$	$\left(1-\frac{n}{N}\right)\sigma_E^2+n\sigma_{RCS}^2$
Repetition	$rcs(n-1)$	σ_E^2

Source: Jerome Cornfield and John W. Tukey (1956). "Average values of mean squares in factorials," *Annals of Mathematical Statistics*, 27, 907–949 (Table 2, p. 929). Reproduced by permission of the Institute of Mathematical Statistics.

closely, even when the distribution of the data is quite far from Gaussian. Even when we have no grounds for helpful distributional assumptions, we can always ask whether the mean square for the "above" line is substantially larger than that for its companion.

Some pairs of lines do not have an "above" relation in either direction; for example, the line for rows and the line for columns do not have such a relation in any of the cases of Table 10-1.

10D. THREE-WAY SPECIAL CASES

As in the two-way analysis, we write out special cases for the average mean squares in the three-way analysis of variance in Tables 10-3, 10-4, and 10-5. The special cases we list are

3 fixed and 0 random factors,
2 fixed and 1 random factors,
1 fixed and 2 random factors,
0 fixed and 3 random factors.

We save a lot of paper if we arbitrarily first let slices, then slices and columns, be the random factors instead of writing out all the possibilities. In an application, the labeling of the factors as rows, columns, and slices would need to match the choices in these tables, so that these special cases cover all the situations that are needed. Table 10-3 gives results for the two cases of (1) all three factors fixed and (2) rows and columns fixed, but slices random. This allows us to begin with the simplest situations, delaying more complicated but more realistic situations briefly.

All three fixed. (The simplest model, and often farthest from reality.) Each variance component in Table 10-3—σ_R^2, σ_C^2, σ_S^2, σ_{RC}^2, σ_{RS}^2, σ_{CS}^2, and σ_{RCS}^2—can be tested separately to see whether it is zero by comparing the observed mean square for the line including it with the observed within-cell mean square.

Rows and columns fixed, slices random. The arrows in the rightmost column of Table 10-3 guide the comparisons. To test whether σ_S^2, σ_{RS}^2, σ_{CS}^2, or σ_{RCS}^2 is zero, we compare the observed mean squares for these lines with the observed within-cell mean square. The observed mean square for *RC* interaction needs to be compared with the observed *RCS* interaction. Similarly, the mean square for the main effect of rows needs to be compared with the *RS* interaction line, and so on.

Table 10-4 gives the corresponding results for the case of one fixed and two random factors. By now the arrows are self-explanatory. The new problem is how to test for row effects alone. We could compare the line for rows with the *RS* interaction line to see whether both σ_{RC}^2 and σ_R^2 are zero

Table 10-3. Average (or Expected) Mean Squares for the Three-Way Analysis When All Factors Are Fixed (Left Panel) and When Rows and Columns Are Fixed but Slices Are Random (Right Panel)[a]

Line of the ANOVA Table	Average Mean Square: R, C, and S Fixed	Average Mean Square: R and C Fixed, S Random
Rows	$\sigma_E^2 + ncs\sigma_R^2$	$\sigma_E^2 + nc\sigma_{RS}^2 + ncs\sigma_R^2$
Columns	$\sigma_E^2 + nrs\sigma_C^2$	$\sigma_E^2 + nr\sigma_{CS}^2 + nrs\sigma_C^2$
Slices	$\sigma_E^2 + ncr\sigma_S^2$	$\sigma_E^2 + ncr\sigma_S^2$
RC interaction	$\sigma_E^2 + ns\sigma_{RC}^2$	$\sigma_E^2 + n\sigma_{RCS}^2 + ns\sigma_{RC}^2$
RS interaction	$\sigma_E^2 + nc\sigma_{RS}^2$	$\sigma_E^2 + nc\sigma_{RS}^2$
CS interaction	$\sigma_E^2 + nr\sigma_{CS}^2$	$\sigma_E^2 + nr\sigma_{CS}^2$
RCS interaction	$\sigma_E^2 + n\sigma_{RCS}^2$	$\sigma_E^2 + n\sigma_{RCS}^2$
Error	σ_E^2	σ_E^2

[a]For each line except Error, the arrow leads to the line that provides the denominator in a test of significance.

Table 10-4. Average Mean Squares for the Three-Way Analysis-of-Variance Table: Rows Fixed, Columns and Slices Random

Line of the ANOVA Table	Average Mean Square
Rows (fixed)	$\sigma_E^2 + n\sigma_{RCS}^2 + nc\sigma_{RS}^2 + ns\sigma_{RC}^2 + ncs\sigma_R^2$
Columns (random)	$\sigma_E^2 + nr\sigma_{CS}^2 + nrs\sigma_C^2$
Slices (random)	$\sigma_E^2 + nr\sigma_{CS}^2 + ncr\sigma_S^2$
RC interaction	$\sigma_E^2 + n\sigma_{RCS}^2 + ns\sigma_{RC}^2$
RS interaction	$\sigma_E^2 + n\sigma_{RCS}^2 + nc\sigma_{RS}^2$
CS interaction	$\sigma_E^2 + nr\sigma_{CS}^2$
RCS interaction	$\sigma_E^2 + n\sigma_{RCS}^2$
Error	σ_E^2

Table 10-5. Average Mean Squares for the Three-Way Analysis with All Factors Random

Line of the ANOVA Table	Average Mean Square
Rows	$\sigma_E^2 + n\sigma_{RCS}^2 + nc\sigma_{RS}^2 + ns\sigma_{RC}^2 + ncs\sigma_R^2$
Columns	$\sigma_E^2 + n\sigma_{RCS}^2 + nr\sigma_{CS}^2 + ns\sigma_{RC}^2 + nrs\sigma_C^2$
Slices	$\sigma_E^2 + n\sigma_{RCS}^2 + nr\sigma_{CS}^2 + nc\sigma_{RS}^2 + ncr\sigma_S^2$
RC interaction	$\sigma_E^2 + n\sigma_{RCS}^2 + ns\sigma_{RC}^2$
RS interaction	$\sigma_E^2 + n\sigma_{RCS}^2 + nc\sigma_{RS}^2$
CS interaction	$\sigma_E^2 + n\sigma_{RCS}^2 + nr\sigma_{CS}^2$
RCS interaction	$\sigma_E^2 + n\sigma_{RCS}^2$
Error	σ_E^2

(explicitly whether $ns\sigma_{RC}^2 + ncs\sigma_R^2$ is zero). But testing which is zero separately requires special effort, as we discuss later.

Table 10-5 gives the average mean squares for the special case when all three factors are random. Again we have a similar problem, this time in assessing any one of the main effects (rows, columns, and slices), and we deal with these next.

10E. CONSTRUCTING AN APPROPRIATE ERROR TERM

In Tables 10-4 and 10-5, some of the lines for main effects do not have single lines beneath them that make assessments of the individual components possible. We turn now to constructing error terms with appropriate average values, error terms that will offer such comparisons.

In Table 10-4, the line for rows, with average mean square

$$\sigma_E^2 + n\sigma_{RCS}^2 + nc\sigma_{RS}^2 + ns\sigma_{RC}^2 + ncs\sigma_R^2,$$

has no direct comparison line beneath it for testing whether $\sigma_R^2 = 0$. Looking at lines that might contribute to a comparison without introducing extraneous effects like σ_C^2 or σ_{CS}^2, we come first to the *RC* interaction line, with average mean square

$$\sigma_E^2 + n\sigma_{RCS}^2 + ns\sigma_{RC}^2,$$

and next to the RS interaction line, with average mean square

$$\sigma_E^2 + n\sigma_{RCS}^2 + nc\sigma_{RS}^2.$$

These two lines have the components we need, and if we add them together we get

$$2\sigma_E^2 + 2n\sigma_{RCS}^2 + nc\sigma_{RS}^2 + ns\sigma_{RC}^2,$$

but the coefficient 2 prevents a match for σ_E^2 and for $n\sigma_{RCS}^2$ with the row line. If we subtracted the RCS interaction line, with average mean square

$$\sigma_E^2 + n\sigma_{RCS}^2,$$

we would get exactly what we need:

$$\sigma_E^2 + n\sigma_{RCS}^2 + nc\sigma_{RS}^2 + ns\sigma_{RC}^2.$$

Thus the observed error term involving three mean squares

$$\text{ET(for } R) = RC \text{ interaction-MS} + RS \text{ interaction-MS} - RCS \text{ interaction-MS} \quad (1)$$

would give us an error term with the appropriate average against which we can compare the mean square for rows.

We need to know more about the behavior of the observed mean squares. One drawback springs out: subtracting an observed mean square exposes us to the possibility of a negative total of mean squares, even though the correct population quantity of interest must be positive. That feature is disturbing, and we hope it does not often arise in practice.

One way around this difficulty is to recall that we anticipate both

$$RC\text{-average mean square} \geq RCS\text{-average mean square}$$

and

$$RS\text{-average mean square} \geq RCS\text{-average mean square}.$$

If the observed mean squares behave oppositely, that is, if either

$$RC \text{ mean square} < RCS \text{ mean square} \quad (2a)$$

or

$$RS \text{ mean square} < RCS \text{ mean square}, \quad (2b)$$

we need to take some special steps.

If we want to make an approximate F test using such a linear combination of the three mean squares as our error term, we need a stand-in for the degrees of freedom in the denominator. If we designate the components as

$$\begin{aligned} a &= \text{Rows MS}, \\ b &= RC \text{ interaction MS}, \\ c &= RS \text{ interaction MS}, \\ d &= RCS \text{ interaction MS}, \end{aligned} \tag{3}$$

then the error term for R is given by

$$b + c - d \quad \text{when } d < b \text{ and } d < c.$$

If inequalities go in the wrong direction, we want to avoid negative quantities. The steps we take are to use as the error term for R

$$b \quad \text{when } c < d < b \text{ or } c < b < d,$$

$$c \quad \text{when } b < d < c \text{ or } b < c < d,$$

after which the appropriate (approximate) F test is given by

$$F_{\text{approx}} = \frac{R \text{ mean square}}{R \text{ error term}}. \tag{4}$$

The degrees of freedom associated with the numerator in this specific case is $r - 1$, whereas the approximate degrees of freedom for the denominator, $b > d, c > d$, is given by

$$\frac{(b + c - d)^2}{\dfrac{b^2}{f_b} + \dfrac{c^2}{f_c} + \dfrac{d^2}{f_d}}, \tag{5}$$

where f_b, f_c, and f_d are the respective degrees for the three mean squares. If we use only b or only c as our error term, we use the corresponding degrees of freedom (f_b or f_c).

The general rule is that, if we weight observed mean squares $M_1, M_2, \ldots, M_k$ by weights $w_1, w_2, \ldots, w_k$, respectively, to get $\Sigma w_i M_i$, and the degrees of freedom for the mean squares are $f_1, f_2, \ldots, f_k$, then the approximating number of degrees of freedom for the weighted sum for the denominator is

$$\text{Estimated d.f.} = \frac{(\Sigma w_i M_i)^2}{\Sigma \dfrac{(w_i M_i)^2}{f_i}}. \tag{6}$$

The degrees of freedom for the numerator of the F ratio will ordinarily be that associated with its line in the analysis-of-variance table.

We note that the pattern of adding the two relevant (two-factor) interaction lines and subtracting the corresponding three-factor interaction line will find appropriate comparison values in all the cases without a single comparison line in Tables 10-4 and 10-5; we need merely to adapt the listing in (3) and the formulas in (4) and (5) to the case at hand.

10F. ESTIMATION OF VARIANCE COMPONENTS IN A TWO-WAY ANALYSIS OF VARIANCE BY EQUATING AVERAGE VALUES

Table 10-1 gives us guidelines for the simplest way of estimating variance components in the balanced two-way analysis of variance. To illustrate, we carry out an example for the mixed case.

Our usual procedure is to begin with a practical example. By now, however, matters have become somewhat complicated, and it seems wise to begin with a synthetic example whose roots we understand so that we can analyze it and see both exactly what we are trying to estimate numerically and how well we do it. Often such an approach is called a simulation, but usually simulations are carried out many times. Here we describe the total population, then the sample population, and then the numerical example. The numbers and formulas are chosen arbitrarily, except that we have allowed ourselves to be guided by numerical convenience while maintaining the full complication of the example.

The First Synthetic Example

The population has grand mean $\mu = 10$. The population has $r = R = 4$ rows with effects -3, -1, $+1$, $+3$. The population variance for rows is $20/3$ using the σ_R^2 definition. The population has infinitely many columns. It will be convenient to index them by a continuous number between 0 and 1 and to let this have a uniform distribution. If we let x_c be the uniformly distributed variable, then the column effect associated with a particular x_c could be—and will be—chosen as $4(x_c - \frac{1}{2})$. The variance of column effects is then $4^2(\frac{1}{12})$, because $\frac{1}{12}$ is the variance of a uniform distribution on a unit interval.

Next we want the interaction term σ_{RC}^2 in the new notation. The essential point is that the population interaction has zero sum for each row and each column. We have chosen for the interaction value at row I and column J

$$\text{Interaction}_{IJ} = 20\left(I - \tfrac{5}{2}\right)\left[\left(x_c - \tfrac{1}{2}\right)^2 - \tfrac{1}{12}\right],$$

where $I = 1, 2, 3, 4$ (with equal weights) and as before x_c is uniform on 0 to 1. For fixed x_c, the sum of $I - \frac{5}{2}$ over the four values of I is zero. For fixed

Table 10-6. Contributions from the Row Effects[a]

Row	Column 1	2	3	4	5
1	−3	−3	−3	−3	−3
2	−1	−1	−1	−1	−1
3	1	1	1	1	1
4	3	3	3	3	3

[a]Each number is used for each of three observations in a cell. Thus −3 is used in each of the three observations in each cell of row 1, and so on.

I, the integral of $(x_c - \frac{1}{2})^2 - \frac{1}{12}$ is zero. This gives us the desired boundary conditions. We also want to know the variance of this interaction effect. It is given by

$$\sigma^2_{\text{int}} = \frac{400}{3} \sum_{I=1}^{4} \left(I - \tfrac{5}{2}\right)^2 \int_0^1 \left\{\left(x_c - \tfrac{1}{2}\right)^2 - \tfrac{1}{12}\right\}^2 dx_c.$$

The result is $\sigma^2_{\text{int}} = 100/27$.

Finally, we regard each pigeonhole as having infinitely many observations in its population (N infinite), and all pigeonhole populations as having equal variances ($\sigma^2 = 4$). And we plan to select 3 observations per cell.

Thus we have a layout with four rows; we plan to choose five columns using a table of uniform random numbers, thus fixing column effects and row-by-column interaction. For the within-cell contribution, we multiply each of $60 = 3 \times (4 \times 5)$ standard random normal deviates by 2 to get deviates with variance $\sigma^2 = 4$.

By adding up all the effects, we get a 4×5 table with three observations per cell. Tables 10-6, 10-7, 10-8, and 10-9 show the contributions to the totals before we get the final table that we analyze (Table 10-10).

Table 10-7. Column Effects Obtained by Sampling x_c from a Uniform Distribution[a]

	x_c:	.43	.51	.60	.85	.88
Row	Column:	1	2	3	4	5
1		−.28	.04	.40	1.40	1.52
2		−.28	.04	.40	1.40	1.52
3		−.28	.04	.40	1.40	1.52
4		−.28	.04	.40	1.40	1.52

[a]The value of the column effect for the chosen sample is $4(x_c - \frac{1}{2})$. Each entry is used three times in the same cell to contribute to the three observations in that cell.

Source: The RAND Corporation (1955). *A Million Random Digits with 100,000 Normal Deviates.* New York: The Free Press. Random numbers x_c, drawn from row 02400, are formed from the first two digits in each of the first five blocks of five digits for uniform random digits.

Table 10-8. The Interaction Component for Each Observed Row and Column is Given by $20(I - \frac{5}{2})\{(x_c - \frac{1}{2})^2 - \frac{1}{12}\}$ for $I = 1, 2, 3, 4$ and the Values of x_c Shown in Table 10-7

x_c	.43	.51	.60	.85	.88
$(x_c - \frac{1}{2})^2$	.005	.000	.010	.122	.144
$(x_c - \frac{1}{2})^2 - \frac{1}{12}$	−.078	−.083	−.073	.039	.061

		Column				
		1	2	3	4	5
Row	$20(I - \frac{5}{2})$	Interaction[a]				
1	−30	2.34	2.49	2.19	−1.17	−1.83
2	−10	.78	.83	.73	−.39	−.61
3	10	−.78	−.83	−.73	.39	.61
4	30	−2.34	−2.49	−2.19	1.17	1.83

[a] For convenience we have rounded $(x_c - \frac{1}{2})^2 - \frac{1}{12}$ to three decimal places before multiplying by $20(I - \frac{5}{2})$.

Table 10-9. Standard Gaussian Deviates Multiplied by 2 to Form the Error Term in the Two-Way Table (Three Observations per Cell)

	Column				
Row	1	2	3	4	5
1	.8	.2	4.8	−.6	.0
	.0	−5.0	−1.0	−.2	1.0
	2.8	−.6	−1.2	1.2	1.8
2	2.0	−.8	2.4	7.0	1.0
	2.6	−1.0	.0	.6	5.8
	1.8	−1.0	−1.0	1.0	1.6
3	2.2	−2.0	.0	1.4	1.8
	−3.0	−.8	−.2	−.2	2.0
	−1.2	1.4	−3.2	−.6	−1.0
4	2.6	.4	.6	1.4	.2
	−.8	3.2	.0	−2.4	−.8
	−2.6	−.2	2.6	−1.0	−.4

Source: The RAND Corporation (1955). *A Million Random Digits with 100,000 Normal Deviates.* New York: The Free Press. Gaussian deviates from lines 0500 through 0511, first five columns cut to one decimal place before multiplication by 2.

Table 10-6 lays out the row effects, it being understood that each number will be used three times, once for each observation from the corresponding cell. Each column adds to zero because we have sampled all the rows, and each row is constant because these are row effects.

Table 10-7 gives the contributions from the five columns sampled. Note that, because no observations (x_c) were smaller than .43, the variance component of columns is likely to be underestimated.

Table 10-8 gives the contributions for the interaction term. As in Table 10-6, the columns add to zero because we used all the rows, but the rows do not add to zero because the columns were sampled. Only the average over the whole population of columns must average to zero in a row in this example.

Table 10-9 gives the within-cell contributions as found from a random number table. Note that we observed a value of 7.0 (a deviation from the mean of more than 3.5 standard deviations) in our sample of 60 observations. Note also the run of eight positive numbers in the final column. Things like this happen in random number tables and in real random experiments. We combine the numbers in Table 10-9 with the other numbers to produce the elements of the final 4 × 5 × 3 data table, Table 10-10.

We form Table 10-10 by adding all the contributions, including the grand mean, which we have taken as 10. For example, in Table 10-10 the upper left

Table 10-10. Observed Values for the Desired Two-Way Table for the Mixed Model Obtained by Summing the Numbers in Tables 10-6 Through 10-9 Plus the Grand Mean 10 (Three Observations in Each Cell)

	Column				
Row	1	2	3	4	5
1	9.86	9.73	14.39	6.63	6.69
	9.06	4.53	8.59	7.03	7.69
	11.86	8.93	8.39	8.43	8.49
2	11.50	9.07	12.53	17.01	10.91
	12.10	8.87	10.13	10.61	15.71
	11.30	8.87	9.13	11.01	11.51
3	12.14	8.21	10.67	14.19	14.93
	6.94	9.41	10.47	12.59	15.13
	8.74	11.61	7.47	12.19	12.13
4	12.98	10.95	11.81	16.97	16.55
	9.58	13.75	11.21	13.17	15.55
	7.78	10.35	13.81	14.57	15.95

Table 10-11. Expected Mean Squares for the Balanced Two-Way ANOVA with Rows Fixed and Columns Random

Source	d.f.	Expected Mean Square
Rows	$r - 1$	$\sigma_E^2 + n\sigma_{RC}^2 + nc\sigma_R^2$
Columns	$c - 1$	$\sigma_E^2 + nr\sigma_C^2$
Interaction	$(r - 1)(c - 1)$	$\sigma_E^2 + n\sigma_{RC}^2$
Within-cell	$(n - 1)rc$	σ_E^2

entry is composed of

Grand mean	10.	
Row effect	−3.	from Table 10-6
Column effect	−.28	from Table 10-7
Interaction effect	2.34	from Table 10-8
Error term	.8	from Table 10-9
Total	9.86	

We wish to carry out an analysis of variance for the data of Table 10-10. We can then use Table 10-1 to guide us to method-of-moments estimates for the various components and compare these with the true values. Of course, in a real problem we do not know the true values. We are doing this example to drive home the nature of the components and their construction and to emphasize what sort of quantities are being estimated by the variance-component analysis.

The rightmost column of Table 10-1 gives the average values for the two-way mixed ANOVA. It has columns fixed and rows random, but the example has rows fixed and columns random. To avoid misreadings, we rewrite that column to suit the example (see Table 10-11).

We have $n = 3$, $r = 4$, $c = 5$. Table 10-12 gives the ANOVA table obtained from the data of the example, already shown in Table 10-10. Let us calculate the method-of-moments estimates, guided by the layout in Table 10-11.

Within-cell. We read the estimate of σ_E^2, 3.97, from the MS column in the error line of Table 10-12. Recall that we constructed the error terms by multiplying a standard normal deviate by 2, making their variance 2^2 or 4. Our estimate 3.97 is extremely close to that value, even considering that it is based on 40 degrees of freedom.

Interaction. For the universe interaction variance component we have

$$\frac{\text{Interaction MS} - \text{Within-cell MS}}{n} = \frac{9.48 - 3.97}{3} = 1.84.$$

Table 10-12. Analysis-of-Variance Table for the Artificial Example with Data Given in Table 10-10; $n = 3$, $r = 4$, $c = 5$

Source	d.f.	SS	MS	ET[a]	DIV[b]	VC[c]
Rows	3	142.17	47.39	9.48	15	$\hat{\sigma}_R^2 = 2.53$
Columns	4	76.25	19.06	3.97	12	$\hat{\sigma}_C^2 = 1.26$
Interaction	12	113.78	9.48	3.97	3	$\hat{\sigma}_{RC}^2 = 1.84$
Within-cell	40	158.67	3.97		1	$\hat{\sigma}_E^2 = 3.97$
Total	59	490.86				

[a]Error term appropriate for the model.
[b]The DIV for a line equals the number of observations for each combination of versions of the factors that define the line. It multiplies the corresponding variance component wherever that variance component appears in an average mean square.
[c]Estimate of the variance component.

The true population value, $100/27 = 3.7$, is roughly double the estimate. (Although the general interaction has $(4 - 1) \times (5 - 1) = 12$ degrees of freedom, the particular kind of interaction we have generated involves only five random numbers and hence no more than 5 degrees of freedom. Thus a factor of 2 seems not unlikely.)

Column variance component. The estimate is

$$\frac{\text{Column MS} - \text{Within-cell MS}}{nr} = \frac{19.06 - 3.97}{3 \times 4} = 1.26.$$

This compares very well with the true value $16/12 = 1.33$, especially with only 4 degrees of freedom.

Row variance component. The estimate is

$$\frac{\text{Row MS} - \text{Interaction MS}}{nc} = \frac{47.39 - 9.48}{3 \times 5} = 2.53$$

compared with the true value $20/3 = 6.67$—not very close, but about as close as we can expect with only 3 degrees of freedom.

We summarize these comparisons in Table 10-13.

Table 10-13. Comparisons Between Population Values and Sample Estimates of Variance Components Obtained by Equating Mean Squares to Their Average Values for the Two-Way Mixed Model in the Artificial Example ($n = 3$, $r = 4$, $c = 5$)

Sampling Model	Theoretical Component	True Value	Estimate	Other Factor
Fixed	σ_R^2	6.67	2.53	Random
Random	σ_C^2	1.33	1.26	Fixed
	σ_{RC}^2	3.70	1.84	—
Random	σ_E^2	4.00	3.97	—

10G. AN ALTERNATIVE MODEL FOR INTERACTION IN TWO-WAY ANALYSIS OF VARIANCE

In Section 10F one factor was a continuous variable, and the interaction was chosen so that it too was continuous. When the values of the continuous variable were chosen at random, the value of the interaction was simultaneously randomly chosen by a formula. If we are dealing with factors that are ordered or continuous variables, such a continuous structure for the interaction seems reasonable (not the specific formula, but the idea of continuity). When a factor is formed from a collection (such as a sample of animals, cities, or firms), the smoothness of the interaction has no good grounding unless associated continuous variables (such as height, size, or total sales, respectively) are taken into account. If they are not, then a random-interaction-effects model may be appropriate. When some factors are random, a common approach takes the interaction elements as random, sampled independently from a distribution. Because this approach is very frequently used, we illustrate it with an artificial example, which we keep as similar to the previous example as possible.

A Second Synthetic Example

Let us recall that the previous example had the following:

1. A set of fixed row effects shown in Table 10-6. The new example retains these unchanged.

2. A random sample of five values x_c drawn from a continuous distribution, uniform on the interval 0 to 1, leading to effects of size $4(x_c - \frac{1}{2})$. In the new example we do not assume the continuity, but we suppose that the objects being sampled have an outcome effect with a population mean of 0 and a variance of $16/12$ ($= 1\frac{1}{3}$), just as the previous population did. The range will be -2 to $+2$ as before, and we shall suppose that the specific values sampled will be the numbers entered in the body of Table 10-7, though we can ignore the x_c line above the body of the table. Thus there has been a slight change in our attitude toward the factor, because we gave up the continuity, but numerically the assumptions are the same, and so are the essential values sampled.

3. The interaction term had a formula that used the row and column values to produce the interactions. We want to preserve some features in the population and some in the sample.
   (a) In the population, the sum across each row should be zero, and the same should hold for the columns.
   (b) In the sample, because rows are fixed (we included all population rows), then for any given column, the interaction sum down the column should be zero because of condition (a), just as in Table 10-8.

(c) In the sample, because columns are random, the sum of interaction terms for a row does not ordinarily turn out to be zero, though the expected value for the row is zero.

To make this example similar to the previous one, we maintain the variance of the interaction $\sigma_{\text{int}}^2 = 100/27$. Although the details are not important, it may be useful to show one way to accomplish it. If we use standard random normal deviates as a basis for the interaction effects, then we can multiply them by $\sigma_{\text{int}} = \sqrt{100/27}$ to make their variance $100/27$. Suppose, for a given column j, that the observed values are $g_{1j}, g_{2j}, \ldots, g_{rj}$, with average $g_{\cdot j}$. Then the differences $g_{ij} - g_{\cdot j}$ add to zero as required by condition (b). The variance is

$$\text{var}(g_{ij} - g_{\cdot j}) = [(r-1)/r]\sigma_{\text{int}}^2,$$

and so we want to replace $g_{ij} - g_{\cdot j}$ by $(g_{ij} - g_{\cdot j})/\sqrt{(r-1)/r}$ to retain the variance of the measurements as σ_{int}^2. Thus the interaction effects for the given column are

$$c_{ij} = \frac{(g_{ij} - g_{\cdot j})}{\sqrt{(r-1)/r}}$$

and we have $\Sigma_i c_{ij} = 0$ and $\text{var}\, c_{ij} = \sigma_{\text{int}}^2$ as desired.

We therefore replace Table 10-8 by Table 10-14 to get the new interaction terms.

4. The within-cell contributions are ε_{ijk} as before, and they can be unchanged. Now we sum the entries in Tables 10-6, 10-7, 10-9, 10-14, and the grand mean, which we have taken as 10, to get a new Table 10-15 that we analyze as before.

Table 10-16 gives the ANOVA table obtained from the data in Table 10-15.

In Table 10-17 we compare the sample estimates of variance components and the population values.

Table 10-14. Interaction Terms for the Second Synthetic Example

	Column				
Row	1	2	3	4	5
1	2.63	1.35	.22	−1.44	.41
2	−1.49	−.26	−.70	2.65	2.28
3	.01	.23	−.19	−.54	−1.95
4	−1.15	−1.32	.67	−.67	−.74

Table 10-15. Observed Values for the Desired Two-Way Table for the Second Synthetic Example of the Mixed Model, Obtained by Summing the Numbers in Tables 10-6, 10-7, 10-9, 10-14, Plus the Grand Mean 10 (Three Replications in Each Cell)

	Column				
Row	1	2	3	4	5
1	10.15	8.59	12.42	6.36	8.93
	9.35	3.39	6.62	6.76	9.93
	12.15	7.79	6.42	8.16	10.73
2	9.23	7.98	11.10	20.05	13.80
	9.83	7.78	8.70	13.65	18.60
	9.03	7.78	7.70	14.05	14.40
3	12.93	9.27	11.21	13.26	12.37
	7.73	10.47	11.01	11.66	12.57
	9.53	12.67	8.01	11.26	9.57
4	14.17	12.12	14.67	15.13	13.98
	10.77	14.92	14.07	11.33	12.98
	8.97	11.52	16.67	12.73	13.38

Table 10-16. Analysis-of-Variance Table for the Data of the Second Synthetic Example, Given in Table 10-15; $n = 3$, $r = 4$, $c = 5$

	d.f.	SS	MS	ET	DIV	VC
Rows	3	167.61	55.87	14.14	15	$\hat{\sigma}_R^2 = 2.78$
Columns	4	76.25	19.06	3.97	12	$\hat{\sigma}_C^2 = 1.26$
Interaction	12	169.65	14.14	3.97	3	$\hat{\sigma}_{RC}^2 = 3.39$
Within-cell	40	158.67	3.97		1	$\hat{\sigma}_E^2 = 3.97$
Total	59	572.17				

Table 10-17. Comparisons Between Sample Estimates of Variance Components Obtained by the Method of Moments and the Population Values

Sampling Model	Theoretical Component	True Value	Estimate	Other Factor
Fixed	σ_R^2	6.67	2.78	Random
Random	σ_C^2	1.33	1.26	Fixed
	σ_{RC}^2	3.70	3.39	—
Random	σ_E^2	4.00	3.97	—

Greater generality. These examples only begin to touch the generality of the pigeonhole model. We have illustrated (1) a formula-based interaction and (2) a separately random interaction. We could equally easily consider a situation involving both kinds of interaction added together. (It will almost always be true that adding up the entries—for the contributions at one or more splittings—will give us another pigeonhole model—for which the same average-mean-square formulas will apply.)

10H. A THREE-WAY EXAMPLE: TUMOR SIZE

In a study designed to discover the reliability of measurements of tumor sizes in cancer patients, Lavin and Flowerdew (1980) set up a simulation. They made objects in two materials, rubber and cork, of three different sizes each that might represent tumors. These objects were placed "upon a folded blanket and arranged at random in two rows of six each and were then covered with a sheet of foam rubber, half an inch thick" (p. 1287). Oncologists, cancer experts experienced in assessing tumor sizes, measured them with their usual equipment, just as they would a human tumor. Pairs of objects identical in size and material were used, and thus each size–material combination was measured twice by each oncologist. (Section 6A gives some further information on this set of data.)

We shall analyze the data as if we had two independent repetitions in each size–material–oncologist cell. (Because each oncologist measured the same two objects of a given size–material combination, repetition is actually nested in size × material, and crossed with oncologists.)

Table 10-18 gives the estimated cross-sectional areas produced by the oncologists. One unusual feature of this study is that we know the actual cross-sectional areas of the objects being measured. We can compare these with the averages produced by the oncologists.

Let us turn now to the sums-of-squares analysis. We have identified rows with the oncologists who made the measurements, columns with the sizes of the objects whose cross-sectional area was being measured, and slices with the materials. Thus $r = 26$, $c = 3$, $s = 2$, and the number of replications in each cell is $n = 2$. Table 10-19 gives the mean squares for this example, including the average mean squares applicable when all three factors are random.

Because the error mean square is larger than the three-way interaction mean square, we estimate the three-way interaction effect as zero. We thus pool the triple interaction with the repetition mean square to get $[50(.0222) + 156(.0274)]/206 = .0261$ with 206 d.f.

We compare each two-way interaction with the pooled mean square. The oncologist × size interaction does not amount to much, reaching a little beyond the 25% level in the F table.

Table 10-18. Actual Versus Estimated Cross-Sectional Areas of Simulated Tumors

	Cork Model						Rubber Model					
Actual Area	Small		Oblong		Large		Small		Oblong		Large	
Oncologist (mm²):	980	980	1199	1199	1467	1467	980	980	1199	1199	1467	1467
	Estimated Area (*mm*²)											
1	667	529	1258	810	1050	960	696	675	875	910	1225	1155
2	625	900	750	1225	1225	1225	625	900	900	1600	1225	1600
3	783	900	986	1184	1089	1024	676	625	884	1200	1330	1260
4	1120	1050	1050	1290	1400	1520	960	840	950	1184	1400	1600
5	1575	1225	1520	2025	1225	2025	1365	1050	1482	1575	2250	1600
6	1600	1400	1710	1750	2236	2500	1485	1400	1500	1575	2100	2310
7	1056	1024	1280	1320	1404	1368	1023	961	1365	1160	1640	1640
8	1188	1184	1287	1295	1332	1638	961	928	1344	1120	1600	1406
9	1225	900	1225	1225	1225	1225	900	750	1050	980	1600	1444
10	1225	900	1600	1050	1600	1800	1225	1480	1600	2000	2025	2500
11	960	924	1470	1400	1800	1225	900	1050	1344	1400	1692	1200
12	1020	784	1160	1120	1480	1400	900	1156	1160	1120	1849	1634
13	928	783	1050	1120	1224	1224	840	400	1110	1292	1369	1369
14	1225	900	1225	1360	1365	1406	992	960	1188	1240	1369	1681
15	1023	1120	1110	1280	1368	1330	1056	1122	1209	1360	1482	1722
16	784	754	1050	980	1225	1155	900	900	1140	1050	1330	1444
17	1400	1225	1400	1400	1800	1400	1225	1225	1350	1350	1600	1400
18	900	1350	1184	1200	1369	1225	900	1225	1120	1200	1600	1444
19	780	638	864	875	1225	1188	840	780	875	665	900	1184
20	900	840	1369	1050	1600	1400	900	960	1050	1000	1260	1225
21	1054	900	1330	1116	1600	1440	1224	1088	1260	1394	1520	2025
22	1080	1089	1295	1575	1800	1400	1225	1326	1440	1280	1813	1936
23	1296	1330	992	1428	1764	1600	1122	1156	1470	1240	1681	1680
24	900	784	784	900	1225	1024	725	960	840	1050	1225	960
25	960	930	1200	1254	1600	1295	1024	1225	1287	1400	1600	1600
26	784	729	1295	1053	1295	1680	1054	1024	899	918	1147	1443
Group mean	1041	965	1209	1242	1443	1411	990	1006	1180	1241	1532	1556

Source: Philip T. Lavin and Gordon Flowerdew (1980). "Studies in variation associated with the measurement of solid tumors," *Cancer*, 46, 1286–1290 (Table 2, p. 1288). Reprinted with permission.

The oncologist × material effect is modest in size though significant at .05; and the size × material (column × slice) effect is substantial, even though we have only two degrees of freedom for the numerator sum of squares. Below we shall find little first-order effect of materials, and so we are tempted to not take this result very seriously.

To assess the first-order effects of oncologists, sizes, and materials, we must construct an error term as described in Section 10E because we have taken all three factors as random. (We use the observed triple interaction rather than the pooled MS, though the latter might be preferable. Exercise 23 requests a comparison.)

We might well expect substantial differences among oncologists, and so we construct an error term for comparison with the row-effect mean square. We add the Row × Column MS and the Row × Slice MS and subtract off the

Table 10-19. Analysis-of-Variance Table for the Simulation of Tumor Measurements (Measurements in Table 10-18 Have Been Divided by 1000)[a]

Source	Expected Mean Square	MS	d.f.
Rows (oncologists)	$\sigma_E^2 + n\sigma_{RCS}^2 + nc\sigma_{RS}^2 + ns\sigma_{RC}^2 + ncs\sigma_R^2$	.5541	25
Columns (sizes)	$\sigma_E^2 + n\sigma_{RCS}^2 + nr\sigma_{CS}^2 + ns\sigma_{RC}^2 + nrs\sigma_C^2$	6.1370	2
Slices (materials)	$\sigma_E^2 + n\sigma_{RCS}^2 + nr\sigma_{CS}^2 + nc\sigma_{RS}^2 + ncr\sigma_S^2$	.0826	1
RC interaction (oncologist × size)	$\sigma_E^2 + n\sigma_{RCS}^2 + ns\sigma_{RC}^2$	.0290	50
RS interaction (oncologist × material)	$\sigma_E^2 + n\sigma_{RCS}^2 + nc\sigma_{RS}^2$	.0471	25
CS interaction (size × material)	$\sigma_E^2 + n\sigma_{RCS}^2 + nr\sigma_{CS}^2$	.1402	2
RCS interaction (triple interaction)	$\sigma_E^2 + n\sigma_{RCS}^2$	.0222	50
Repetition	σ_E^2	.0274	156
(pooled MS)		(.0261)	(206)
Total			311

[a]The dimensions of the data are $r = 26$ rows (oncologists), $c = 3$ columns (sizes: small, oblong, large), $s = 2$ slices (materials), and $n = 2$ replications. Each arrow indicates the line that provides the error term for the line from which the arrow originates (rows, columns, and slices require constructed error terms).

Row × Column × Slice MS to get .0290 + .0471 − .0222 = .0539. The approximate number of degrees of freedom for the denominator is 25.2, from equation (5) in Section 10E. The pseudo F ratio for oncologists is then .5541/.0539 = 10.3 with 25 and 25.2 degrees of freedom—very significant, just as we expected. Oncologists were not doing the same thing.

The corresponding pseudo F ratios for sizes and for materials are

$$\text{Sizes:} \quad F_{2,2.2} = 6.137/(.0290 + .1402 - .0222) = 41.7;$$

$$\text{Materials:} \quad F_{1,2.7} = .0826/(.0471 + .1402 - .0222) = 0.50.$$

Thus sizes matter a good deal, as they should, because size is what the measurement is all about and the investigators deliberately made them quite

Table 10-20. Useful Summaries Involving Sizes and Materials in the Simulated Tumor Experiment

(*a*) *Means for Size × Material Combinations*

	Small	Medium	Large
Cork	1003	1225.5	1427
Rubber	998	1210.5	1544

(*b*) *Bordering Tables* (*Rounded*)

19	24	−42	−16
−19	−24	42	+16

1000	1218	1486

(*c*) *Size × Material and Material Combined*

3	8	−58
−3	−8	58

different. Materials have made no difference, on average. Because we found an appreciable size × material interaction, though, we should look to see what appears to be going on. To do this, we turn to the corresponding overlays or, more simply, to the corresponding bordering tables. Table 10-20 shows some selected bordering tables and the result of adding two of them together. Panel C tells the story: material appears to matter only for the large simulated tumors. (If we calculate the standard errors of our entries, we find that this appearance is to be taken seriously.)

What we have just seen is fairly common. If factor B matters only within one version of factor A, the two mean squares—for factor B and for interaction AB—are often of similar size. Particularly when the interaction is (or is in) the error term for factor B, this is likely to make the effects of B appear to be mostly noise, whereas the AB interaction effects appear to be more than noise.

We see that the oncologists were responsive to differences in tumor size, but that they varied a good deal among themselves. This means that if we wanted to get good measures of absolute tumor size, as we might in a study of treatments designed to control the growth of tumors, the investigators would need to consider having more than one oncologist measure the size of the tumors. This is troublesome because patients are likely to find the process disagreeable. A possible alternative would have the same assessor make all measurements on any one patient, thus confounding oncologists and patients (actually, nesting patients in oncologists), something that will often be acceptable.

Calculations from Overlay Entries

Sometimes it is helpful to be able to compute the sums of squares from first principles. These sums of squares are essentially based on the value-splitting approach described in Chapters 5 and 6. We use the notation y_{ijkl} for the observation at row i, column j, slice k, and replication l. Dots in subscripts mean that we have summed over the dotted subscripts and then averaged. Thus $y_{....}$ is the grand mean. The subscripts run $i = 1, \ldots, r$; $j = 1, \ldots, c$; $k = 1, \ldots, s$; and $l = 1, \ldots, n$. Then the analysis-of-variance outline for calculating the three-way layout with replication is as shown in Table 10-21.

It may be worth noting that the quantities that are being summed can be rewritten to show how they arise from value splitting. Thus the table of values

$$y_{.j..} - y_{....}, \qquad j = 1, \ldots, c,$$

Table 10-21. Calculations for the Three-Way Analysis-of-Variance Table with Replications

Source	Sum of Squares	Degrees of Freedom
Rows	$ncs\sum_{i}(y_{i...} - y_{....})^2$	$r - 1$
Columns	$nrs\sum_{j}(y_{.j..} - y_{....})^2$	$c - 1$
Slices	$nrc\sum_{k}(y_{..k.} - y_{....})^2$	$s - 1$
Row × Column	$ns\sum_{ij}(y_{ij..} - y_{i...} - y_{.j..} + y_{....})^2$	$(r - 1)(c - 1)$
Row × Slice	$nc\sum_{ik}(y_{i.k.} - y_{i...} - y_{..k.} + y_{....})^2$	$(r - 1)(s - 1)$
Column × Slice	$nr\sum_{jk}(y_{.jk.} - y_{.j..} - y_{..k.} + y_{....})^2$	$(c - 1)(s - 1)$
Row × Column × Slice	$n\sum_{ijk}(y_{ijk.} - y_{ij..} - y_{i.k.} - y_{.jk.} + y_{i...} + y_{.j..} + y_{..k.} - y_{....})^2$	$(r - 1)(c - 1)(s - 1)$
Within-cell	$\sum_{ijkl}(y_{ijkl} - y_{ijk.})^2$	$(n - 1)rcs$
Total sum of squares	$\sum_{ijkl}(y_{ijkl} - y_{....})^2$	$nrcs - 1$

is the marginal table of column means with the grand mean removed. Also,

$$y_{ij..} - y_{i...} - y_{.j..} + y_{....}$$

can be rewritten as

$$(y_{ij..} - y_{....}) - (y_{i...} - y_{....}) - (y_{.j..} - y_{....}),$$

showing it more clearly to be the result of removing the grand mean from row mean and column mean and then removing these from the joint marginal means for rows and columns to get the interaction effects.

Estimating the Variance Components

The approach of combining mean squares from Table 10-5 leads to the estimates of the sizes of variance components shown in Table 10-22. In two instances, materials, σ_S^2, and the triple interaction, σ_{RCS}^2, the estimates were actually negative, and we have therefore assigned the value 0 to them.

We notice that the variance component for sizes is estimated as .0575. This is most encouraging, because the variance of the true sizes of the cross sections of the simulated tumors is just about .06.

Logarithmic Transformation

The original measurements were of area of cross section. We might expect that the errors of measurement made on the larger objects would be

Table 10-22. Estimates of the Variance Components in the Tumor Simulation Investigation

	Variance Component	Estimate	(DIV)	ET[a]
Oncologists	σ_R^2	.0412	(12)	.0500
Sizes	σ_C^2	.0575	(104)	.1431
Materials	σ_S^2	.0000	(156)	.1612
	σ_{RC}^2	.0017	(4)	.0261
	σ_{RS}^2	.0042	(6)	.0261
	σ_{CS}^2	.0023	(52)	.0261
	σ_{RCS}^2	.0000	(2)	
	σ_E^2	.0274	(1)	

[a]Three-way interaction and within-cell pooled.

Table 10-23. Analysis-of-Variance Table for the Logarithms of the Simulated Tumor Data

Source		MS	F
R	Oncologists	.0681	14.1
C	Sizes	.792	65.0
S	Materials	.00573	Small
$R \times C$		.00354	Small
$R \times S$		.00484	Small
$C \times S$		.0122	3.5
$R \times C \times S$		.00356	Small
Within-cell		.00348	

proportionately larger than the errors made on the smaller objects. If this happened, we would benefit by taking logarithms, because that would translate multiplication into addition, and similar ratios into similar differences. Much more important, the oncologists seem somewhat more likely to differ by a constant ratio than by a constant difference (and the same could happen for materials). Thus we might expect to get more pointed results from an analysis of the logarithms than from the raw measurements.

When we carried out such an analysis, that turned out to be true. Table 10-23 shows the mean squares for the logarithmic analysis. It shows a big effect for sizes, which we are pleased to have, and a substantial effect for oncologists, as we expected. As in the analysis of the raw measurements (Tables 10-19 and 10-20), it also shows some joint effects for sizes with materials, a not-so-welcome finding, especially when materials seem not to matter at the level of the main effects.

We notice that the within-cell mean square and the three-factor interaction mean square are about of equal size, as they were in the analysis of the raw data. Thus again the variance component for the three-factor interaction is estimated as about zero. The F ratios, except for sizes by materials, are less than 2 when we compare the mean squares for two-way interactions with that for the three-way. To get an appropriate denominator for the main effects, we combine as usual and get huge pseudo F ratios for oncologists and sizes, but not for materials. And so without a much more detailed analysis we can sum up by saying that sizes are the principal effect, with oncologists second, and some interaction between sizes and materials. We shall examine these in a little more detail with value splitting.

10I. EXAMINING THE EFFECTS IN THE TUMOR SIMULATION EXPERIMENT

Because the substantial effects in the tumor example are limited to sizes, oncologists, and sizes × materials, we can readily look at their effects (or

Table 10-24. Selected Effects from a Logarithmic Analysis of the Tumor Simulation Data

Common = grand mean = 3.076

	Small	Oblong	Large
Size Effects	−.088	.001	.087

Oncologist effects (26 oncologists)

−.136	.170	.021	−.059	.038	−.043
−.068	.021	.003	.067	.074	
−.089	.023	−.071	.007	.063	
−.006	−.026	.012	−.130	−.105	
.112	.107	.021	−.032	.025	

	Cork	Rubber
Material effects	−.0043	.0043

Material × size interactions

	Small	Oblong	Large
Cork	.0053	.0072	−.0125
Rubber	−.0053	−.0072	.0125

overlays). From the analysis of the logarithmic scores, we find the results shown in Table 10-24.

We see the broad distribution of effects for both sizes and oncologists, the very small differences in materials, and the modest effects for materials × sizes. A fairly good summary would ignore the material × size interaction and the materials and use only oncologists and sizes and the grand mean.

10J. THE TWO-WAY UNBALANCED DESIGN

As executed, investigations often turn out unbalanced even when planned to be balanced. Although our main emphasis is on balanced designs, we consider here a common unbalanced situation.

A two-way design with the same number of replications in each cell becomes an unbalanced design when one or more observations are missing. We write the model

$$y_{ijk} = \mu + \alpha_i + \beta_j + \gamma_{ij} + \varepsilon_{ijk}, \qquad i = 1, \ldots, I$$
$$j = 1, \ldots, J$$
$$k = 1, \ldots, n_{ij}. \qquad (7)$$

Throughout this section we study only the case where the two factors are fixed. We refer to the factor associated with the α_i as factor A and that associated with the β_j as factor B. Because observations are replicated within a cell, we can estimate an interaction term, denoted by γ_{ij}. We assume that no cell is empty (i.e., we assume $n_{ij} \geq 1$), as this simplifies the problem greatly and holds in many practical situations.

Unbalanced designs are a headache for the data analyst because the imbalance introduces many complications into the analysis. For example, if we wish to calculate a sum of squares for the main effects for one factor, the result usually depends on whether we have already swept out the main effects for the other factor. Thus the sums of squares that are most appropriate for judging the overall contribution of the two sets of main effects may not go together in an ANOVA table in which the individual sums of squares add up to the total sum of squares.

We can sidestep some of the complications by not trying to work with a single set of main effects and interactions, as equation (7) might tempt us to do. Instead, one standard approach (discussed, for example, by Searle 1987, Chap. 4) adopts the *cell means model*,

$$y_{ijk} = \mu_{ij} + \varepsilon_{ijk}. \tag{8}$$

Thus we would sweep out the cell means,

$$y_{ij\cdot} = \frac{1}{n_{ij}} \sum_{k=1}^{n_{ij}} y_{ijk},$$

and stop, leaving the residuals, $y_{ijk} - y_{ij\cdot}$, in each cell.

In this framework we explore the effects of factor A or the effects of factor B by working with the cell means, $y_{ij\cdot}$. For example, the hypothesis of no main effects of A, ordinarily associated with the F test based on MS_A/MS_e in a balanced layout, would take the form

$$\frac{1}{J}\sum_j \mu_{1j} = \frac{1}{J}\sum_j \mu_{2j} = \cdots = \frac{1}{J}\sum_j \mu_{Ij}. \tag{9}$$

That is, we simply compare the population marginal means for the rows, which we define by averaging the cell means in each row. We might consider other (weighted) averages, but the unweighted averages in equation (9) are particularly appropriate in situations where factor A is fixed and differences among the actual n_{ij} represent accidental departures from a planned balanced design.

To judge whether the population marginal means for factor A are all equal (and similarly for factor B), we may still use an F test. The denominator is still MS_e, but now the numerator combines the information from the rows in a way that reflects their respective numbers of observations. The procedure is part of the weighted-squares-of-means analysis originally proposed by Yates (1934). The words "weighted mean" capture the idea behind the technique: the analysis uses sums of squares of cell means, $y_{ij\cdot}$, where the terms in those sums of squares are *weighted* in inverse proportion to their variance according to the number of observations in the cell. We need weights w_{Ai} associated with factor A, and w_{Bj} associated with factor B. We define these weights

$$w_{Ai} = J^2 \bigg/ \left(\sum_{j=1}^{J} \frac{1}{n_{ij}} \right)$$

and

$$w_{Bj} = I^2 \bigg/ \left(\sum_{i=1}^{I} \frac{1}{n_{ij}} \right)$$

and the corresponding weighted means

$$y_{...}(A) = \sum_{i=1}^{I} w_{Ai} \tilde{y}_{i..} \bigg/ \sum_{i=1}^{I} w_{Ai},$$

$$y_{...}(B) = \sum_{j=1}^{J} w_{Bj} \tilde{y}_{.j.} \bigg/ \sum_{j=1}^{J} w_{Bj},$$

where

$$\tilde{y}_{i..} = \sum_{j=1}^{J} y_{ij.} \bigg/ J$$

$$\tilde{y}_{.j.} = \sum_{i=1}^{I} y_{ij.} \bigg/ I.$$

Then the sum of squares for A and the sum of squares for B are, respectively,

$$\mathrm{SS}_A = \sum_{i=1}^{I} w_{Ai} [\tilde{y}_{i..} - y_{...}(A)]^2,$$

$$\mathrm{SS}_B = \sum_{j=1}^{J} w_{Bj} [\tilde{y}_{.j.} - y_{...}(B)]^2.$$

We do not approach the sum of squares for interaction, SS_{AB}, in this way because we cannot in general express it as a simple function of the data. Instead, we rely on the ANOVA capabilities of statistical software such as SAS or BMDP to handle the details numerically.

The error sum of squares, however, is straightforward. It pools the within-cell variation from all the cells:

$$SS_e = \sum_{i=1}^{I} \sum_{j=1}^{J} \sum_{k=1}^{n_{ij}} (y_{ijk} - y_{ij\cdot})^2.$$

Each cell contributes $n_{ij} - 1$ degrees of freedom, and so SS_e has $\Sigma\Sigma(n_{ij} - 1) = n - IJ$ degrees of freedom, where $n = \Sigma\Sigma n_{ij}$. Under the customary assumption that the ε_{ijk} in equation (7) came independently from a Gaussian distribution with mean 0 and variance σ^2, SS_e is distributed as σ^2 times a variable that has a chi-squared distribution on $n - IJ$ degrees of freedom. Thus the error mean square, $MS_e = SS_e/(n - IJ)$ provides a suitable estimate of σ^2.

For its part, when the population marginal means for A are all equal, SS_A behaves like σ^2 times a variable that has a chi-squared distribution on $I - 1$ degrees of freedom. Similarly, when the population marginal means for B are all equal, $SS_B \sim \sigma^2\chi^2_{J-1}$. Thus we may use the F statistics $F_A = MS_A/MS_e$ and $F_B = MS_B/MS_e$ to test these overall hypotheses about the effects of A and the effects of B. When the data are balanced, these hypotheses translate into simple statements about the α_i and the β_j in equation (7). For example, H_0: $\alpha_i = 0$ for all i. When the data are unbalanced and may involve interactions, however, the row and column averages of those interactions,

$$\gamma_{i\cdot} = \frac{1}{J} \sum_j \gamma_{ij} \quad \text{and} \quad \gamma_{\cdot j} = \frac{1}{I} \sum_i \gamma_{ij},$$

come into play. Thus the hypothesis in equation (9) would be equivalent to H_0: $\alpha_i + \gamma_{i\cdot}$ are all equal, and the corresponding statement about effects of B would be H_0: $\beta_j + \gamma_{\cdot j}$ are all equal. Thus only with the restriction $\gamma_{i\cdot} = 0$ for $i = 1, \ldots, I$ can equality of the α_i be tested. Similarly for the β_j with $\gamma_{\cdot j} = 0$.

To illustrate the weighted-squares-of-means analysis, we return to a balanced example introduced in Chapter 5 (see Section 5G). The data came from an experiment to test the effectiveness of six feeding treatments. The treatments were beef, cereal, and pork, each at high and low amounts of protein. Each combination was assigned to 10 male rats in a 2 × 3 design with 10 replications per cell. In Table 10-25, the ANOVA table for the full data, we see that only protein level is significant at the 5% level.

We have made the two-way layout of Table 5-12 unbalanced by omitting eight observations in a deliberate way. Table 10-26 shows the resulting data layout, along with tables of the cell counts and the cell means. We prepare for the weighted-squares-of-means analysis by calculating the weights w_{Ai} and w_{Bj}, the observed marginal means $\tilde{y}_{i\cdot\cdot}$ and $\tilde{y}_{\cdot j\cdot}$, and the corresponding

Table 10-25. ANOVA Table for the Full 60 Observations of the Rat Feeding Data (see Table 5-12 and Figure 5-9)

Source	Sum of Squares	d.f.	Mean Square	F	Tail Probability
Mean (common)	463,233.07	1	463,233.07	2,159.04	0.0000
Protein level	3,168.27	1	3,168.27	14.77	0.0003
Type	266.53	2	133.27	0.62	0.5411
Interaction	1,178.13	2	589.07	2.75	0.0732
Residual	11,586.00	54	214.55		

weighted averages $y_{\dots}(A)$ and $y_{\dots}(B)$, shown in Table 10-27. For example,

$$w_{A1} = J^2 \Bigg/ \left(\frac{1}{n_{11}} + \frac{1}{n_{12}} + \frac{1}{n_{13}} \right) = 9 \Bigg/ \left(\frac{1}{8} + \frac{1}{8} + \frac{1}{10} \right) = 25.7143,$$

$$\tilde{y}_{1\cdot\cdot} = \tfrac{1}{3}(y_{11\cdot} + y_{12\cdot} + y_{13\cdot}) = \tfrac{1}{3}(105.00 + 96.25 + 85.90) = 95.7167,$$

and

$$\begin{aligned} y_{\dots}(A) &= (w_{A1}\tilde{y}_{1\cdot\cdot} + w_{A2}\tilde{y}_{2\cdot\cdot})/(w_{A1} + w_{A2}) \\ &= (25.71 \times 95.72 + 25.43 \times 79.71)/51.14 = 87.7598. \end{aligned}$$

Table 10-26. A Deliberately Unbalanced Version of the Rat Feeding Data[a]

	Beef	Pork	Cereal
		Data, y_{ijk}	
High	81, 100, 102, 104, 107, 111, 117, 118	79, 91, 94, 96, 98, 102, 102, 108	56, 74, 77, 82, 86, 88, 92, 95, 98, 111
Low	51, 64, 72, 76, 78, 86, 95	49, 70, 73, 81, 82, 82, 86, 97, 106	58, 67, 74, 74, 80, 89, 95, 97, 98, 107
		Cell counts, n_{ij}	
High	8	8	10
Low	7	9	10
		Cell means, $y_{ij\cdot}$	
High	105.00	96.25	85.90
Low	74.57	80.67	83.90

[a]The omitted observations are as follows: 73 and 87 in the (High, Beef) cell; 105 and 120 in the (High, Pork) cell; 90, 90, and 90 in the (Low, Beef) cell; and 61 in the (Low, Pork) cell.

Table 10-27. Weights, Marginal Means, and Corresponding Weighted Averages for the Weighted-Squares-of-Means Analysis of the Unbalanced Rat Feeding Data

i	w_{Ai}	$\tilde{y}_{i\cdot\cdot}$
	Factor A (Protein Level)	
1	25.7143	95.7167
2	25.4260	79.7127
		$y_{\cdots}(A) = 87.7598$
j	w_{Bj}	$\tilde{y}_{\cdot j\cdot}$
	Factor B (Type)	
1	14.9333	89.7857
2	16.9412	88.4583
3	20.0000	84.9000
		$y_{\cdots}(B) = 87.4686$

Then the (weighted) sum of squares for A (protein level) is

$$\text{SS}_A = w_{A1}[\tilde{y}_{1\cdot\cdot} - y_{\cdots}(A)]^2 + w_{A2}[\tilde{y}_{2\cdot\cdot} - y_{\cdots}(A)]^2 = 3274.4985.$$

Similarly, for B (type)

$$\text{SS}_B = 228.7266.$$

Because we are treating both factors as fixed, the error term for both of these is the error mean square, $\text{MS}_e = 9105.0143/46 = 197.9351$. Thus we have

$$F_A = \text{MS}_A/\text{MS}_e = (3274.4985/1)/197.9351 = 16.54$$

and

$$F_B = \text{MS}_B/\text{MS}_e = (228.7266/2)/197.9351 = 0.58.$$

As an observation from an F distribution on 1 and 46 d.f., F_A has a P value of .0002. Thus the population marginal means for high and low levels of protein, $(\mu_{11} + \mu_{12} + \mu_{13})/3$ and $(\mu_{21} + \mu_{22} + \mu_{23})/3$, are significantly different at a comfortably extreme level. By comparison, an F distribution on 2 and 46 d.f. assigns F_B a large P value of .57. When we average over levels, the population marginal means for the three types of protein—$(\mu_{11} + \mu_{21})/2$, $(\mu_{12} + \mu_{22})/2$, and $(\mu_{13} + \mu_{23})/2$—do not differ significantly.

As in balanced layouts, interpretation of such statements about main effects is easier if the interactions are mostly noise. We approach this question in an overall way via the sum of squares for interaction, whose details we leave to the statistical software. (Briefly, the interaction sum of squares equals the difference between the residual sum of squares for the main-effects-only model, $y_{ijk} = \mu + \alpha_i + \beta_j + \varepsilon_{ijk}$, and the residual sum squares for the cell means model, $y_{ijk} = \mu_{ij} + \varepsilon_{ijk}$.) In this example $SS_{AB} =$ 1733.2532 on 2 degrees of freedom, so that

$$F_{AB} = MS_{AB}/MS_e = (1733.2532/2)/197.9351 = 4.38.$$

From the F distribution on 2 and 46 d.f. the P value is .0182. Thus the interactions between protein type and level seem to be more than just noise.

This outcome does not invalidate the tests that we based on SS_A and SS_B. The corresponding hypotheses involve only the cell means, μ_{ij}. In the presence of interaction, however, we cannot interpret those hypotheses in terms of only the α_i and the β_j. In this particular unbalanced example we have reached a conclusion about the interactions that differs from the message of the balanced data (Table 10-25 shows a P value of .0732). The actual choice of the eight observations that we omitted plays a major role in this shift, mainly by producing a larger sum of squares for interaction.

As a repository for the various sums of squares that we have calculated in this unbalanced example, we set up the ANOVA table in Table 10-28. The key feature to note about this ANOVA table is that the sums of squares do not add up to the (corrected) total sum of squares. We could develop a set of sums of squares with this property (essentially by choosing an order for the factors and adopting the multiple-regression approach of assigning to each line the incremental sum of squares as it enters the fit), but the ones for the main effects would not bear on the main questions that we ordinarily want to answer.

For the general unbalanced two-way layout with some data in each cell, the cell means model provides a suitable framework for flexible analysis. The limitations of the natural sums of squares do not constitute a major obstacle.

Table 10-28. ANOVA Table Based on the Weighted-Squares-of-Means Analysis of the Unbalanced Rat Feeding Data[a]

Source	Sum of Squares	d.f.	Mean Square	F	P
Protein level	3,274.50	1	3,274.50	16.54	.0002
Type	228.73	2	114.36	0.58	.5652
Interaction	1,733.25	2	866.63	4.38	.0182
Residual	9,105.01	46	197.94		
Total	13,912.31	51			

[a]The entries in this table come from the general linear models procedure (PROC GLM) in SAS. The appropriate sums of squares for protein level and type are the "Type III" sums of squares.

10K. SUMMARY

After the introductory Section 10A, Section 10B describes a pigeonhole model, for sets of populations organized by factors, that is the basis for the sampling that leads to the usual ANOVA tables. This model leads to fixed effects for a factor when every level of a factor is drawn into the sample to be analyzed and to random effects for a factor when a subset of the levels of a many-leveled factor is chosen. Although we give formulas for the average mean squares when finite samples of levels are drawn randomly from finite collections of levels in the population, in practice we almost always treat problems as if we have either fixed effects or random effects associated with a factor, and we concentrate on these special cases for the two-factor and the three-factor layouts.

These mean squares (Tables 10-1 through 10-5) guide us in figuring out what lines of the analysis of variance to compare with what other lines when we want to assess the magnitude of an effect. When the average mean square for a line contains all the components of another line, the first line is said to be above the second (Section 10C). And so when ratios of observed mean squares are computed to get an idea of the size of the variance components, and especially to see whether they might reasonably be neglected, the special tables of average mean squares in the three-way layout not only guide the choice of ratios in Section 10D but also suggest in Section 10E how to construct artificial mean squares for providing error terms for single variance components that do not have lines that automatically lead to comparisons. Section 10E also offers an assessment of the degrees of freedom or degrees of firmness for the artificially constructed mean squares used for ratios of mean squares (formula (5)).

Because of the complications involving factors and populations and samples, first from the factor levels and then from the populations, leading to the ANOVA table, Section 10F begins with an artificially constructed example, so that we are able to illustrate concretely the populations and the samples we are drawing and check how well the variance components are being estimated in the specific numerical example. Two versions of interaction effects are offered, one in Section 10F, the other in Section 10G.

To get experience with the three-way analysis, we reanalyze in Sections 10H and 10I a simulated investigation of measuring the sizes of tumors in cancer patients involving many oncologists, two kinds of materials, and three sizes of simulated tumors. We found that sizes mattered a great deal, as one would hope, but that variation among oncologists was large enough that good measurement might require the same oncologist to do all the measuring if one wanted to assess growth or reduction of a tumor in a patient. We also found that using a logarithmic transformation sharpened the results of the analysis.

In Section 10J, we treated the analysis of unbalanced two-way layouts. We introduced the cell means model and applied a weighted-squares-of-means analysis.

REFERENCES

Cornfield, J. and Tukey, J. W. (1956). "Average values of mean squares in factorials," *Annals of Mathematical Statistics*, 27, 907–949.

Lavin, P. T. and Flowerdew, G. (1980). "Studies in variation associated with the measurement of solid tumors," *Cancer*, 46, 1286–1290.

The RAND Corporation (1955). *A Million Random Digits with 100,000 Normal Deviates*. New York: The Free Press.

Searle, S. R. (1971). *Linear Models*. New York: Wiley.

Searle, S. R. (1987). *Linear Models for Unbalanced Data*. New York: Wiley.

Yates, F. (1934). "The analysis of multiple classifications with unequal numbers in the different classes," *Journal of the American Statistical Association*, 29, 51–66.

EXERCISES

1. Give an example of a factor that may have infinitely many levels.

2. In the notation of this chapter what are the distinctions between the following:

(a) Y_{IJK} and y_{ijk}?
(b) Y_{IJ-} and $y_{ij\cdot}$?
(c) Y_{I--} and $y_{i\cdot\cdot}$?
(d) Y_{-J-} and $y_{\cdot j\cdot}$?
(e) Y_{---} and $y_{\cdot\cdot\cdot}$?

You may be able to respond to all five parts in one statement.

3. **(a)** Does Y_{-K-} have a meaning?
(b) Does $y_{\cdot K\cdot}$ have a meaning?

4. Given the table representing numerical samples drawn from the pigeonholes of a two-factor model,

		B		
	j	1	2	3
	i			
A	1	12 14 13	10 8	14 6
	2	8 10	9 13 11	11 15

fill in the following blanks:

y_{112} = ______ . y_{231} = ______ .
$y_{11\cdot}$ = ______ . $y_{12\cdot}$ = ______ . $y_{13\cdot}$ = ______ . $y_{21\cdot}$ = ______ .
$y_{22\cdot}$ = ______ . $y_{23\cdot}$ = ______ .
$y_{1\cdot\cdot}$ = ______ . $y_{2\cdot\cdot}$ = ______ . $y_{\cdot 1\cdot}$ = ______ . $y_{\cdot 2\cdot}$ = ______ . $y_{\cdot 3\cdot}$ = ______ .
$y_{\cdots}$ = ______ .

5. (Continuation) There are 14 observations in the table. If you sum the 14 observations and divide by 14, is this the same as $y_{\cdots}$? Explain.

6. (Continuation) If each pigeonhole (cell) had 2 observations, would the mean of the 12 observations be the same as $y_{\cdots}$?

7. In Table 10-1 show that if $c = C$, $r = R$, and N is infinite with n finite, the general case reduces to the corresponding lines for the fixed case.

8. (Continuation) Show that, when N, C, and R are infinite, the general case, when n, r, and c are finite, reduces to the corresponding lines for the random case.

9. (Continuation) Show that, when $c = C$ with N and R infinite, the general case reduces to the corresponding lines for the mixed case.

10. (Continuation) If $r = R$ and N and C are infinite, is the case fixed, random, or mixed?

11. For Table 10-1, write the lines for the mixed case when $r = R$, c finite, N and C infinite.

12. In Table 10-2, sum the degrees of freedom to get the total degrees of freedom. (Note that this table has a degree of freedom for the general mean, whereas many analysis-of-variance summaries omit this line, as we often do.)

13. Extend Table 10-3 by writing out lines for rows and slices fixed, columns random, and draw appropriate arrows.

14. (Continuation) Write lines for columns fixed, rows and slices random.

15. (Continuation) In the result for Exercise 14, what line would you use to test $\sigma_C^2 = 0$?

16. If in a table having rows, columns, and slices fixed, you found the *RC* interaction MS to be 3.00 and the Error MS to be 2, how would you test whether $\sigma_{RC}^2 = 0$?

17. (Continuation) If rows and columns were fixed and slices were random, to what would you want to compare the *RC* interaction?

18. (Continuation) If rows were fixed and columns and slices were random, to what would you want to compare the *RC* interaction?

19. (Continuation) If all three factors were random, to what would you want to compare the *RC* interaction?

20. If there are two rows, using the notation of expressions (3), (4), and (5), if $a = 6$, $b = 4$, $c = 3$, $d = 1$, and $f_b = 6$, $f_c = 8$, and $f_d = 24$, use expressions (4) and (5) to get the approximate degrees of freedom and approximate *F* ratio for rows.

21. Sections 10F and 10G develop two distinct ideas for interaction terms. Explain the general idea in each instance, not the details, and indicate the important distinction between them.

22. Sections 10F and 10G provide exact information about the various components of variance and then carry out a single example in each section to estimate the components. Why, when we know the true values, are we estimating them?

23. For the tumor simulation study of Table 10-19 create a suitable pooled mean square with approximate degrees of freedom for assessing:

(a) The effects of oncologists, and carry out the *F* test.

(b) The effects of sizes, and carry out the *F* test.

(c) The effects of materials, and carry out the *F* test.

24. For Table 10-20, explain the relation of the bordering tables (Panel (b)) to the means table (Panel (a)).

25. For Table 10-20, explain the meaning of Panel (c).

26. Why do we take the logarithms of the tumor data for the analysis of Table 10-23?

27. Compare the *F* ratio for sizes × materials from Table 10-23 with the corresponding *F* ratio from Table 10-19 to see the effect of taking logarithms on that interaction. Note also the corresponding *F* ratios for oncologists × materials.

28. In the unbalanced two-way layout of Section 10J, note that $\text{var}(y_{ij\cdot}) = \sigma^2/n_{ij}$. Verify that w_{Ai} is inversely proportional to the variance of $\tilde{y}_{i\cdot\cdot}$ and that w_{Bj} is inversely proportional to the variance of $\tilde{y}_{\cdot j\cdot}$.

Data for Exercises 29-32.

The following data set was generated from the model

$$y_{ijk} = \mu + \alpha_i + \beta_j + \varepsilon_{ijk}$$

for certain fixed values of the α's and β's and for $i, j = 1, 2$ and $k = 1, 2, 3$.

i \ j	1	2
1	2.2 2.4 2.3	−2.0 0.2 0.8
2	5.2 7.6 6.8	3.8 3.5 4.3

29. Construct the ANOVA table for this layout and test the hypotheses H_0: $\alpha_1 = \alpha_2$ and H_0: $\beta_1 = \beta_2$.

30. Suppose that $y_{121} = -2.0$ is missing, so that the (1, 2) cell has only two observations (0.2 and 0.8). Using the method described in Section 10J, test the hypotheses H_0: $(\mu_{11} + \mu_{12})/2 = (\mu_{21} + \mu_{22})/2$ and H_0: $(\mu_{11} + \mu_{21})/2 = (\mu_{12} + \mu_{22})/2$ and compare the results to those you got in Exercise 29.

31. Suppose that $y_{122} = 0.2$ were missing instead of y_{121} (so that the (1, 2) cell has observations −2.0 and 0.8). Without doing any calculations, would the results of the analysis be similar to those in Exercise 30 or more like those in Exercise 29? Why?

32. Suppose that both $y_{122} = 0.2$ and $y_{121} = -2.0$ are missing. Cell (1, 2) now contains only one observation, $y_{123} = 0.8$. Again using Section 10J analyze this unbalanced design. Can the hypotheses H_0: $(\mu_{11} + \mu_{12})/2 = (\mu_{21} + \mu_{22})/2$ and H_0: $(\mu_{11} + \mu_{21})/2 = (\mu_{12} + \mu_{22})/2$ be tested here? Why?

33. If Table 10-2 had only one slice, then all interaction components and variances containing S would be zero, and the actual situation would have only two factors. Check that Rows in Table 10-2 will then reduce to the general case in Table 10-1 and that the interaction RC in Table 10-2 reduces to the interaction in Table 10-1.

CHAPTER 11

Assessing Changes

John W. Tukey
Princeton University

Frederick Mosteller and Cleo Youtz
Harvard University

After taking measurements apart into components related to various factors and interactions (as in Chapters 5 and 6), we may find ourselves with more numbers than we can comprehend. To find the messages, some masked by deafening noise, we simplify by recombining portions. As we explained in Chapter 2, simplifying by setting aside factors aids analysis. When we set a component aside or pool it with other sources of variability, this action does not mean that we believe that the factor has no effect, but only that the noise in the background is drowning out the message or distorting it enough that we expect to learn more by focusing elsewhere.

In this chapter we develop some approaches to assessing changes and illustrate them on examples.

After describing a one-factor model in Section 11A, we explain when noise in the sampling and measurement system impedes us in estimating differences between true means. In Section 11B, we set up a variety of targets for estimation of effects in two-factor situations to show that different circumstances in the second factor affect what we estimate for the first factor. In Section 11C we give details of the two-factor model with interaction. In Section 11D we introduce the Rule of 2 for deciding whether to pool certain effects with others further down in an analysis-of-variance table. To get experience with this rule, we apply it to a collection of hypothetical two-factor examples in Sections 11E, F, and G. Then we extend the two-factor situation to include replication in Section 11H, and illustrate with extensive hypothetical examples in Section 11I, followed by a brief discussion of the Rule of 2 in Section 11J.

In Section 11K, we turn from hypothetical examples to a complex situation from dentistry, having four factors. This requires special analyses, given in

Section 11L. Having learned a good deal about the example, we are able to make a new, more pointed analysis in Section 11M.

At the close of the chapter, Section 11N explains why fixed-factor analyses may not adequately represent the genuine sources of variation.

11A. LOOKING AT DIFFERENCES OF MEANS THROUGH NOISE

Suppose that one factor occurs in two levels, labeled $i = 1, 2$, and that we have J measurements of a response at each level. The usual model would use j as an index for replication, y_{ij} for the response, μ_i for the unknown population mean at level i, and ε_{ij} for the unknown disturbances, so that

$$y_{1j} = \mu_1 + \varepsilon_{1j} \quad \text{and} \quad y_{2j} = \mu_2 + \varepsilon_{2j}.$$

We may use the sample means $y_{i\cdot}$ as estimates of μ_i, so that $\hat{\mu}_i = y_{i\cdot}$, and we can express a measurement as

$$y_{ij} = \hat{\mu}_i + \left(y_{ij} - \hat{\mu}_i\right) = y_{i\cdot} + \left(y_{ij} - y_{i\cdot}\right).$$

If we believe that $y_{1j} - y_{1\cdot}$ and $y_{2j} - y_{2\cdot}$ represent noise, then we ask also whether $y_{1\cdot} - y_{2\cdot}$ seems to be noise. If so, we would estimate $\mu_1 - \mu_2$ as near zero. If, instead, only a little of $y_{1\cdot} - y_{2\cdot}$ is noise, then $y_{1\cdot} - y_{2\cdot}$ can estimate $\mu_1 - \mu_2$.

We aim to make such choices when the data have more complicated arrangements than two samples.

11B. QUESTIONS AND TARGETS: MODELS OF TWO-FACTOR SITUATIONS

Let us look at the two-way situations—two factors, called A and B—at first with no replication in the cells, later with replication. We focus on one or more changes of level in factor A. The levels of factor B provide additional circumstances, in most investigations giving similar results.

We approach the definition of our targets for estimation by asking the general question, "What is the estimate of the effect of changing factor A from level i to level k?" The answer depends on how we take into account the levels of factor B. We describe five different situations:

a. Specified level of B. When factor B will be at a fixed level, j.

Example. In Chapter 4, Table 4-2 gives results for an experiment where differing numbers of shelves display a product at six supermarkets. The outcome is sales volume. Regard the number of shelves as factor A and the supermarkets as factor B. One of these supermarkets will have interest in its own sales and much less interest in those of other supermarkets. Thus that supermarket would be regarded as a fixed level of B with concern for differences in sales according to numbers of shelves.

b. Random level of B. When the level of factor B will be chosen, randomly or equally often, from a set of versions that already appears in the study we are analyzing.

Example. In a state welfare department a new way of processing cases is to be compared with the standard method. These two methods form the levels of factor A. Three separate units of comparable size in the welfare department process cases, and the cases go to these units essentially randomly. No other units process cases. These units form the versions of factor B. All the versions of factor B are present in the investigation. This study would be regarded as **b-all**. If a random pair of units were chosen for following through the investigation, that study would be labeled **b-some**. (It will help below to distinguish **b-all**, where the equally weighted set includes all levels in the data, from **b-some**, where only some of those available levels are to be included.)

c. Random sample of levels of B**, equally likely.** When the level of factor B will be chosen, randomly or equally often, from a population of levels, and a random sample of these levels appears in the study we are analyzing (in this situation our standard error will ordinarily refer to predicting the average change, averaged over the population of levels).

Example. Continuation of the welfare example. Imagine that factor B has many units, but that only a small random set of these is chosen for the experiment, even though cases are distributed evenly over all of them.

d. Sample of levels of B **from an unknown population.** When the levels of factor B will be chosen, in some way we cannot foresee, from a population of levels, and a sample of those levels appears in the study we are analyzing (we are often stuck with proceeding as if we were using a target of type **c**).

Example. Continuation of the welfare example. The welfare system is being reorganized, and it will have units of various sizes. The study of the two methods of processing will be carried out on a set of units that are currently in place and will be present after the reorganization.

e. Illustrative levels. When the level of factor B will be chosen, in some way we cannot foresee, from a population of levels, and the levels that appear in the study we are assessing are, we believe, "illustrative" of all the versions we might care about.

Example. Burden on the courts. We desire to find the burden on the courts (number of cases) associated with new criminal laws. We have no population of laws (past *and* future) from which to sample. We pick a few state laws, perhaps one from the area of fraud, one from traffic violations, one from the area of assault on persons, and so on. These form factor B. Factor A will be the socioeconomic class of the victim.

When the levels of our factor are quantitative, we have further alternatives.

Target **e** may seem too loose to be realistic, yet it occurs in nearly every agricultural experiment, where experience has shown (1) that more than a single year's trial is essential; (2) that the last two, three, or five years do *not* behave like a sample of the future years for which a forecast is needed; and (3) that usually the best we can do treats the years studied like a sample of the years that concern us.

Investigations in many fields fall into the "illustrative" class. When they come from the social sciences or the humanities, we are not very surprised; but some life sciences such as biology or paleontology, physical sciences such as astronomy, and technology provide rich sources as well.

In working with our possible targets, we prefer assessments that at least include both an estimated value and a standard error for that value.

11C. RELATION TO THE USUAL "PROBABILITY MODELS"

The usual approach, very limited as we shall see, to distinguishing different targets is to speak of fixed or random factors, as discussed in Chapter 10. We need to relate this distinction to the distinctions we make throughout this chapter. We find ourselves, in particular, adopting a clearer and more specific attitude toward the actual meaning of "fixed factors." In dealing with "probability models," we are planning, as in most inferential applications of probability ideas to data, to assess the results we have in hand in the light of the results we might have had. The probability models that are most usually thought of as guiding our analysis treat factors as either fixed or random.

Factorial Structure

Let us consider an $A \times B$ table of data, initially unreplicated. If it pays us to think of the data in factorial terms, a natural formal model has

$$y_{ij} = \mu + \alpha_i + \beta_j + (\alpha\beta)_{ij} + \varepsilon_{ij} \tag{1}$$

with error expectations

$$E(\varepsilon_{ij}) = 0$$

so that the table of population means, whose terms we expect to define our targets, is given by

$$\mu_{ij} = \mu + \alpha_i + \beta_j + (\alpha\beta)_{ij}. \tag{2}$$

Parallel to (1) and (2), which refer to the quantities that interest us—but that we cannot touch—is a breakdown of the observations

$$y_{ij} = \hat{\mu} + \hat{\alpha}_i + \hat{\beta}_j + \widehat{(\alpha\beta)}_{ij} \tag{3}$$

conventionally defined so that $\Sigma\hat{\alpha}_i = 0$, $\Sigma\hat{\beta}_j = 0$, and $\Sigma_i\widehat{(\alpha\beta)}_{ij} = 0$ for each j and $\Sigma_j\widehat{(\alpha\beta)}_{ij} = 0$ for each i. These definitions mean, for example, that

$$\hat{\alpha}_i = \frac{1}{J}\sum_j y_{ij} - \frac{1}{IJ}\sum_i\sum_j y_{ij}$$

so that

$$E(\hat{\mu} + \hat{\alpha}_i) = [\mu_{i1} + \mu_{i2} + \cdots + \mu_{iJ}]/J$$

and

$$E(\hat{\alpha}_i - \hat{\alpha}_k) = [(\mu_{i1} - \mu_{k1}) + (\mu_{i2} - \mu_{k2}) + \cdots + (\mu_{iJ} - \mu_{kJ})]/J.$$

Thus $\hat{\alpha}_i - \hat{\alpha}_k$ tells us about the equally weighted average of the $\mu_{ij} - \mu_{kj}$ —the average over j of what happens, within the jth version of the second factor, when we change the first factor from i to k.

This means that a "fixed factor" is one for which, across all our background of alternatives, not only are the versions exactly the same as those in the actual data, but also *we are willing to confine our attention to means taken across this fixed set of versions with equal weights*. To have "fixed effects" means to accept a target of type **b-all**.

A **b-all** summary is not bad. We would use the same summary value if B were treated as a random factor. The discomfort comes from assuming that our next use (and then all future uses) will be restricted to this one target. What "random factor" really does is to allow for a flexible relation between observed circumstances and target circumstances!

If we need to consider a target of type **b-some** or **b-unequal**, the fixed analysis is not adequate unless we make a corresponding change in the value splitting, which is equivalent to a change in what we call a main effect.

11D. LEAVING OUT OR POOLING NOISY ESTIMATES: THE $F > 2$ PRINCIPLE

Value splitting takes the data apart into parcels described by overlays. We want to use this splitting by putting back together some of these parcels. If a parcel is "all noise," we do well to forget it when finding some value that interests us. More important, if a parcel is only "mostly noise," we still do better to leave it out or sweep it down.

In our present context, this translates into "a parcel that is more than half noise is better left out of the assessed value, but it is likely to contribute to the size of our estimate of the assessed value's variability (specifically its standard error)." We return to this important principle later and use it often. This whole chapter illustrates and describes the use of this principle, first in simple instances and then in more complicated ones.

We need to begin by seeing that such a rule makes sense. We plan to do this by thinking through some simple numerical examples, in some of which very informal thinking will show us what we ought to do. Not only will these examples help us appreciate this principle, but they should also help to give us a clearer understanding of what is going on and why a principle is needed.

The rule is to downsweep a line (overlay) into a line (overlay) just below it when the former's mean square is less than twice that of the latter.

Anscombe (1967) asked how to make residuals, in any regression, close to the corresponding ϵ's in the model. The resulting procedure proves to be close to including a final group of terms in the regression only if MS(group)/MS(residuals) is greater than 2. This rule corresponds to downsweeping an overlay whenever its mean square is less than twice the mean square with which downsweeping would combine it. This analogy is not always appropriate in analysis of variance, but it supports use of the Rule of 2 from the view of the lower overlay involved.

Readers concerned with the application of this rule may wish to turn to Section 11J. (The intervening sections discuss illustrative synthethic examples.)

11E. A HYPOTHETICAL EXAMPLE ILLUSTRATING TWO-FACTOR SITUATIONS

For illustration Table 11-1 shows three examples of a 3×4 set of data. The 3×4 interaction box is the same for each example. The factor A effects vary from trivial to borderline to large and are shown in reduced form in the leftmost column. For example, in Panel $T(A)$, the factor A effects would be a 3×4 box having the values $(0, -4, 4)$ appearing in each of the four columns. The two effects (factor A and interaction) are added to give the data in the rightmost column. We have not considered the common value C and the main effects of factor B with versions (levels) B_1, B_2, B_3, and B_4 because their actual values do not concern us at this time. Compared with the interaction, the factor A effects in Panel $T(A)$, $(0, -4, 4)$, are trivial, those in Panel $S(A)$, $(0, -8, 8)$, are borderline, and those in Panel $L(A)$, $(0, -40, 40)$, are large.

We can see that the 3×1 box of A effects in Panel $T(A)$ is largely noise —plausibly very close to "all noise."

To spell out an argument, the range of the values in the interaction box of Panel $T(A)$ is $17 - (-12) = 29$, and a rough estimate of the standard deviation based on the range of n measurements is $29/\sqrt{n} = 29/\sqrt{12} \approx 8.4$.

Table 11-1. Three 3 × 4 Hypothetical Examples with Three Levels of Factor *A* and Four Levels of Factor *B*[a]

Panel T(A)

$C + [B_1 \; B_2 \; B_3 \; B_4]$ + (A effects) + (interaction) = $C + [B_1 \; B_2 \; B_3 \; B_4]$ +

A		Interaction						Sum			
0	+	10	−5	−5	0	=		10	−5	−5	0
−4		2	10	−12	0			−2	6	−16	−4
4		−12	−5	17	0			−8	−1	21	4

Panel S(A)

$C + [B_1 \; B_2 \; B_3 \; B_4]$ + (A effects) + (interaction) = $C + [B_1 \; B_2 \; B_3 \; B_4]$ +

A		Interaction						Sum			
0	+	10	−5	−5	0	=		10	−5	−5	0
−8		2	10	−12	0			−6	2	−20	−8
8		−12	−5	17	0			−4	3	25	8

Panel L(A)

$C + [B_1 \; B_2 \; B_3 \; B_4]$ + (A effects) + (interaction) = $C + [B_1 \; B_2 \; B_3 \; B_4]$ +

A		Interaction						Sum			
0	+	10	−5	−5	0	=		10	−5	−5	0
−40		2	10	−12	0			−38	−30	−52	−40
40		−12	−5	17	0			28	35	57	40

[a]The panels are ordered according to the size of the A effects: $T(A)$ (for trivial), $S(A)$ (for small), and $L(A)$ (for large) instances of a factor A apparent main effect combined with the same apparent interaction. (C = common term = $\hat{\mu}$, B_1 to B_4 = factor B main effects—whose values do not matter in the present context.)

If we had an average of four measurements, the standard error would be roughly estimated as $8.4/\sqrt{4} \approx 4.2$, and the factor A effects observed in Panel $T(A)$ are no larger than this rough estimate of standard error.

Turning to Panel $S(A)$, the observed factor A effects run as large as about twice the rough standard error; and, when we consider degrees of freedom and unreliability of estimates, the sizes of these factor A effects seem borderline. Finally, the effects shown in Panel $L(A)$ leave little room for doubt about reality and magnitude of effects.

Analysis of Panel $T(A)$

If we want in Panel $T(A)$ to compare the 3rd version of factor A with the 2nd version (taking no account of factor B), we could choose between the following differences

$$C - C = 0 \quad \text{(common only)},$$
$$(C - C) + (4 - (-4)) = 8 \quad \text{(common and main effect 3rd MINUS 2nd)}.$$

Because $4 - (-4) = 8$ is largely noise, as we know from our earlier analysis, we should not include it; thus in the specified situation, 0 is a preferable assessment. This means that we gain nothing by separating

0
−4
4

from the interaction, and we may as well put these two back together, as they are shown on the right-hand side of Panel $T(A)$.

So far we have been "taking no account of factor B." We need to do better. Let us consider what we might do if each of targets **a** to **e** listed in Section 11B were, in turn, our target.

*Target **a**. Specified Level of B*

If a specified version, say the 2nd, of factor B were our target, we would have the same two choices and a third, namely,

$$(C - C) = 0 \quad \text{(common)},$$
$$(C - C) + (4 - (-4)) = 8 \quad \text{(common + main)},$$
$$(C - C) + (4 - (-4)) + (-5 - (10)) = -7 \quad \text{(common + main + interaction)}.$$

Because the right-hand box of $T(A)$ is the sum of the factor A overlay and the interaction overlay, the third choice can also be written

$$(C - C) + (-1 - 6) = -7.$$

As before, we believe $4 - (-4)$ is largely noise and is best not included. The next question is: Is $(-1 - 6)$ "mostly noise" or not? The data we have so far supposed to be at hand do not let us answer that question directly. We return to it later because the definition of "noise" for interaction needs to depend on something "below" interaction, such as replication. All we can say here is: "Maybe we should take 0, maybe -7." In either choice the right-hand (unified) breakdown of the $T(A)$ case is inadequate.

*Target **b**. Random Level of B*

If we know that all four versions of B, and no others, occurring with *equal* weights, are our target, then we would convert the two breakdowns of the $T(A)$ case into the results of averaging. Because each row of the interaction box of Table 11-1 adds to zero, their averages are zero. These averages give

$$
\begin{array}{ccccccc}
C & + & B(=0) & & C & + & 0 \\
 & & + & & & & + \\
0 & & 0 & = & & & 0 \\
-4 & + & 0 & & & & -4 \\
4 & & 0 & & & & 4
\end{array}
$$

The choices for our assessment of the 3rd version of factor A versus the 2nd are again

$$C - C = 0,$$
$$(C - C) + (4 - (-4)) = 8,$$

and we now have to ask whether 8 is "mostly noise." We need not expect the same answer as before, because the sort of variation shown in the interaction table now does affect our assessment. It can be quite proper, and we later illustrate circumstances, to regard 8 as mostly noise for target **a** but definitely not "mostly noise" for target **b**.

In any event, the left- and right-hand splittings offer the same two alternatives—though possibly with different labels. For this target and this set of data, we do not need the more detailed (left-hand) splitting, though when we come to assessing variability there may well be advantage to having it.

*Target **b-some**. Random Levels of B, Not All Available*

If we have a version of target **b** where equally weighted averaging is to be applied to only some of the available versions of the second factor B, things are a little different. Such situations might arise if something about the investigation ruled out our using some of the B outcomes. Let us suppose that we are to combine versions 1, 2, and 4 of factor B. The averages for the interactions become

Factor *A* Effects		Average Interaction		Sum
0		1.7		1.7
−4	+	4	=	0
4		−5.7		−1.7

For this target, we have no excuse for paying attention to the original A main effect alone. The mean of factor A over four versions of factor B seems irrelevant when a mean over three versions is the target! Thus, here too, a tiny A main effect leads us to want to combine the main effect of factor A with the interaction.

In the specific instance at hand, the averaged-over-three-versions-of-B combined box has small enough entries that we may well believe the box contains only noise.

Another way to describe the situation is to say that, when we have **b-some** as a target, we need a new, redefined A main effect. Such a redefined main effect is given by the combined average interaction and the old factor A effects, both limited to the versions of B we are considering. Taking it out of the entire combined box by subtracting 1.7 from each term in the first row of the interaction table, 4 from terms in the second row, and -5.7 from those in the third row gives

$$\begin{array}{|c|}\hline 1.7\\ 0\\ -1.7\\ \hline\end{array} + \begin{array}{|cccc|}\hline 8.3 & -6.7 & (-6.7) & -1.7\\ -2 & 6 & (-16) & -4\\ -6.3 & .7 & (22.7) & 5.7\\ \hline\end{array} = \begin{array}{|cccc|}\hline 10 & -5 & (-5) & 0\\ -2 & 6 & (-16) & -4\\ -8 & -1 & (21) & 4\\ \hline\end{array},$$

where we have put certain entries in parentheses to remind us that they are not included in the definition of the target. Note that averaging over the 1st, 2nd, and 4th versions of B leaves us with

$$\begin{array}{c}A\\ \text{main effect}\\ \begin{array}{|c|}\hline 1.7\\ 0\\ -1.7\\ \hline\end{array}\end{array} + \begin{array}{c}\text{Average}\\ \text{interaction}\\ \begin{array}{|c|}\hline 0\\ 0\\ 0\\ \hline\end{array}\end{array} = \begin{array}{c}\\ \\ \begin{array}{|c|}\hline 1.7\\ 0\\ -1.7\\ \hline\end{array}\end{array}$$

in close parallel to the case for **b-all** itself.

Target ***b-unequal****. Versions of B Not Equally Likely*

It is worth considering another type of target of the same general nature. Suppose we are to weight the versions *un*equally. Then including all versions is not enough to make this target behave like target **b-all**. Thus if weights 1, 3, 3, 1 are to be applied to versions 1, 2, 3, 4 of B, the weighted entries are (for row 1 we have $(10 + 3(-5) + 3(-5) + 0)/8 = -2.5$, etc.)

A main effect		Average interaction*		
0		−2.5		−2.5
−4	+	−.5	=	−4.5
4		3.0		7.0

*weighted 1, 3, 3, 1 for the versions of B.

Again it is hard to see why we should use the leftmost (A main effect) box. Here also, we are better off to combine the A main effect and the AB interaction. (We could go on to redefine and split off a corresponding new A main effect.)

Target c. Version of B Drawn Randomly from a Population of Levels

If we are aiming at a version of B drawn at random from a population of which our levels are a random sample, we have to accept the interaction as noise, at least so far as comparisons (differences) between versions of A go. In the $T(A)$ case

0
−4
4

appears to be almost all noise, and we have no need to split this parcel out.

The same is true and applies with even more force for targets **d** and **e**.

Summary

For the $T(A)$ case (trivial A main effect), we found no need for preserving a distinction between the A main effect and the AB interaction. The only situation where there might be an advantage to maintaining separation was the special case of target **b**, where we restricted ourselves to looking at equally weighted means over all the available versions of factor B— both for our arithmetic and our intended targets. In the course of this analysis we have two steps. The first is to decide whether to pool the A effects with the AB interaction; the portion of our exposition just completed deals with this. In due course we will introduce replications and ask whether the AB interaction should be combined with the replications (Section 11I). This is a slight oversimplification when we come to real data because we may want to examine the pool of $(A + AB)$ effects as compared with replications, but for now let us ignore that. Thus the second effort decides whether to pool the

AB interaction with replications. If *A* effects have already been pooled with *AB* interaction, they join *AB* before we decide whether to pool with the replications.

11F. EXAMPLES WITH LARGER FACTOR *A* EFFECTS

The *S*(*A*) Panel

The second example, Panel *S*(*A*) of Table 11-1, has been carefully chosen to make visual and mental calculations difficult. Is

A
0
−8
8

mostly noise, or is it not? We approach this from a sum-of-squares point of view, rather than through the rough estimation procedure used in Section 11C. With 2 d.f., $0^2 + (-8)^2 + 8^2 = 128$ gives for the four columns a mean square of $4 \times 128/2 = 512/2 = 256$. The interaction sum of squares, with $2 \times 3 = 6$ d.f., is $10^2 + (-5)^2 + \cdots + 17^2 + 0^2 = 856$, and the corresponding mean square is $856/6 = 143$. Twice 143 is 286, somewhat larger than 256, so our rule of thumb (requiring a ratio of at least 2 to keep a factor separate) tells us to treat the *S*(*A*) case like the *T*(*A*) case and therefore to combine the *A* effect with the *AB* interaction. (For slightly larger *A* effects $0, -9, 9$ instead of $0, -8, 8$ this would no longer be so because the *A* effect mean square would be $4 \times 162/2 = 324$ and 286 is smaller. Thus our situation is near borderline, as we said.)

The *L*(*A*) Case Is Different!

So let us turn to the *L*(*A*) case (*L*(*A*) for large *A* main effect). For target **a**, specifically for the second version of *B*, we choose among

$$(C - C) = 0,$$

$$(C - C) + (40 - (-40)) = 80,$$

$$(C - C) + (40 - (-40)) + (-5 - (10)) = 65.$$

Clearly $40 - (-40)$ is *not* mostly noise (we already know that $0, -9, 9$ is not mostly noise, so that $0, -40, 40$ is a clear signal), and we want to include this parcel. Whether or not $-5 - (10) = -15$ is mostly noise is unclear, because we have not yet introduced the appropriate hypothetical evidence.

We could get 65 from the right-hand breakdown for the $L(A)$ case as

$$(C - C) + (35 - (-30)) = 65,$$

but there is no natural way to get 80 from this less complete breakdown (without reanalysis). Thus we greatly prefer the left-hand (more complete) breakdown, giving an A main effect and interactions, for this case and this target.

For targets of type **b** and the $L(A)$ case, things are much as before. For **b-all** we can live with either combining or not combining (because averaging this interaction in exactly this way gives all zeros). For **b-some** or **b-unequal**, however, we want to recombine before averaging, not because the two parcels "belong together," but because the A main effect, as conventionally defined, is inappropriate.

For all forms of target **b**, as in the last section, we can live with either breakdown, because averaging over the versions of B, after sweeping the A main effect and the AB interaction together, will reconstitute the A main effect.

For targets **c**, **d**, and **e** we certainly want to move from

$$(C - C) = 0$$

to

$$(C - C) + (40 - (-40)) = 80.$$

Equally certainly we have no excuse for going further (and will need to include interaction-revealed variability in our assessed standard error). For these targets, like target **a**, we need the more detailed left-hand breakdown when the A main effect is large.

Overall Summary

When the A main effect is "small"—when MS(for A) $\leq$ 2(MS(for AB))—the rule of thumb says that we can usually sweep A and AB effects together for all targets. When the A main effect is "large"—when MS(for A) $>$ 2(MS (for AB))—the rule of thumb says that we do not want to sweep together A and AB for targets **a**, **c**, **d**, and **e**. For **b-some** and **b-unequal** we shall want to sweep these two parcels together always, but only as a step in redefining the "main effect," which will automatically redefine the "interaction." For target **b-all** it makes, so far as our present calculations go, little difference whether or not we sweep things together.

Conclusions from the Rule of Thumb

1. Sweep together if MS(for A) $\leq$ 2(MS(for AB)).
2. Sweep together (as a step in redefining the breakdown) for targets of type **b** other than **b-all**.
3. Wait for further discussion for target **b-all**.

11G. SWEEPING THE COMMON TERM DOWN?

Most discussions of analysis of variance pass by the common term (when we are splitting with means, the grand mean) without a word. The common term often deserves separate treatment, because we seldom want to compare it with anything. But we may want to look at its value, say $\bar{\bar{y}}$, and its mean square (= its sum of squares), either $(\bar{\bar{y}})^2 \times$ (number of observations) or $(\bar{\bar{y}} - y_{ref})^2 \times$ (number of observations). Here y_{ref} is some chosen reference value whose use corresponds to taking $y - y_{ref}$ as the response. For example, in using a Fahrenheit temperature scale, we might want the deviation from 32°F.

Unless either (1) the y's themselves are changes (differences), not all of the same sign, or (2) they are differences of observed from standard, also of varying sign, there is little chance that the common-term mean square will be nearly as small as the other mean squares. But exceptions occasionally occur.

If MS(for common term) is less than twice one or both of MS(for A) and MS(for B), we can sweep it down, presumably to join whichever parcel, A or B, has the larger mean square. If we contemplate sweeping "common" in with B, we can have any one of the alternative targets with which we opened this chapter. Everything goes much the same.

11H. LOOKING AT TWO FACTORS WITH REPLICATION

Suppose now that our 3×4 table arose as the 12 means of 2 replicates, one pair in each of the 12 cells. This adds another component, the one for replication within cells. Figure 11-1 schematizes the overlays that we would then have. Previously we asked about downsweeping A into AB, in which circumstances AB was taken as "noise." By analogy, we can now consider downsweeping AB into replication when replication is taken as "noise."

We use the analogous rule of thumb for mean squares:

$$\text{downsweep if MS(for } AB) \leq 2\text{MS(replication)}, \qquad (4)$$

but we often cannot use the same justification, because we are likely to be

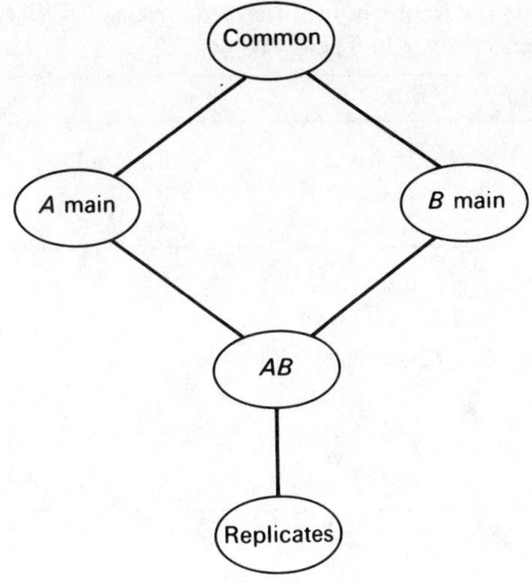

Figure 11-1. The five overlays for a two-way table that has replication within the cells. The lines represent possible paths for downsweeping.

concerned with either

the *AB* parcel

or

the combined *AB* and replication parcel,

as a basis for assessing variability of our values for main effects and simple changes (differences from one version to another of a chosen factor), rather than as a part of assessing the simple changes themselves. (When we assess the values appearing in the interaction, we can, of course, use the previous justification.)

In practice, we use the rule of thumb (4), but with a different justification. The question of when to "pool" *AB* with replications as a basis for judging the main effects was studied by A. E. Paull (1950), who found that (4)—or a rule that used a multiplier somewhat less than 2—gave good results for the cases he studied.

Let us return to our simple numerical example, expanded by alternative overlays for replication in each cell.

11I. REPLICATED CELLS: POOLING INTERACTION WITH REPLICATION

Suppose that our 3×4 data patterns are the result of analyzing 24 values, 2 in each of $12 = 3 \times 4$ cells. Table 11-2 offers some alternatives for the

Table 11-2. Four Possible Sets of Residuals for Replication, Intended for Use (in All Combinations) with the Three Sets of Higher Boxes in Table 11-1

Replication $T(R)$ Box

+2, −2 0, 0 −2, +2	+1, −1 0, 0 +1, −1	−2, +2 +2, −2 +2, −2	−1, +1 +3, −3 −4, +4

Sum of squares = 96, d.f. = 12, MS = 8

Replication $S(R)$ Box

+6, −6 0, 0 −6, +6	+3, −3 0, 0 +3, −3	−6, +6 +6, −6 +6, −6	−3, +3 +9, −9 −12, +12

Sum of squares = 9 × 96, d.f. = 12, MS = 72

Replication $L(R)$ Box

+8, −8 0, 0 −8, +8	+4, −4 0, 0 +4, −4	−8, +8 +8, −8 +8, −8	−4, +4 +12, −12 −16, +16

Sum of squares = 16 × 96, d.f. = 12, MS = 128

Replication $V(R)$ Box

+20, −20 0, 0 −20, +20	+10, −10 0, 0 +10, −10	−20, +20 +20, −20 +20, −20	−10, +10 +30, −30 −40, +40

Sum of squares = 100 × 96, d.f. = 12, MS = 800

overlay for replicates involving two residuals, necessarily equal in magnitude but opposite in sign, in each cell.

If we compare the interaction box from Table 11-1 with the replication $T(R)$ box from Table 11-2, it is quite clear that, if the variation in these replications is typical "noise," then the interaction in Table 11-1 is *not* "mostly noise." A mean square of 8 or a standard deviation of $\sqrt{8} = 2.8$ from $T(R)$ is not compatible with "noise" interactions of size 17 and −12. At the other extreme, if we compare the interaction box from Table 11-1 with the replication $V(R)$ box from Table 11-2, it is clear that, if the $V(R)$ replication residuals are typical "noise," then the interaction *is mostly noise.*

So we would want to sweep the interaction down into the replication $V(R)$ but leave interaction separate for the replication $T(R)$. (We can, of course, confirm these judgments by calculating and comparing appropriate mean squares using our rule of thumb.)

The $S(R)$ box and $L(R)$ box for replication are likely to leave us somewhat confused: they were chosen to be marginal. To overcome such confusion, we now systematically develop the sums of squares that we need in order to compare the interaction box in Table 11-1 with each of the replication boxes in Table 11-2. Then the resulting mean squares can guide our choice of whether to sweep the interaction down into the replication.

Working Out the Mean Squares

We want to calculate our mean squares on the basis of each replication of a response. The presence of 2 replications per cell implies that the original data layout would contain $3 \times 4 \times 2$ observations on the response. If we were calculating sums of squares (and mean squares) from full overlays, we would have to expand the (reduced) A overlay and the AB overlay of Table 11-1 accordingly. That is, we would use 8 copies of each of the 3 A effects and 2 copies of each of the 12 AB interactions. Table 11-3 illustrates two versions of this expansion process for the $T(A)$ box from Table 11-1 and the $T(R)$ box from Table 11-2. (That is, we have tiny factor A effects and tiny residual variance.) One version adds the replications to the expanded AB interaction effects, and the other version also expands the A effects by adding them into the resulting layout, cell by cell. Although the amount of arithmetic is substantial, it is a comfort to be able to verify directly that the sums of squares add up as anticipated.

Note that in the new layout in the lower part of Table 11-3, the sum of squares for factor A is twice that of the layout having one observation per cell, that is, $2(128) = 256$; the 128 comes from the full overlay of the factor A effects in Table 11-1, and the 2 comes from the twoness of replications. The sum of squares for the new AB effects overlay is twice that of the overlay having one observation per cell, plus the sum of squares of the replication table: $2(856) + 96 = 1808$. For the new $(A + AB)$ effects overlay, the SS is twice the SS for the $(A + AB)$ overlay having one observation per cell plus the SS for the replications, all told giving $2(984) + 96 = 2064$; that is, in full detail $2(128 + 856) + 96 = 2064$.

Now that we have introduced replication as another indication of random "noise," we face two possibilities for sweeping down. First, should we sweep the A effect down into the interaction? Second, should we sweep the interaction down into the replication? (We may be working with the combination of the interaction and the A effect, if we have already decided to sweep these two together.) We must consider these two choices for each of the 12 data sets that we get by combining each of the 3 sets of A effects (and the AB interaction effects) from Table 11-1 and each of the 4 sets of residuals (replication) from Table 11-2. As a basis for these hypothetical analyses, Table 11-4 gives the sums of squares for the effects from the

Table 11-3. Expanding the $T(A)$ Box and the Interaction Box of Table 11-1 to Combine with the Replications from the $T(R)$ Box from Table 11-2

From Table 11-1:

A effects

0
−4
4

SS = 128

AB effects

10	−5	−5	0
2	10	−12	0
−12	−5	17	0

SS = 856
= SS(for *AB*)

$(A + AB)$ effects

10	−5	−5	0
−2	6	−16	−4
−8	−1	21	4

SS = 984
= SS(for $(A + AB)$)

Replications from $T(R)$ box of Table 11-2:

+2, −2	+1, −1	−2, +2	−1, +1
0, 0	0, 0	+2, −2	+3, −3
−2, +2	+1, −1	+2, −2	−4, +4

SS = 96 d.f. = 12 MS = 96/12 = 8
= SS(for replication) = MS(for replication)

We expand the *AB* effects above by adding the first replication in a cell to the corresponding entry in the *AB* effects box and then adding the second replication to the same cell in the *AB* effects box, thus

10 + 2, 10 − 2	−5 + 1, −5 − 1	−5 − 2, −5 + 2	0 − 1, 0 + 1
2 + 0, 2 + 0	10 + 0, 10 + 0	−12 + 2, −12 − 2	0 + 3, 0 − 3
−12 − 2, −12 + 2	−5 + 1, −5 − 1	17 + 2, 17 − 2	0 − 4, 0 + 4

to give a new layout, having two observations in each cell:

A effects

0
−4
4

SS = 2(128)

AB effects with replications

12, 8	−4, −6	−7, −3	−1, 1
2, 2	10, 10	−10, −14	3, −3
−14, −10	−4, −6	19, 15	−4, 4

SS = 1808 = 2(856) + 96
= 2(SS(for *AB*)) + SS(for replication)

$(A + AB)$ effects with replications

12, 8	−4, −6	−7, −3	−1, 1
−2, −2	6, 6	−14, −18	−1, −7
−10, −6	0, −2	23, 19	0, 8

SS = 2064 = 2(128 + 856) + 96
= 2(SS(for *A*) + SS(for *AB*)) + SS(for replication)

Table 11-4. Sums of Squares[a] and Degrees of Freedom from Tables 11-1 and 11-2

From Table 11-1

	A Effects (2 d.f.)	*AB* Effects (6 d.f.)	$(A + AB)$ Effects (8 d.f.)
$T(A)$	128	856	984
$S(A)$	$2^2 \times 128 = 512$	856	1,368
$L(A)$	$10^2 \times 128 = 12{,}800$	856	13,656

From Table 11-2

	Replication (12 d.f.)
$T(R)$	$96 = 96$
$S(R)$	$3^2 \times 96 = 864$
$L(R)$	$4^2 \times 96 = 1{,}536$
$V(R)$	$10^2 \times 96 = 9{,}600$

[a]When the elements to be squared are multiplied by a constant k, the sum of squares is multiplied by k^2.

three panels of Table 11-1 and the sums of squares for the four replication boxes in Table 11-2. Table 11-5 then uses these to construct the sums of squares for the 12 expanded tables.

For example, each of the 4 enlarged data sets derived from $T(A)$ by combining with $T(R)$, $S(R)$, $L(R)$, and $V(R)$ from Table 11-2 would have SS for factor A equal to 2(128), and its SS for interaction would be 2(856) + SS(for replication) from Table 11-2. That is,

	Interaction SS
$T(A)$ with $T(R)$	$2(856) + 96 = 1{,}808$
$T(A)$ with $S(R)$	$2(856) + 864 = 2{,}576$
$T(A)$ with $L(R)$	$2(856) + 1536 = 3{,}248$
$T(A)$ with $V(R)$	$2(856) + 9600 = 11{,}312$

as shown in the first three lines of Table 11-5.

Comparing Interaction with Replication

In Sections 11E and 11F we have already considered whether to sweep the A effects down into the AB interaction in each of the three cases—$T(A)$, $S(A)$, and $L(A)$. We now turn to the question of whether to sweep the interaction down into the replication in each of the four cases—$T(R)$, $S(R)$, $L(R)$, and $V(R)$. In all instances MS(for AB) $= 2(856)/6 = 285$ (from Tables 11-4 and 11-5), and we get MS(for replication) from Table 11-2.

Table 11-5. Sums of Squares for Expanded Tables[a]

	A Effects		AB Effects		$(A + AB)$ Effects	
$T(A), T(R)$	2(128) =	256		96 = 1,808		96 = 2,064
$T(A), S(R)$		256	2(856) +	864 = 2,576	2(128 + 856) +	864 = 2,832
$T(A), L(R)$		256		1,536 = 3,248		1,536 = 3,504
$T(A), V(R)$		256		9,600 = 11,312		9,600 = 11,568
$S(A), T(R)$	2(512) =	1,024		96 = 1,808		96 = 2,832
$S(A), S(R)$		1,024	2(856) +	864 = 2,576	2(512 + 856) +	864 = 3,600
$S(A), L(R)$		1,024		1,536 = 3,248		1,536 = 4,272
$S(A), V(R)$		1,024		9,600 = 11,312		9,600 = 12,336
$L(A), T(R)$	2(12,800) =	25,600		96 = 1,808		96 = 27,408
$L(A), S(R)$		25,600	2(856) +	864 = 2,576	2(12,800 + 856) +	864 = 28,176
$L(A), L(R)$		25,600		1,536 = 3,248		1,536 = 28,848
$L(A), V(R)$		25,600		9,600 = 11,312		9,600 = 36,912

[a]The multiplications by 2 come from the twoness of the replications.

$T(R)$: If the replication $T(R)$ box is a fair indication of what is "noise," then any quick look at the interaction boxes of Table 11-1 (all the same) shows that they are *not* "mostly noise." Accordingly, if there were replication, of a size corresponding to the $T(R)$ box of Table 11-2, we would conclude that we ought *not* to sweep the interaction down into the replication.

$S(R)$: If we have replication with variation of size corresponding to Table 11-2's $S(R)$ box, we have MS(for replication) = 72 from Table 11-2. Thus, because 285 > 2(72), we shall again *not* want to combine *AB* and replication.

$L(R)$: If we have replication variation of size corresponding to Table 11-2's $L(R)$ box, we have MS(for replication) = 128 from Table 11-2. Thus MS(for *AB*) = 285 is more than twice the MS(for replication). Therefore we shall still not want to combine *AB* with replication.

$V(R)$: The MS(for replication) for $V(R)$ is 800. Because 285 is less than MS(for replication), and hence less than twice this MS, we shall want to combine *AB* with replication.

If we went systematically through both steps of possible downsweeping for each of the 12 expanded data sets represented in Table 11-5, the details would be a little more complicated because in some instances we would be combining *A* and *AB* with replication. Without yet illustrating the process at

Table 11-6. The Swept-down Breakdowns Based on Combinations of $T(A)$, $S(A)$, and $L(A)$ with $T(R)$, $S(R)$, $L(R)$, and $V(R)$

A and *AB* combined; replication separate

- $T(A)$ and $T(R)$
- $T(A)$ and $S(R)$
- $T(A)$ and $L(R)$
- $S(A)$ and $T(R)$
- $S(A)$ and $S(R)$
- $S(A)$ and $L(R)$

A, *AB*, and replication all combined

- $T(A)$ and $V(R)$
- $S(A)$ and $V(R)$

A, *AB*, and replication all separate

- $L(A)$ and $T(R)$
- $L(A)$ and $S(R)$
- $L(A)$ and $L(R)$

A separate but *AB* and replication combined

- $L(A)$ and $V(R)$

Table 11-7. Some Final Breakdowns[a] in the Replicated Situation, Selected Illustrations from the Summary in Table 11-6

Panel 1

A and *AB*, combined					Replication							
10	−5	−5	0		+2,	−2	+1,	−1	−2,	+2	−1,	+1
−2	6	−16	−4	+	0,	0	0,	0	+2,	−2	+3,	−3
−8	−1	21	4		−2,	+2	1,	−1	+2,	−2	−4,	+4

(for *A* as *T*(*A*) box and replication as *T*(*R*) box)

Panel 2

A, *AB*, and replication combined							
30,	−10	5,	−15	−25,	15	−10,	+10
−2,	−2	6,	6	+4,	−36	+26,	−34
−28,	12	9,	−11	41,	1	−36,	+44

(for *A* as *T*(*A*) box and replication as *V*(*R*) box)

Panel 3

A		*AB*					Replication							
0		10	−5	−5	0		+2,	−2	+1,	−1	−2,	+2	−1,	+1
−40	+	2	10	−12	0	+	0,	0	0,	0	+2,	−2	3,	−3
+40		−12	−5	17	0		−2,	+2	1,	−1	2,	−2	−4,	4

(for *A* as *L*(*A*) box and replication as *T*(*R*) box)

Panel 4

A		*AB* and replication combined							
0		30,	−10	5,	−15	−25,	15	−10,	10
−40	+	2,	2	10,	10	8,	−32	30,	−30
+40		−32,	8	5,	−15	37,	−3	−40,	40

(for *A* as *L*(*A*) box and replication as *V*(*R*) box)

[a]Omitting the boxes for the common term and the *B* effects. T, tiny; L, large; and V, very large.

this level of detail, we note that it has four possible outcomes. Table 11-6 itemizes these and lists the data sets to which each applies.

The Breakdowns

In Sections 11E and 11F we already looked into the question of whether to combine A and AB. We concluded that with tiny A effects, $T(A)$, we should combine A effects and AB effects. We made the same decision, though it was borderline, for the small effects, $S(A)$; but we found that the large A effects, $L(A)$, should not be combined with AB.

We now consider examples of deciding whether to sweep AB into the replication sum of squares. We again use the Rule of 2. If the MS(for AB) is more than twice the MS(for replications), we do not combine. Otherwise we do. Close calls are left up to the investigator.

Table 11-7 collects one instance of each of the different breakdowns specified in Table 11-6. If we examine each in turn, we can be qualitatively happy about the outcome in each case. In the top breakdown, we have no desire to separate an A box from the 3×4 combined box, and no desire to mix up the tiny replication entries with the other larger ones.

In the second panel, where all has been left together, we find no suggestion of anything but noise.

In the third panel, with its 3 parts of very different size, we have no urge to combine anything.

In the fourth panel, we see that the square root of the mean square for AB and replication is about 25, and this would be an estimate of the standard deviation for single measurements. The 40s of the A effect, based on averages of 4 measurements, are large compared to noise and appropriately separated from the pool.

Thus Table 11-7 illustrates the kinds of consequences we expect when we have two factors with replication. We would have to interpret the resulting table in the specific problem being studied.

11J. COMMENTARY ON THE RULE OF 2

We plan to use the rule of thumb—the Rule of 2—freely, in almost any context. Before we think about what that means in more complex cases, it is important to be sure what it does *not* mean in the simple cases. A decision to sweep the A effect down into the AB interaction (in simplified terms, to sweep the A effect down into the AB effect):

Does not mean that we have decided that the underlying A effect (defined in terms of subpopulation means, perhaps) is really zero.

Does mean that we find the amount of contamination of the underlying A effect by "noise" (defined by the underlying AB effect, possibly combined with replicate variation) is large enough to make it unprofitable to keep the A effect separate.

Not to view the underlying effect as zero, even when we sweep the corresponding packet down, is important. If we thought, erroneously, that we were forcing things to be zero, we might find ourselves worrying unnecessarily about whether we forced something to be zero in one place while attending to it elsewhere. Deciding that something is drowned in one kind of noise may still be consistent with deciding that it is not drowned in some other, weaker, kind of noise.

As a practical matter, as we have often noted, when different treatments are applied, we expect some effect (not absolutely none); the issue we have been attending to is whether it is small compared with various sources of noise.

In the next section we use the Rule of 2 freely, comparing a mean square at one level with the largest appropriate mean square at the next level down. If the ratio of 2 is not reached, we sweep the upper mean square down. The comparisons being made—ratios of individual mean squares—are not specially constructed as described in Chapter 10, where combinations of individual mean squares are used to match expected mean squares. Our view is that the simple ratio works rather well in practice and seemingly avoids downsweeps that would rather too often lead us to fail to explore effects that deserve exploration. The choice of 2 is a rule of thumb of a type that we have sometimes found useful in exploratory data analysis. The rule has some grounding in special cases and analogies with other types of analysis, but we shall not try to report those arguments here. Instead, we want to get on with the downsweeping and its consequences.

11K. MORE COMPLEX SITUATIONS WITH AN EXAMPLE HAVING FOUR FACTORS

If we adhere to "start by treating all factors as random" as another rule of thumb, we have a reasonably easy time analyzing any (balanced) factorial design. Suppose that we have several factors $A, B, \ldots, I$. If we ask about downsweeping the interaction AB, we can consider sweeping into any of the three-way interactions $ABC, ABD, \ldots, ABI$ that exist for our data. Paull's rule of 2 suggests that we sweep into whichever of $ABC, \ldots, ABI$ has the largest mean square, as long as twice this 3-factor mean square exceeds the AB mean square. Note that A and B must be two of the factors in the 3-factor mean square to be compared. Other two-factor interactions, or main effects, or higher-order interactions are to be treated similarly.

We recommend, at this point, the following rules:

1. Start at the top and work down.
2. Do not sweep a parcel down unless its mean square is less than twice the mean square for the parcel into which we are sweeping.
3. Rule 2 applies to sweeping down compound parcels produced earlier (after they have been combined).
4. Sweep down as much as Rules 2 and 3 allow.

(It may occasionally be necessary to make a second cycle of downsweeping.)

We use the dental-gold data (Table 11-8) to illustrate sweeping down. As Section 6B explains, this data set has four factors: M (Method), A (Alloy), T (Temperature), and D (Dentist). From these four factors, we get overlays of five orders:

One zeroth-order or common term;
Four first-order or main effects: M, A, T, and D;
Six second-order or two-factor interactions: M × A, M × T, M × D, A × T, A × D, and T × D;
Four third-order or three-factor interactions: M × A × T, M × A × D, M × T × D, and A × T × D;
One fourth-order (M × A × T × D is the residual term, because there is no replication).

In sweeping down, we compare the MS for the zeroth-order term with the MS for each of the four first-order terms; the MS for each first-order term with each of the 3 appropriate second-order terms; the MS of each second-order term with the 2 appropriate third-order terms; and the MS of each third-order term with the MS of the fourth-order term. We compare a term with the term of the next order containing it; for example, we compare M × D with M × T × D.

To make the comparisons easier, Table 11-9 arranges the terms within each order from the largest MS to the smallest. We now compare each mean square in one order with the highest (and largest) relevant one of the next higher order. Once we have made a comparison of the item in an order with *the largest appropriate item* in the next order, we are finished with the comparison of the item from above. It is either combined or not, as far as that comparison for the pair of orders is concerned. Once a combination is made, that combination is compared with the next order down, after we have completed the sweeping from one level to the next.

Recall that, by our rule of thumb, we combine if the MS of the lower-order term is less than twice the MS of the one we are comparing in the next higher order. For example, in Table 11-9, the MS of D (first-order term) is 39,449

Table 11-8. Dental-Gold Example: Two Alloys, Three Temperatures (T1 = 1500, T2 = 1600, T3 = 1700°F), Three Methods, and Five Dentists (Raw Data)

		A1			A2		
D	M	T1	T2	T3	T1	T2	T3
1	1	813	792	792	907	792	835
	2	782	698	665	1115	835	870
	3	752	620	835	847	560	585
2	1	715	803	813	858	907	882
	2	772	782	743	933	792	824
	3	835	715	673	698	734	681
3	1	743	627	752	858	762	724
	2	813	743	613	824	847	782
	3	743	681	743	715	824	681
4	1	792	743	762	894	792	649
	2	690	882	772	813	870	858
	3	493	707	289	715	813	312
5	1	707	698	715	772	1048	870
	2	803	665	752	824	933	835
	3	421	483	405	536	405	312

Source: Morton B. Brown (1975). "Exploring interaction effects in the ANOVA," *Applied Statistics*, 24, 288–298 (data from Table 1, p. 290). Data reproduced by permission of the Royal Statistical Society.

and the MS of M × D is 38,309. Because 39,449 < 2(38,309), we combine. Put another way, we combine when the F ratio is less than 2, and not otherwise.

Another way to examine the mean squares using this rule of thumb is to take logs to base 2, and if the $\log_2$(MS of lower-order term) minus $\log_2$(MS of the largest appropriate one in the next higher order) is less than 1, then we combine. For example,

$$\log_2(\text{MS for D}) = \log_2(39{,}449) = 15.27,$$

$$\log_2(\text{MS for M} \times \text{D}) = \log_2(38{,}309) = 15.23.$$

Because 15.27 − 15.22 < 1, we would combine. In Table 11-9 we also show the logs. Even if you prefer not taking logs to the base 2, you will find these logs very convenient in following this example.

We plan to revise Table 11-9 by working in steps.

Step 1. We compare the MS of the common term with the largest first-order term. Because the common term is very large, it is obvious that we would not combine. As discussed in Section 11G, this is the usual situation.

Table 11-9. First Downsweep for the Dental-Gold Example, Order by Order

	MS	$\log_2$(MS)	d.f.	Downsweeping
Zeroth-Order				
Common term	Very large		1	
First-order				
Method, M	296,714	18.18	2	
Alloy, A	105,816	16.69	1	
Temperature, T	41,089	15.33	2	
Dentist, D	39,449	15.27	4	sweep
Second-order				
M × D	38,309	15.23	8 ←	D
M × A	27,343	14.74	2	
T × D	16,803	14.04	8	
A × T	10,863	13.41	2	
M × T	7,663	12.90	4	
A × D	1,422	10.47	4	
Third-order				
M × T × D	11,345	13.47	16	
M × A × D	9,616	13.23	8	
A × T × D	5,930	12.53	8	
M × A × T	4,143	12.02	4	
Fourth-order				
M × A × T × D	4,736	12.21	16	

Because nothing is combined so far, Table 11-9 remains unchanged at the end of Step 1.

Step 2. Comparing first-order terms with second-order terms: M does not combine with any second-order term because the difference $\log_2$ MS(M) − $\log_2$ MS(M × D) is greater than 1 for M × D, the largest appropriate term. Therefore, because of the ordering within groups, it is greater than 1 for all other appropriate second-order terms. Similarly for A and for T.

Next we come to D, with $\log_2$ MS(D) = 15.27. The largest appropriate second-order term is M × D, with $\log_2$ MS(M × D) = 15.23. The difference 15.27 − 15.23 is less than 1; therefore we combine, as already noted. We write a "D" on the line for M × D in the rightmost column of Table 11-9 to indicate that D and M × D are to be combined. (We make no further comparisons of first-order terms with second-order ones because we would combine D with one and only one second-order term. Also, because D is the last of the first-order terms, we make no further comparisons of first-order with second-order.)

We pause now to calculate the combinations so far. We compute the mean square for the combination D with M × D:

$$[4(39{,}449) + 8(38{,}309)]/12 = 38{,}689, \quad \text{with 12 d.f.}$$

We revise Table 11-9 to get Table 11-10, in which the old MS for M × D, 38,309 with 8 d.f., has been replaced by 38,689 with 12 d.f. Note that Table 11-10 does not have D among the first-order terms. When we compute a pooled mean square, the line may change its position within its order because of the changed mean square.

Step 3. Comparing second-order terms with third-order ones. We use Table 11-10 to make these comparisons.

M × D,D with M × T × D — 15.24 − 13.47 > 1, do not combine. (Nor would we combine with any other third-order term.)

M × A with M × A × D — 14.74 − 13.23 > 1, do not combine. (Nor with any other third-order term.)

T × D with M × T × D — 14.04 − 13.47 < 1, therefore combine. In Table 11-10, we write T × D on the M × T × D line. (Having combined, we make no further comparisons for T × D.)

A × T with A × T × D — 13.41 − 12.53 < 1, therefore combine. In Table 11-10 we write A × T on the line A × T × D. (We make no further comparisons for A × T.)

M × T with M × T × D — 12.90 − 13.47 < 1, therefore combine. In Table 11-10 we write M × T on the line M × T × D. (We make no further comparisons for M × T.)

A × D with M × A × D — 10.47 − 13.23 < 1, therefore combine. In Table 11-10 we write A × D on the M × A × D line. (We make no further comparisons for A × D.)

We pause again to make the calculations for the combinations in Step 3.

M × T × D: $[8(16{,}803) + 4(7{,}663) + 16(11{,}345)]/28 = 12{,}378$, with 28 d.f.

M × A × D: $[4(1{,}422) + 8(9{,}616)]/12 = 6{,}885$, with 12 d.f.

A × T × D: $[2(10{,}863) + 8(5{,}930)]/10 = 6{,}917$, with 10 d.f.

Table 11-10. Downsweep for the Dental-Gold Example, After Any Resulting Combinations of First-Order Terms with Second-Order Ones

	MS	$\log_2$(MS)	d.f.	New Downsweeping	
Zeroth-order					
Common term	Very large		1		
First-order					
M	296,714	18.18	2		
A	105,816	16.69	1		
T	41,089	15.33	2		
Second-order					
M × D, D	38,689	15.24	12		
M × A	27,343	14.74	2		
T × D	16,803	14.04	8		
A × T	10,863	13.41	2		
M × T	7,663	12.90	4		
A × D	1,422	10.47	4		
Third-order					
M × T × D	11,345	13.47	16	T × D	M × T
M × A × D	9,616	13.23	8	A × D	
A × T × D	5,930	12.53	8	A × T	
M × A × T	4,143	12.02	4		
Fourth-order					
M × A × T × D	4,736	12.21	16		

We now revise Table 11-10 to get Table 11-11, replacing the old mean squares for M × T × D, M × A × D, and A × T × D with our new combined ones. Note that we now have only two second-order terms, M × D (actually M × D,D) and M × A.

Step 4. Comparing mean squares for third-order terms in Table 11-11 with the M × A × T × D mean square. When a line consists of more than one source, we now use the label of the highest-order source (most factors).

M × T × D $13.60 - 12.21 > 1$, do not combine
M × A × D $12.75 - 12.21 < 1$, therefore combine. We write "M × A × D, A × D" on the M × A × T × D line.

Similarly for A × T × D and M × A × T. We now calculate a new fourth-order term.

$$[12(6{,}885) + 10(6{,}917) + 4(4{,}143) + 16(4{,}736)]/42 = 5{,}813, \quad \text{with 42 d.f.}$$

Making these substitutions, we have our final table, Table 11-12.

Table 11-11. Downsweep for the Dental-Gold Example After Any Appropriate Combinations of Second-Order Terms with an Appropriate Third-Order One

	MS	$\log_2(\text{MS})$	d.f.	New Downsweeping
Zeroth-order				
Common term	Very large		1	
First-order				
M	296,714	18.18	2	
A	105,816	16.69	1	
T	41,089	15.33	2	
Second-order				
M × D, D	38,689	15.24	12	
M × A	27,343	14.74	2	
Third-order				
M × T × D, T × D, M × T	12,378	13.60	28	
M × A × D, A × D	6,885	12.75	12	
A × T × D, A × T	6,917	12.76	10	
M × A × T	4,143	12.02	4	
Fourth-order				
M × A × T × D	4,736	12.21	16	M × A × D, A × D; A × T × D, A × T; M × A × T

Table 11-12. Final Downsweep for the Dental-Gold Example After the Appropriate Combinations of Terms

	MS	$\log_2(\text{MS})$	d.f.
Zeroth-order			
Common term	Very large		1
First-order			
M	296,714	18.18	2
A	105,816	16.69	1
T	41,089	15.33	2
Second-order			
M × D, D	38,689	15.24	12
M × A	27,343	14.74	2
Third-order			
M × T × D, T × D, M × T	12,378	13.60	28
Fourth-order			
M × A × T × D + 5 terms	5,813	12.50	42

Table 11-13. Effects for M, A, T, M × D, and M × A

Panel 1

	M
M1	52
M2	63
M3	− 115

A1	A2
− 34	34

T1	T2	T3
31	10	− 41

Panel 2

	D1	D2	D3	D4	D5
M1	− 13	− 9	− 56	6	73
M2	− 18	− 42	− 41	38	63
M3	31	51	97	− 44	− 135

	A1	A2
M1	− 8	8
M2	− 25	25
M3	34	− 34

From Table 11-12, we see that a second cycle is not needed. That is, comparing first-order terms with second-order ones, M would not combine with M × D nor with M × A; and A would not combine with M × A. Comparing second-order terms with third-order ones, pooled M × D,D would not combine with M × T × D. And the third-order term pooled M × T × D would not combine with M × A × T × D.

Having made the sweeps, we focus attention on the means for the one-factor and two-factor terms of Table 11-12 (i.e., M, A, T, M × D, and M × A). For these terms we construct Table 11-13. Auxiliary computations, starting with the means of M, A, T, D, M × D, and M × A, help to get these results. Recall that M has three levels or versions, A two, T three, D five, M × D fifteen, and M × A six. For one entry in each of the five terms, Table 11-14 illustrates the calculation by giving the number of observations and the mean.

We begin with the raw data for the dental-gold example in Table 11-8 and compute the grand mean of the 90 observations:

$$\text{grand mean} = 741.78.$$

In Panel 1 of Table 11-14 for the illustrative one-factor entries we give the mean and the deviation from the grand mean. These deviations appear in the cells of Panel 1 in Table 11-13. Panel 2 of Table 11-14 gives, for two two-factor entries, the number of observations and the mean. From these means we subtract the grand mean, and then we subtract the deviation given in Panel 1 for each component. For example, the mean of the six observa-

Table 11-14. Sample Computations Leading to First- and Second-Order Effects for the Dental-Gold Data

Computations for Panel 1 of Table 11-13

	Number of Observations	Mean	Mean MINUS Grand Mean (741.78)
M1	30	793.90	52.12
A1	45	707.49	−34.29
T1	30	772.77	30.99
(D1	18	783.06	41.28)

Computations for Panel 2 of Table 11-13

	Number of Observations	Mean	Mean MINUS Grand Mean		M		D		Second-Order Effect
M1D1	6	821.83	80.05	−	52.12	−	41.28	=	−13.35
M1A1	15	751.13	9.35	−	52.12	−	(−34.29)	=	−8.48

tions for M1D1 is 821.83. From this we subtract the grand mean, 741.78, the deviation 52.12 for M1 in Panel 1, and the deviation 41.28 for D1 in Panel 1, to give

$$821.83 - 741.78 - 52.12 - 41.28 = -13.35.$$

Rounding off to whole numbers, we have −13 in the M1D1 cell in Panel 2 of Table 11-13.

If we attend only to the largest effects, we see that

- Methods 1 and 2 appear to give higher results—harder fillings—than Method 3.
- Alloy 2 seems to give higher results than Alloy 1.
- Temperatures 1 and 2 appear to give higher results than Temperature 3.
- Methods 1 and 2, relative to Method 3, seem to give higher results for Alloy 2 rather than Alloy 1; and Method 3 with Alloy 1 gives higher results.
- Methods 1 and 2, relative to Method 3, seem to do worse for Dentists 1, 2, and 3 and better for Dentists 4 and 5.

We can clarify some of these statements by temporarily downsweeping relevant main effects into two-factor interactions, as shown in Table 11-15,

Table 11-15. Temporary Further Recombination of Boxes[a]

Panel 1: Method and Alloy

A1	A2
−34	34

+

M1	52
M2	63
M3	−115

+

	A1	A2
M1	−8	8
M2	−25	25
M3	34	−34

=

	A1	A2
M1	9	95
M2	3	122
M3	−115	−114

Panel 2: Method and Dentist

D1	D2	D3	D4	D5
41	45	7	−28	−65

+

M1	52
M2	63
M3	−115

+

	D1	D2	D3	D4	D5
M1	−13	−9	−56	6	73
M2	−18	−42	−41	38	63
M3	31	51	97	−44	−135

=

	D1	D2	D3	D4	D5
M1	80	88	3	30	60
M2	86	66	29	72	60
M3	−42	−19	−11	−187	−315

Panel 3: Translations

- Method 3 gives lower results than Methods 1 and 2 for either alloy.
- Alloy 2 gives considerably higher results than does Alloy 1 for Methods 1 and 2.
- Methods 1 and 2 do better than Method 3 for each of the dentists separately.
- This difference (Methods 1 and 2 above 3) is enhanced for Dentist 4, and especially for Dentist 5.

[a]Entries come from Table 11-13.

where Panel 3 offers translations into words of the appearances in these additionally downswept boxes. The statements above, supplemented by those at the foot of Table 11-15, tell us most things of an overall quantitative nature that these data seem likely to reveal.

We can arrive at Table 11-15 in another way. We look now at the cell mean (raw data) MINUS grand mean. For the M1A1 entry in the M × A box at the right of Panel 1 in Table 11-15, for example, we have

$$751.13 - 741.78 = 9.36.$$

Similarly, for the M1D1 entry of the M × D box in Panel 2

$$821.83 - 741.78 = 80.05.$$

Parallel calculations for the other entries complete the boxes in Table 11-15.

11L. DISPLAYING A TWO-WAY PACKET

In Table 11-13 we did not display the three-way packet M × T × D nor the residual, because we had not yet faced up to the display of multiway boxes. Because this seems likely to call for some form of condensation, it is natural to begin by trying to condense one of the two-way boxes that we have already displayed in full. Because M × D,D has the larger MS of the second-order terms in Table 11-12, we start with it.

Table 11-16. The Condensation Process for Three- or More-Way Boxes Applied to a Two-Way Box for Illustration[a]

	D1	D2	D3	D4	D5
M1	80	88	3	30	60
M2	86	66	29	72	60
M3	−42	−19	−11	−187	−315

Stem-and-leaf

8	860
7	2
6	600
5	
4	
3	0
2	9
1	
0	3
−0	
−1	19
−2	
−3	
−4	2
LO	−187, −315

Median = 30
Fourths = −15 and 69
Cutoffs at −141 and 195

[a]From Table 11-15.

What might we want to detect in a large, possibly complicated overlay? Perhaps some of the following:

Values that are very extreme (by comparison with most entries);

Concentration of extreme values in a subbox;

Other forms of size–structure.

A good start on the first and last item would be to make a stem-and-leaf display (Section 3E) of the overlay and look for outliers and other patterns. Table 11-16 gives the swept-down table, which we could represent symbolically as the sum of overlays M + D + M × D, and also its stem-and-leaf.

The upper cutoff has no points near or outside it, and the outliers are −187 and −315. This assigns M3D4 and M3D5 as outliers. We have therefore decided that Dentists 4 and 5 using Method 3 got unusually low numbers.

A second feature of the stem-and-leaf worth noting is the clumping of observations at the upper end of the scale, almost as if there is a ceiling on the alloy hardness averages. Stem-and-leafs of the original data would give more information about such a conjecture.

When we are experimenting, with the aim of reaching as high a value as we can find, we may get this kind of picture because the maximum is nearly flat. Sometimes, however, we must choose among several isolated maxima.

11M. TAKING THE EXAMPLE FURTHER BY USING WHAT WE HAVE ALREADY LEARNED

In the dental-gold example, the goal is to achieve high values of hardness. We have seen three notable points: (1) that Methods 1 and 2 produce high values, (2) that Alloy 2 produces high values, and (3) that we have two low outliers in the downswept method-by-dentist overlay.

Several lines of attack suggest themselves:

1. Set Method 3 aside and analyze anew the remaining data (60 observations).
2. Set both Method 3 and Alloy 1 aside and analyze the remaining data (30 observations).
3. Do something about the extreme observations we have observed and reanalyze.

Both approaches 1 and 2 would already do something drastic about the extreme values. Both approaches will surely be instructive. Therefore we carry out approach 1 in the text and leave approach 2 to the exercises.

Starting over with the 60 observations omitting Method 3, Table 11-17 gives a new table of mean squares and their logarithms to the base 2.

Table 11-17. Dental-Gold Example: Two Alloys, Three Temperatures, Five Dentists, Methods 1 and 2 Only (60 Observations)

Parcel	MS	$\log_2$ MS	d.f.
Zeroth-order			
Common	Very large		1
First-order			
A	156,366	17.25	1
T	10,611	13.37	2
D	8,474	13.05	4
M	1,633	10.67	1
Second-order			
A × D	6,337	12.63	4
T × D	6,263	12.61	8
A × M	4,117	12.01	1
A × T	2,761	11.43	2
T × M	1,918	10.91	2
D × M	1,819	10.83	4
Third-order			
T × D × M	8,260	13.01	8
A × T × D	7,285	12.83	8
A × D × M	6,510	12.67	4
A × T × M	4,993	12.29	2
Fourth-order			
A × T × D × M	1,846	10.85	8

Table 11-18. Final Table Resulting from Downsweeping the Dental-Gold Example: Two Alloys, Three Temperatures, Five Dentists, Methods 1 and 2 Only (60 Observations)

Parcel	MS	$\log_2$ MS	d.f.
Zeroth-order			
Common	Very large		1
First-order			
A	156,366	17.25	1
Third-order			
T × D × M; T × D; T; T × M; D × M	6,188	12.60	24
A × T × D; A × D; D; A × T	6,836	12.74	18
A × D × M; A × M; M	5,298	12.37	6
A × T × M	4,993	12.29	2
Fourth-order			
A × T × D × M	1,846	10.85	8

None of the third-order terms will combine with the fourth-order term.

Table 11-18 exhibits the final results of the downsweeping (we omit the intermediate details).

We see that dentist, temperature, and method have been swept down, as have all the second-order interactions. Further sweeping is not called for because these effects and interactions are distinguishable from the residual.

We therefore find that in the subgroup of Methods 1 and 2, alloy still shows a difference, but everything else is in triple interactions or in the residual.

11N. REASONABLENESS OF FIXED-FACTOR ANALYSES

Why should we want to confine our attention to such a narrow target as **b-all**—or a specific instance of **b-some**? All the targets discussed in this chapter lead us, for better or worse, to a "random factor" state of analysis, which clearly has much greater flexibility. Fixed factors would have faded away, long ago, unless their use offered some apparent advantage.

The apparent advantage is simple. Treating B as a fixed factor, rather than a random one, tends to produce a smaller standard error for the estimated effects of changing the first factor (and may considerably increase the number of degrees of freedom associated with this standard error). (The fixed-factor error term is lower in the ANOVA table and thus, usually, has a smaller mean square.) This apparent advantage is illusory when we do not want **b-all** as our exclusive target and therefore do not deserve the smaller standard error.

Arguments for b-all

It proves to be hard to identify cases where we can be sure that the narrow focus corresponding to **b-all** is appropriate—particularly as a basis for standard errors. With discrete versions, the most plausible instances seem to involve factors that can have only two versions—such as sex (= male or female). Even in such cases it is often far from clear why equal weights are important: shouldn't we reflect the needs we are trying to fill more closely with some carefully chosen form of **b-unequal**, which implies the use of the corresponding re-definition of main effect? If we are working with nurses, for example, is there currently any situation where 50–50 male and female is appropriate?

Undoubtedly in a few cases discrete versions deserve to be weighted—once and for all—in a specific, well-chosen way, but not nearly as many as one might at first suppose. Therefore, when we use them without justification, we usually have to regard the analysis as an example with an optimistic view of the standard errors.

REFERENCES

Anscombe, F. J. (1967). "Topics in the investigation of linear relations fitted by the method of least squares," *Journal of the Royal Statistical Society, Series B*, 29, 1–52 (with discussion). See particularly Tukey's comments (pp. 47–48).

Paull, A. E. (1950). "On a preliminary test for pooling mean squares in the analysis of variance," *Annals of Mathematical Statistics*, 21, 539–556.

EXERCISES

1. In the light of the approach presented by the Rule of 2, how would you decide in Section 11A whether $y_{1\cdot} - y_{2\cdot}$ is mostly noise?

2. In a two-factor investigation, we plan to examine differences in outcomes for two levels of factor B. The 10 possible levels of B occur about equally often. Is it therefore a random factor? How do you know?

3. (Continuation) If only one of the 10 possible levels of B is going to occur, how does that affect the means for the levels of factor A?

4. Give an example from a field familiar to you, such as transportation, involving two factors, at least one of which is an illustrative factor rather than random or fixed.

5. If you change a two-factor situation from the type we call **b-all** to type **b-some**, what is the consequence for the main effects for factor A?

6. The text uses a rough argument based on ranges to conclude that the effects of A in Panel $T(A)$ of Table 11-1 are nearly all noise. Use an argument based on standard deviations (or variances) to make this same point.

7. (Continuation) About how big would the A effects need to be in order to be regarded as real?

8. In an investigation related to Panel $T(A)$ of Table 11-1, only B_4 is going to occur. How does this affect our attitude toward the sizes of the A effects?

9. In Panel $S(A)$ of Table 11-1, the levels B_2 and B_3 will occur equally often, and the levels B_1 and B_4 will not occur. Recompute the A main effects.

10. (Continuation) In Exercise 9, how will C be affected?

11. The common term is rarely swept down. Why not?

12. In Table 11-2, Replication $S(R)$ (and Table 11-1, Panel $S(A)$), check whether the A effects should be swept down into the AB interaction.

13. (Continuation) After the decision in Exercise 12 has been made, check whether the AB interaction (plus any pooling) should be swept down into replication.

14. Work through the steps of Table 11-3, and check that the upper left-hand corner cell in the final table has the correct entries. Then compute the mean square for the final table.

15. In Table 11-6 check that the last entry in each of the four kinds of breakdown has been correctly placed. Show your work.

16. In Table 11-7, verify Panel 2. Show your work.

17. In downsweeping as in Tables 11-9, 11-10, 11-11, and 11-12, what is the benefit of the entry $\log_2$(MS)? Find $\log_2(10)$. How do you use your calculator to compute $\log_2(x)$?

18. In sweeping down, as from Table 11-10 to 11-11, what happens to the degrees of freedom of the swept-down line?

19. In sweeping down from one order to the next, how do you combine the mean squares?

20. If MS(for A) with d.f. f_A and MS(for AB) with d.f. f_{AB} are combined using the Rule of 2, what is an upper bound on the MS the combination can produce, in terms of MS(for AB)?

21. Explain why the tables of effects given in Table 11-13 are the only ones considered. What do the dashed lines imply?

22. Considering our various discussions of fixed and random effects, what is your view about the decision to treat all the effects as random? (That is, consider each factor in the dental-gold analysis and say why that factor might reasonably be taken as random (or why not).)

Introduction to Exercises 23–26. Section 11M looks for high values of hardness by carrying out approach 1. Instead, use approach 2 to the same end with the aid of Tables 11-19, 11-20, and 11-21 as follows.

Table 11-19. Downsweeping the Table of Mean Squares of the Dental-Gold Example: Three Temperatures, Five Dentists, Methods 1 and 2, Alloy 2 (30 Observations)

Parcel	MS	$\log_2$ MS	d.f.	New Downsweeping
Zeroth-order				
Common	Very large		1	
First-order				
T	11,626	13.51	2	
D	10,368	13.34	4	
M	5,468	12.42	1	
Second-order				
T × D	10,168	13.31	8	T,D
D × M	5,105	12.32	4	M
T × M	1,899	10.89	2	
Third-order				
T × D × M	5,900	12.53	8	

Table 11-20. Table Resulting from Downsweeping First-Order Terms into Second-Order Terms for Table 11-19 (30 Observations)

Parcel	MS	$\log_2$ MS	d.f.	New Downsweeping
Zeroth-order				
Common	Very large		1	
Second-order				
T × D, T, D	10,433	13.35	14	
D × M, M	5,178	12.34	5	
T × M	1,899	10.89	2	
Third-order				
T × D × M	5,900	12.53	8	T × D, T, D; D × M, M; T × M

Note that all second-order terms are swept into the third-order term.

Table 11-21. Final Table After Downsweeping the Table of Mean Squares for 30 Observations—A2 Only, Excluding M3—from Table 11-20

Parcel	MS	$\log_2$ MS	d.f.
Common	Very large		1
T × D × M + 6 terms	7688	12.91	29

Note that everything except the common term has been swept into the third-order term.

23. Check the proposed steps of downsweeping in Table 11-19.

24. Compute the T × D,T,D mean square line in Table 11-20. Show your work.

25. Check the proposed downsweeping indicated in Table 11-20.

26. Explain the meaning of Table 11-21 from the point of view of obtaining hard fillings.

CHAPTER 12

Qualitative and Quantitative Confidence

John W. Tukey
Princeton University

David C. Hoaglin
Harvard University

By applying the techniques from the preceding chapters, especially value splitting and downsweeping, we are able to advance from a set of data to a variety of overlays. We usually summarize them numerically, display them graphically, and describe any apparent patterns. In this chapter we convert some of the possible patterns into statements that reflect confidence, in the technical probability-based sense of that word. Thus we are concerned with formal statistical inference from data to underlying reality. We may choose to state our confidence quantitatively or qualitatively, but we shall stick to relatively easily interpretable statements. Our quantitative statements give the sizes of differences (or ranges) of response; we use the term "interval statement" or the familiar "confidence interval." Qualitative statements indicate the direction of differences in response; we may use the term "directional statement" or "confident direction."

Familiar examples of formal inferences include the use of Student's t—for example, in a two-treatment experiment—to set confidence limits on the difference between the underlying (population) mean for one treatment and the underlying mean for the other. The meaning of "underlying" depends on the specific situation. For example, the interpretation cannot extend beyond the circumstances sampled, represented, or illustrated in the data before us. Because we almost always need to go beyond the circumstances actually illustrated, we almost always need to weaken further the conclusions we reach by formal inference. Methods for doing this are difficult to formulate, however, and we do not try to discuss them here. But all analysts of data are still responsible for such further weakening.

In this chapter we focus on inferences for the entries in the most common types of overlays: main effects, two-factor interactions, and nested effects. Conveniently, all three of these types arise in a single set of data that we describe in Section 12A. Subsequent sections (12B through 12E) discuss comparisons that arise in analyses of those data and present inference procedures to handle them. The family of distributions known as the Studentized range provides a unifying framework for all the specific techniques.

12A. AN ELECTION EXAMPLE

Section 1C sketched an example drawn from the state-by-state presidential election results for the four Franklin D. Roosevelt elections (1932, 1936, 1940, 1944). We take up this example here in somewhat more detail. The data appear in Table 1-5. For the overlays in the value splitting, we rely on the numerical detail in the six panels of Table 1-6.

As we prepare to draw conclusions from these data, we consider several general types of question. Against the background of the variation in the data, we ask which differences between elections are large enough that we should regard them as more than noise. Similarly, we look at differences among states. Because voting often follows regional patterns, however, this example introduces further structure by placing nearby states in groupings. We then compare the voting behavior of the groupings, and we look at differences between states that belong to the same grouping. More detailed questions take into account both elections and groupings. For a chosen grouping, which differences between elections seem substantial? Or for a particular election, how do the election results in the groupings compare? Furthermore, how does the difference between a particular pair of elections compare among groupings? As the chapter progresses, we relate such questions to the appropriate overlays and discuss procedures for making confidence statements about them.

We have divided 39 of the "lower 48" states into 13 groupings of three states each. These groupings respect the 9 geographic divisions used by the U.S. Bureau of the Census to divide the states (in 1960), and, within these divisions, they favor geographic contiguity. We have given no explicit attention to voting patterns in choosing these groupings, though such attention would have produced groupings with more homogeneous voting behavior. Even when three states take up the whole of a division, we have treated the states as illustrative of what a state in that division or grouping might have been like. As a result, the standard errors we calculate correspond to "random factors." Figure 12-1 maps the groupings, marking omitted states with an X. (We might have used Michigan, Wisconsin, and Iowa as a further grouping, or even Florida, Louisiana, and Texas; but both would have breached census division boundaries, so we did not.)

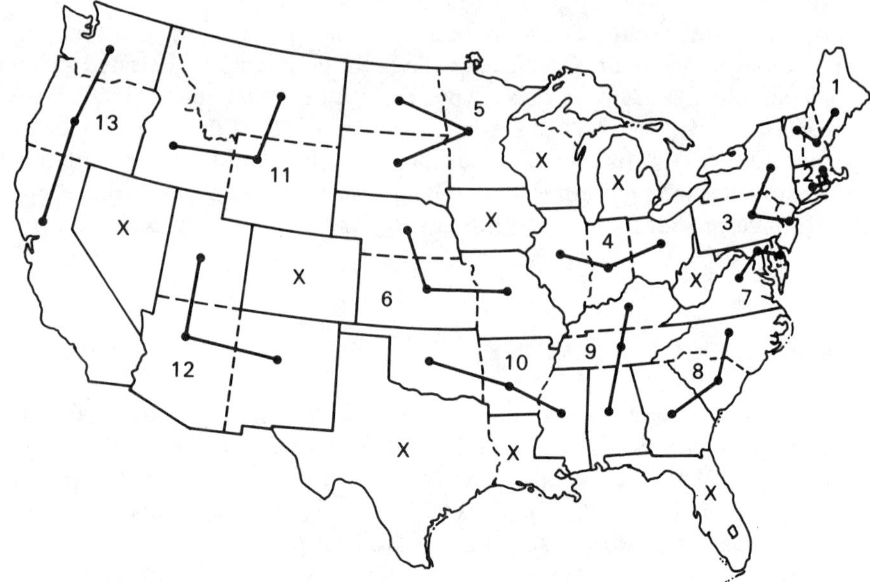

Figure 12-1. The 13 groupings, each of three states. (X marks a state that is not included in any grouping.) Line segments tie together the states of each grouping.

In the data array (Table 1-5) we have expressed the election results in the form

$$1000 \times \frac{\text{Roosevelt vote}}{\text{Roosevelt vote PLUS vote for Republican opponent}}.$$

In most states and years this measure is 10 times the percent Democratic of the major party vote. The actual wording, however, covers several of the quirks that have enlivened U.S. elections. In New York in 1936, 1940, and 1944, Roosevelt was the candidate not only of the Democratic Party but also of the American Labor Party; and in 1944 he was also the candidate of the Liberal Party. In those same three elections the Republican Party in a few states presented two or even three slates of electors. We focus on how the voters divided, rather than on technical questions (which could affect who became a state's electors).

These election data involve three factors: election, grouping, and state. Election and state are crossed, and state is nested within grouping. Thus, in addition to the common term and the residuals, the value splitting (Table 1-6) has four (reduced) overlays:

Main effects for election (E for short);
Main effects for grouping (G);
Two-factor interaction effects for election by grouping (EG or GE);
Effects for state within grouping (S(in G)).

In each of these overlays we may want to make some inferences. We may study differences among elections or differences among groupings. The two-factor interactions may shed light on three main features: differences among elections specific to a particular grouping, differences among groupings specific to a particular election, and comparison of a difference between elections in one grouping to the corresponding difference in another grouping (a double difference or bicomparison). The sections that follow take up these objectives in turn. We describe a procedure for handling each type of comparison.

In preparation for those further analyses, Table 12-1 presents the degrees of freedom, mean squares, and related quantities for the overlays in Table 1-6. The mean squares decrease rather rapidly (from top to bottom), and no downsweeping is required. The error terms for E, S(in G), and EG involve only a single mean square lower in the table. As an error term for groupings we look for a line whose expected mean square is $\sigma^2 + 3\sigma^2_{EG} + 4\sigma^2_{S(G)}$. No line in the table fits this description, but the line for EG has $\sigma^2 + 3\sigma^2_{EG}$, and that for S(in G) has $\sigma^2 + 4\sigma^2_{S(G)}$. Thus, as explained in Section 10E, we form the desired error term from these two mean squares and the residual mean square, according to

$$(\text{MS for S(in G)}) + (\text{MS for EG}) - (\text{MS for residual}).$$

Section 10E also discusses the calculation of d.f. for such a composite error term.

Table 12-1. Mean Squares and Related Quantities for the Election Example[a]

n	d.f.	Overlay	MS	DIV	Expected MS—All Factors Random
1	1	Common	58,001,249	156	
4	3	Elections	60,940	39	$\sigma^2 + 3\sigma^2_{EG} + 39\sigma^2_E$
13	12	Groupings	145,442	12	$\sigma^2 + 3\sigma^2_{EG} + 4\sigma^2_{S(G)} + 12\sigma^2_G$
39	26	States (in G)	23,533	4	$\sigma^2 + 4\sigma^2_{S(G)}$
52	36	EG	4,212	3	$\sigma^2 + 3\sigma^2_{EG}$
156	78	Residual	747	1	σ^2

Overlay	Error Term (ET)	d.f. for Error Term	SE[b]
E	4,212	36	10.39
G	26,998	33.4	47.43
S(in G)	747	78	13.66
EG	747	78	15.78

[a] n is the number of entries in the reduced overlay. An overlay's DIV equals the number of observations for each combination of versions of the factors that define the overlay (page 272).
[b] For each overlay, $\text{SE} = \sqrt{\text{ET}/\text{DIV}}$.

12B. COMPARING MAIN EFFECTS

The Roosevelt election example involves two sets of main effects: elections and groupings. From them we hope to learn how much (averaging over all the groupings) Roosevelt's share of the vote changed from one election to another and also how much stronger (averaging over the elections) his support was in one grouping than in another. Formally, we may compare each pair of elections, and we may compare each pair of groupings.

Among inferences that focus on the entries in an overlay, the most common involve comparisons (i.e., differences) of main effects. When the factor has two versions, as in a one-way layout with two groups, we ordinarily use Student's t as the basis for inferences about the difference between the average response for the two versions. For a difference, Student's t takes the form

$$\frac{\text{observed difference} - \text{contemplated difference}}{\sqrt{\text{estimated variance of observed difference}}}. \tag{1}$$

Often the contemplated difference is zero. If the appropriate (two-sided) critical value of Student's t is, say, 2.2 for an announced error rate of 5%, then we assert that the magnitude of this ratio is ≤ 2.2 with the announced error rate. Simple algebra converts this assertion into

$$\text{obs'd difference} - 2.2\sqrt{\text{est'd var}} \leq \text{contemplated difference}$$
$$\leq \text{obs'd difference} + 2.2\sqrt{\text{est'd var}}\,,$$

a statement pointing out an interval of contemplated difference with 95% confidence of covering the "true" or underlying difference. If we write "underlying difference" in the middle of this double inequality, in the place of "contemplated difference," we then make a *quantitative confidence statement*—asserting a *confidence interval*—with an announced error rate of 5%. If we look for instances where both limits have the same sign (i.e., the left-hand side is ≥ 0 or the right-hand side is ≤ 0), and assert that the "underlying difference" has the corresponding sign, we make a *directional confidence statement*—asserting a *confident direction*.

Sometimes confident directions will meet our needs. It will be enough to know that one version of a factor produces better results than another version, or the opposite, or that we do not yet know enough to say which produces better results. When the factor has three of more versions, these outcomes may combine in various ways. For example, a specific set of data and a chosen confidence level could easily lead to (among others) the following three statements:

We do not have enough data to distinguish the average results for version 2 from the average results for version 1.

We do not have enough data to distinguish the average results for version 3 from the average results for version 2.

We are confident that the average results for version 3 are better than those for version 1.

For example, average results of 467 for version 1, 584 for version 2, and 796 for version 3 would lead to the statements above if a difference had to exceed 238 to be significant. Versions 1 and 3 would differ by 329; but versions 1 and 2 would differ by only 117, and versions 2 and 3, by 212. Because we deal with data that involve noise, such situations happen all the time. We accept them as a consequence of trying to make directional confidence statements about three or more versions.

Two reasons for confidence intervals, rather than mere confident directions, deserve special attention:

1. A theory predicts what the difference between our two results should be, on average, and we want to check the theory quantitatively.
2. We have two or more studies reporting on the same comparison, and we want to judge the compatibility of their results.

The second reason is probably more important. Figure 12-2 illustrates two apparently paradoxical situations by representing a confidence interval as a line segment with an arrow at each end. In the first situation, Study A reports a key comparison as confidently positive, Study B reports the same comparison as without confident direction (the difference is not significant), and Studies A and B are entirely compatible. In the second, almost opposite, situation Study E and Study F both report the key comparison as confidently positive, but Studies E and F are *not* compatible.

Because science depends on the results of repeated study, questions of compatibility among studies that report on the same comparison often take on considerable importance. Figure 12-2 shows how confidence intervals respond to such questions and, by implication, why confident directions cannot be enough.

Rates of Error

In summarizing the familiar Student's t procedure, we referred to "an announced error rate." Any probability-based way of drawing conclusions involves an accepted risk of error. Here the risk is that the confidence

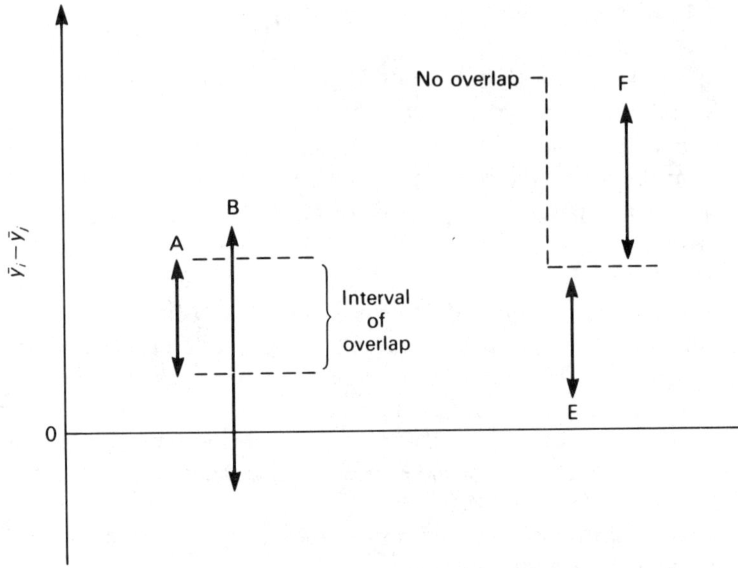

Figure 12-2. Example of compatibility or incompatibility of pairs of studies whose verbal descriptions might seem paradoxical. Each double-ended arrow represents a confidence interval for a true difference. Study A produced a significant difference, Study B did not, and yet A and B are compatible, because their confidence intervals overlap. Studies E and F both produced significant differences in the *same* direction, but E and F are still not compatible, because their confidence intervals do not overlap.

interval will fail to cover the underlying difference, and we express that risk numerically as an error rate. A formal procedure involves an announced error rate.

When the factor has more than two versions, the announced risk of error can be of different kinds. If, for example, we look at 190 differences—whether from 190 different two-treatment experiments or from one experiment with 20 treatments—we may announce an *individual* error rate of 5% for each difference separately, or we may announce a *simultaneous* error rate of 5% for all differences simultaneously. (The latter statement means that, in repetitions of the whole data collection, only one repetition in 20 will involve one or more incorrect statements.)

Collectively, we are used to making statements at different error rates—perhaps 5% and 1% (a ratio of 5) or even 5% and 0.1% (a ratio of 50). In our example of 190 differences, the ratio from individual to simultaneous is somewhat less than 190. Different bases for announcing error rates can matter more than any difference among customary error rates.

In this chapter we emphasize the consequences of choosing a *kind* of error rate by sticking, almost all the time, to 5%. (Those who wish to use, say, 1% throughout can modify all the numbers accordingly.) We discuss different kinds of 5%.

In practice, people's attitudes about kinds of error rates differ. Some are comfortable with using an individual error rate of 5% for each comparison (particularly when the comparisons were planned before seeing the data) and checking whether the number of significant comparisons is more than might reasonably occur by chance. Others prefer to identify a collection of comparisons (such as all pairwise comparisons among the grouping effects in the election example) and adopt a simultaneous error rate of 5% for that collection. Either way, others who read about the results may have different standards for strength of evidence. Thus it is important to announce the choice of error rate and the number of comparisons and to provide enough information that others can make the calculations for alternative choices. (To take a simple example of reporting, a statement that a particular comparison is significant at the 5% level gives no information about significance at levels lower than 5%, whereas a P value, such as $P = .0081$, provides freedom of choice.)

We now turn from this general discussion to consider how we can extend the use of Student's t from one difference to several differences. We describe two approaches: application of Bonferroni's inequality (introduced in Section 8E) and the Studentized range. For the Roosevelt election example, these will put us in position to compare election effects and grouping effects.

Bonferroni *t*

If we use a t test at an error rate of $100p\%$ on each of n differences, the average number of errors we make (per set of n differences) is np, because the average number of errors we make for only one difference is p. If we use an error rate of $100(p/n)\%$ for each difference, the average number of errors will be p (per *set* of n differences). If we make any error, we must make at least one error. Thus, since

$$\begin{aligned}&(\text{chance of any error or errors})\\&\qquad\times(\text{average number of errors, if there is an error}) = p\end{aligned}$$

and the second parenthesized quantity is at least 1, we must have

$$\text{chance of any error or errors} \leq p. \tag{2}$$

This approach applies as long as we can replace $100p\%$ critical values by $100(p/n)\%$ critical values. The only special feature of t is the easy availability of diverse critical values.

When the chance of error in inequality (2) is strictly less than p, the procedure is *conservative*. That is, it takes less risk of error than the nominal error rate announces.

If we are looking at 190 differences, as we supposed above, then we can do individual t tests at $100(p/190)\%$, which for $100p\% = 5\%$ will be 0.0263%.

The tables in the appendix make it easy to find unusual one-sided critical values of Student's t by interpolation. For a one-sided tail area of 0.0132% they yield 5.50 for 10 d.f., 4.42 for 20 d.f., and 3.65 for ∞ d.f. These apply to any set of 190 differences, no matter how interrelated. This Bonferroni t procedure is conservative, from the point of view underlying the Studentized range, but the extent of the conservatism depends on the nature of the interrelationship. Once we have the necessary critical values, we can use the Bonferroni t procedure for any number of differences (or to control the simultaneous error rate for any number of t tests).

Studentized Range

In the special case of all differences of two out of k values (such as the 4 elections and the 13 groupings), we can construct a critical ratio slightly different from that in expression (1), leading to a different set of tables. This approach, the *Studentized range*, controls the simultaneous error rate by working with the range of the differences. From the k values $z_1, z_2, \ldots, z_k$ we can form $k(k-1)/2$ pairs, disregarding order (i.e., z_2 and z_1 are the same pair as z_1 and z_2). When we work with pairwise differences, however, we often find it convenient to include all $k(k-1)$ differences, counting $z_2 - z_1$ and $z_1 - z_2$ separately. For any i and j with $j \neq i$, one of the two differences $z_i - z_j$ and $z_j - z_i$ is positive (or both are zero). Thus, when we base our inferences on the range of the differences, we may approach the task of constructing a set of simultaneous confidence intervals by working with the positive differences, determining a critical value for the largest of these, and then applying it with the opposite sign to the negative differences.

If we are looking at $z_i - z_j$ for all i and j with $1 \leq i, j \leq k$ and $j \neq i$, we can look at the largest value among the $k(k-1)/2$ positive differences, namely,

$$\max_{i,j}\{z_i - z_j\} = \max_i\{z_i\} - \min_j\{z_j\} = \text{range}\{z\}.$$

Then

$$-\text{range}\{z\} \leq z_i - z_j \leq \text{range}\{z\}$$

simultaneously for all i and j, both $\leq k$. To apply this derivation to main effects, we recall from expression (1) and the accompanying discussion that Student's t works with

$$\text{observed difference} - \text{underlying difference}.$$

We denote the observed mean for version i of the factor by $\bar{y}_i$ and the corresponding underlying true mean by μ_i. Then we substitute $\bar{y}_i - \mu_i$ for z_i

and $\bar{y}_j - \mu_j$ for z_j and note that

$$(\bar{y}_i - \mu_i) - (\bar{y}_j - \mu_j) = (\bar{y}_i - \bar{y}_j) - (\mu_i - \mu_j).$$

That is, each difference between the observed $\bar{y}_i - \bar{y}_j$ and the underlying $\mu_i - \mu_j$ must lie between $-\text{range}\{\bar{y}_i - \mu_i\}$ and $\text{range}\{\bar{y}_i - \mu_i\}$. (The difference $\bar{y}_i - \bar{y}_j$ is the same as the difference between the corresponding main effects, because the common value subtracts out.)

Theoretically, the Studentized range distribution arises from taking the range of k mutually independent standard Gaussian variables Z_i and dividing by the square root of V/ν, where V is distributed as chi-squared on ν degrees of freedom and is independent of all of $Z_1, \ldots, Z_k$. If we denote the Studentized range variable by Q, we have

$$Q = \frac{\text{range}\{Z_1, \ldots, Z_k\}}{\sqrt{V/\nu}}. \tag{3}$$

This way of obtaining the Studentized range distribution parallels the definition of Student's t distribution. Here the numerator is a range of k instead of a single standard Gaussian variable, but the denominator is the same.

The distribution of Q has two parameters: k, the number of variables whose range forms the numerator, and ν, the number of degrees of freedom associated with the denominator. For the critical value or percentage point at upper tail area α of the Studentized range distribution, we use the notation $q_{k,\nu}^{(\alpha)}$; that is, $\Pr\{Q \le q_{k,\nu}^{(\alpha)}\} = 1 - \alpha$.

In practice, the numerator of Q comes from variables that have an unknown common variance, and the denominator is intended to estimate the square root of that variance, as in Student's t. Thus, if $\text{SE}(\bar{y})$ is the square root of an estimate (with ν degrees of freedom) of the variance of each $\bar{y}_i$ (about μ_i), we can make the statement

$$\text{range}\{\bar{y}_i - \mu_i\} \le q_{k,\nu}^{(\alpha)}\text{SE}(\bar{y})$$

and rearrange it to get

$$-q_{k,\nu}^{(\alpha)}\text{SE}(\bar{y}) + (\bar{y}_i - \bar{y}_j) \le \mu_i - \mu_j \le q_{k,\nu}^{(\alpha)}\text{SE}(\bar{y}) + (\bar{y}_i - \bar{y}_j) \tag{4}$$

simultaneously for all i and j, both $\le k$. That is, allowing $\pm q_{k,\nu}^{(\alpha)}\text{SE}(\bar{y})$ about each $\bar{y}_i - \bar{y}_j$ gives simultaneous confidence intervals for all the $\mu_i - \mu_j$.

One detail requires care: $q_{k,\nu}^{(\alpha)}$ is defined to multiply the square root of the variance of a single $\bar{y}_i$, which is one-half the variance of a difference $\bar{y}_i - \bar{y}_j$. Thus the natural correspondence between Student's t and the Studentized range is between $t\sqrt{2}$ and Q (and *not* between t and Q). Confidence

Table 12-2. Bonferroni Factors that Correspond to Use of the Studentized Range for 95% Simultaneous Confidence Intervals on All 190 Pairwise Differences Among 20 Means ($k = 20$)

Degrees of Freedom[a] (ν)	Studentized Range (5%) (q)	Corresponding t value ($q/\sqrt{2}$)	Tail Area for $\lvert t \rvert$	Bonferroni Factor, (.05)/(Tail Area)
10	6.47	4.57	.00101	49
20	5.71	4.04	.00064	77
∞	5.01	3.54	.00040	125

[a]The denominator degrees of freedom (ν) takes on three illustrative values.

intervals based on Bonferroni t would take this $\sqrt{2}$ into account, because their half-length would involve $\text{SE}(\bar{y}_i - \bar{y}_j)$, which is $\sqrt{2}\,\text{SE}(\bar{y}_i)$.

For the example of 20 treatment means with their 190 differences, a table of percentage points of the Studentized range, also provided in the appendix, gives 5% values of 6.47 for 10 d.f., 5.71 for 20 d.f., and 5.01 for ∞ d.f. We use the three values of ν (10, 20, and ∞) for illustration. We compare these percentage points (and hence the lengths of the confidence intervals) to the corresponding percentage points for Bonferroni t, which takes no advantage of the mutual independence of the $\bar{y}_i$. To do this, Table 12-2 shows the three values of $q_{20,\nu}^{(.05)}/\sqrt{2}$. Their ratios to the corresponding values for Bonferroni t (not shown) are, respectively, .83, .91, and .97. The extent of conservatism of Bonferroni, which decreases as more degrees of freedom are available for $\text{SE}(\bar{y})$, is substantial.

We may also use the tables in the appendix to go from the corresponding t values to one-sided tail areas. The last two columns of Table 12-2 double these to get two-sided tail areas and then divide them into .05 to get the Bonferroni factor. For these three choices of ν, then, using the Studentized range (for 190 differences) corresponds to using Bonferroni with factors of 49, 77, and 125, respectively, in place of 190. Thus, although we may use Bonferroni's inequality in some situations, we usually base our inferences on the Studentized range, so as to take advantage of the shorter confidence intervals. We prefer to avoid conservatism where we can.

Main Effects for Election

We now use the Studentized range procedure to compare the election main effects. From Tables 12-1 we get the standard error for each main effect, $\text{SE} = 10.39$ with 36 degrees of freedom. Thus $\nu = 36$, and from the number of elections we have $k = 4$. Interpolation on ν in the table for the Studentized range gives $q_{4,36}^{(.05)} = 3.81$; and hence our confidence intervals, paralleling equation (4), use $q_{4,36}^{(.05)}\text{SE} = (3.81)(10.39) = 39.6$.

Table 12-3. Analysis of the Election Main Effects in the Election Example

(a) Ordered Effects

Effect	Election
43	'36
21	'32
−20	'40
−44	'44

(b) Confidence Intervals ($q^{(.05)}$SE = 39.6)

Comparison	Simultaneous 95% Interval
'32 minus '36	−62 to 17
'32 minus '40	2 to 81
'32 minus '44	25 to 105
'36 minus '40	24 to 104
'36 minus '44	48 to 127
'40 minus '44	−16 to 63

Table 12-3 presents the election main effects (in decreasing order) and uses the 5% simultaneous allowance $q^{(.05)}$SE to develop the six pairwise confidence intervals. By shifting the intervals for the four confident directions to the right, the listing in Table 12-3b makes them clearly distinct from the two comparisons that lack confident direction. We shift the four intervals to the right because they have both endpoints positive. This style of listing can also handle intervals that have both endpoints negative. We just shift them to the left.

If we think in terms of an underlying share of the vote, then the data allow us to state that Roosevelt had a higher share in 1932 and 1936 than in 1940 and 1944, but we cannot say whether 1936 was truly higher than 1932 or whether 1940 was truly higher than 1944. Together, these statements cover all six pairwise differences. The main finding is that FDR had greater popularity in the first two elections than in the last two.

Main Effects for Grouping

We now compare the strength of support for FDR among the groupings, as summarized in the main effects for grouping. Table 12-4 shows the effects in decreasing order. Now $k = 13$ and (from Table 12-1) $\nu = 33.4$. Again, interpolation on ν in the table for the Studentized range gives $q^{(.05)}_{13,33.4} = 5.02$. From Table 12-1 we also get SE = 47.43 as the standard error of each main effect. Thus the 5% simultaneous allowance for pairwise confidence intervals among groupings is $q^{(.05)}_{13,33.4}\text{SE} = 238.3$.

Table 12-4. Ordered Main Effects for Grouping in the Election Example

Effect	Grouping
239	8
186	10
87	9
16	12
−3	13
−9	7
−25	11
−52	5
−64	2
−72	3
−74	4
−87	6
−142	1

With 78 pairwise comparisons, a list in the style of Table 12-3b would be too long to have adequate visual impact; we could only refer to it. Instead, we use a graphical display known as an *interference plot*, which conveys information about both confidence intervals and confident directions. Figure 12-3 presents the grouping main effects in this way. Each effect appears as a dot at the center of a bow-tie-shaped icon whose ends are half the simultaneous allowance (i.e., $\frac{1}{2}q^{(.05)}\text{SE}$) above and below the effect. To compare a pair of effects, such as those for Grouping 10 and Grouping 11, we ask whether their two icons overlap. Absence of an overlap is equivalent to having a confident direction for the difference between the two effects (at a simultaneous error rate of 5%). That is, if the two effects are $\bar{y}_i - m$ and $\bar{y}_j - m$ with $\bar{y}_i - m > \bar{y}_j - m$, we check whether

$$\bar{y}_i - m - \tfrac{1}{2}q^{(.05)}\text{SE} > \bar{y}_j - m + \tfrac{1}{2}q^{(.05)}\text{SE},$$

which is the same as

$$\bar{y}_i - \bar{y}_j - q^{(.05)}\text{SE} > 0.$$

Thus, if the two icons do not overlap, the lower confidence limit in equation (4) is greater than zero. (The intervals depicted by the bow ties are sometimes known as *partial intervals* because their combination offers confidence statements.)

For the grouping main effects, $\frac{1}{2}q^{(.05)}\text{SE} = 119$, so that Groupings 10 and 11 (for example) are represented in the plot by the partial intervals $[67, 305]$ and $[-144, 94]$. Because these partial intervals overlap $(67 < 94)$, we do not have a confident direction for the difference between Grouping 10 and Grouping 11.

Looking more closely at Figure 12-3, we see that the leftmost nine groupings have no overlap with Grouping 8 and that the leftmost five have no overlap with Grouping 10. Only these 14 pairwise comparisons give confident directions. As one summary we could say that, when we set aside Groupings 8 (the Carolinas and Georgia) and 10 (Arkansas, Mississippi, and Oklahoma), we are not confident of the order of the remaining groupings. The data allow us to be confident about only the most substantial of the apparent regional differences: Roosevelt's support was stronger in the South than elsewhere.

In a more traditional vein, some displays of main effects show each effect as a point at the center of a line segment whose half-length is one standard-

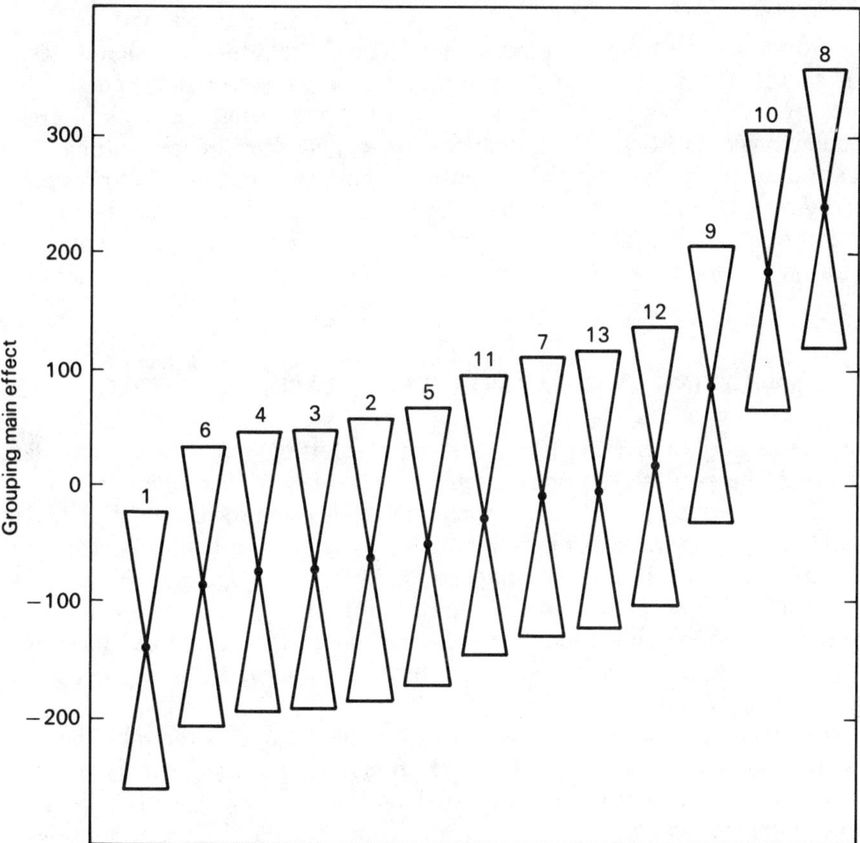

Figure 12-3. Interference plot for the 13 grouping main effects. The half-height of each bow-tie-shaped icon is $\frac{1}{2}q^{(.05)}\text{SE} = 119$. The bottom of icon 8 is above the tops of the left-hand nine icons, and the bottom of icon 10 is above the tops of the left-hand five icons. Absence of overlap implies confident direction of difference.

error. A variation on the same theme shows each effect at the center of a confidence interval. Even if such intervals provide simultaneous coverage of the underlying effects with the proper error rate, however, the failure of two of them to overlap does not exactly correspond to "these two effects are confidently different." We mention three main reasons. First, simultaneous confidence statements for all pairwise differences usually require a different allocation of the error rate (e.g., .05/10 for each pair among five effects, rather than .05/5 for each effect). Second, the standard error of a difference is $\sqrt{2}$ times the standard error of an effect. Third, for overlap or its absence to have the correct interpretation, the half-length of the interval must equal one-half the allowance for a difference. Interference plotting meets these requirements in a specialized display.

Summary

In order to move from inferences about a single difference to simultaneous inferences about all pairwise differences among a set of main effects, the present section has introduced a variety of topics, including rates of error, Bonferroni t statistics, the Studentized range, and interference plotting. This machinery has given us useful conclusions about the main effects for election and grouping, and it will allow us to deal with main effects in other analyses. We also apply most of the same ideas in making inferences about entries in two-way bordering tables.

12C. ANALYZING TWO-WAY BORDERING TABLES CONDITIONALLY

As mentioned in Section 12A, the value splitting for the election data includes the two-way bordering table for two-factor interaction effects of election by grouping. In studying such a table we often focus on the effects of one factor when the other factor remains fixed at a particular version. For example, how much did Grouping 11 differ from Grouping 12 in the 1932 election? Or how much did 1936 differ from 1944 in Grouping 8? At a further level of detail we sometimes compare two such differences, forming a "double difference," as in, "How much did the difference between Groupings 11 and 12 change from 1932 to 1936?"

When we compare the effects at two versions of one factor while holding the version of a second factor fixed, we are making a *conditional comparison*, conditional on the version of the second factor. The present section discusses procedures for conditional comparisons, and Section 12D continues with double differences, also called bicomparisons. Again we base inferences on the Studentized range.

To apply the Studentized range, we need (as before) an estimate of the common standard error of the effects that we are comparing. Thus we now approach that standard error by calculating how much the interaction effects are less variable than the unswept entries in the corresponding two-way bordering table. It suffices to focus only on the two factors involved, so we suppose that the unswept table would contain the entries y_{ij} ($i = 1, \ldots, I$; $j = 1, \ldots, J$), each with variance σ^2 and independent of the others. The observed two-factor interactions are then

$$y_{ij} - y_{i\cdot} - y_{\cdot j} + y_{\cdot\cdot}.$$

If we compare two entries in column j, say the first two, we have

$$(y_{1j} - y_{1\cdot} - y_{\cdot j} + y_{\cdot\cdot}) - (y_{2j} - y_{2\cdot} - y_{\cdot j} + y_{\cdot\cdot}) = y_{1j} - y_{1\cdot} - (y_{2j} - y_{2\cdot}). \tag{5}$$

Thus, for comparisons within column j, the Studentized range would have as its numerator the range of the I quantities $y_{1j} - y_{1\cdot}$, $y_{2j} - y_{2\cdot}, \ldots, y_{Ij} - y_{I\cdot}$. To obtain the variance of any one of these, we write $y_{1j} - y_{1\cdot}$ as a linear combination of the independent quantities y_{1k}—that is, $\Sigma c_k y_{1k}$—and apply the formula

$$\operatorname{var}\left(\sum c_k y_{1k}\right) = \sum c_k^2 \operatorname{var}(y_{1k}), \tag{6}$$

which uses the assumed independence. By substituting $y_{1\cdot} = (1/J)\Sigma y_{1k}$ into $y_{1j} - y_{1\cdot}$, we determine the coefficients c_k in the linear combination: $c_j = 1 - 1/J$, and $c_k = -1/J$ for $k \neq j$. Then equation (6) yields

$$\operatorname{var}(y_{1j} - y_{1\cdot}) = (1 - 1/J)^2\sigma^2 + (J - 1)(1/J)^2\sigma^2 = (1 - 1/J)\sigma^2,$$

so that the desired standard error for comparisons within a column is $\sqrt{1 - 1/J}$ times an estimate of σ.

Similarly, if we compare two entries in the same row, the standard error is $\sqrt{1 - 1/I}$ times an estimate of σ.

Now, to determine the simultaneous allowance for each set of conditional comparisons, we need critical values for the Studentized range. Ordinarily we explore the interaction effects by looking at all pairwise comparisons within each column *and* at all pairwise comparisons within each row. We adopt a 5% simultaneous error rate for this phase of exploration and apportion it equally between the two sets of conditional comparisons: 2.5% for columns and 2.5% for rows. Furthermore, because we are applying the Studentized range separately to each row and each column, we again use Bonferroni's inequality to divide one 2.5% equally among the columns and the other 2.5% equally among the rows.

In the election example we see from Table 12-1 that the error term for the EG bordering table is 747 with 78 degrees of freedom and that the standard error appropriate to an individual entry (y_{ij}) is 15.78. Thus, for comparisons within a column, we have

$$\text{SE}_{\text{within-col}} = \sqrt{1 - 1/4}\,(15.78) = 13.66.$$

The appropriate parameters for the Studentized range distribution are k = 13 and $\nu = 78$. By interpolation we get $q_{13,78}^{(.025/4)} = 5.79$ and then $q_{13,78}^{(.025/4)}\text{SE}_{\text{within-col}} = (5.79)(13.66) = 79.2$. For comparisons within a row we have

$$\text{SE}_{\text{within-row}} = \sqrt{1 - 1/13}\,(15.78) = 15.16,$$

$q_{4,78}^{(.025/13)} = 5.29$, and $q_{4,78}^{(.025/13)}\text{SE}_{\text{within-row}} = (5.29)(15.16) = 80.1$.

We turn now to the interaction effects in Table 1-6e and locate the comparisons that have confident direction. As a shortcut we put all 52 interaction effects in order as a batch (Table 12-5) and work downward, concentrating on the columns and rows that contain at least one comparison large enough to yield a confident direction. Table 12-6 records this search, starting with the entry 83 for Grouping 5 in 1932. For example, we compare the other entries for Grouping 5 to 83 − 80 = 3 and thus declare confident differences for 1940 (at −74) and 1944 (at −46). We compare the other

Table 12-5. Stem-and-Leaf Display of the Two-Factor Interactions for Election by Grouping from Table 1-6e

Stem	Leaf
8	3
7	
6	7
5	1
4	63
3	9854
2	66210
1	5432
0	777443
−0	114667899
−1	136778
−2	11348
−3	8
−4	02336
−5	4
−6	
−7	4

Table 12-6. Conditional Comparisons That Have Confident Direction Among the Election-by-Grouping Interaction Effects[a] in the Election Example

Entry	Label		Confident Directions[b]
83	(5, '32)	In Grouping 5:	'32 election ≫ '40, '44
		In '32 election:	Grouping 5 ≫ 1, 2, 3, 4, 7, 8, 9, 11, 12, 13
67	(6, '32)	In Grouping 6:	'32 election ≫ '40, '44
		In '32 election:	Grouping 6 ≫ 1, 2, 3, 7, 12
51	(1, '44)	In Grouping 1:	'44 election ≫ '32, '36
		In '44 election:	Grouping 1 ≫ 5, 6
46	(2, '44)	In Grouping 2:	'44 election ≫ '32
		In '44 election:	Grouping 2 ≫ 5, 6
43	(1, '40)	In Grouping 1:	'40 election ≫ '32, '36
		In '40 election:	Grouping 1 ≫ 5, 6
39	(10, '32)	In '32 election:	Grouping 10 ≫ 1, 2, 3
38	(5, '36)	In Grouping 5:	'36 election ≫ '40, '44
		In '36 election:	Grouping 5 ≫ 1
35	(11, '36)	In '36 election:	Grouping 11 ≫ 1
34	(13, '36)	In '36 election:	Grouping 13 ≫ 1
26	(7, '40)	In '40 election:	Grouping 7 ≫ 5
26	(9, '40)	In '40 election:	Grouping 9 ≫ 5
21	(8, '40)	In '40 election:	Grouping 8 ≫ 5
20	(2, '40)	In '40 election:	Grouping 2 ≫ 5
7	(3, '40)	In '40 election:	Grouping 3 ≫ 5

[a]The interaction effects appear as a two-way bordering table in Table 1-6e.
[b]Read " ≫ " as "confidently greater than."
Note: Eleven confident directions within groupings and 32 within elections.

entries for 1932 to 83 − 79 = 4 and declare confident differences for Groupings 1, 2, 3, 4, 7, 8, 9, 11, 12, and 13. The search need not consider entries smaller than −74 + 79 = 5, because they cannot be confidently different from the smallest entry, −74, according to the smaller of the two allowances. Table 12-6 does not include entries that did not lead to any confident differences. Although this table lists 14 entries and records a total of 43 comparisons with confident directions (11 within rows and 32 within columns), it still is simpler than enumerating all 390 comparisons, $4(13 \times 12/2) = 312$ within columns and $13(4 \times 3/2) = 78$ within rows.

To summarize what we have found, Table 12-7 reorganizes the comparisons with confident direction according to the grouping or election within

Table 12-7. Summary of Confident Directions Among Conditional Comparisons of Election-by-Grouping Effects[a]

(a) Within Grouping

Grouping 1: '40, '44 $\gg$ '32, '36
Grouping 2: '44 $\gg$ '32
Grouping 5: '32, '36 $\gg$ '40, '44
Grouping 6: '32 $\gg$ '40, '44

(b) Within Election

'32 election:	5, 6 $\gg$ 1, 2, 3, 7, 12
	5 $\gg$ 4, 8, 9, 11, 13
	10 $\gg$ 1, 2, 3
'36 election:	5, 11, 13 $\gg$ 1
'40 election:	1, 2, 3, 7, 8, 9 $\gg$ 5
	1 $\gg$ 6
'44 election:	1, 2 $\gg$ 5, 6

[a]Read "$\gg$" as "confidently greater than."

which they arose. A number of the patterns are easy to describe:

All election-to-election comparisons with confident direction compare an earlier election with a later one. (The outer years, 1932 and 1944, seem to differ more than the inner years, unlike the pattern of the election main effects.)

Election-to-election comparisons are strongest within Groupings 1 and 5, with Grouping 6 not far behind.

More than half of the grouping-to-grouping comparisons with confident direction (18 of 32) are in the 1932 election.

Each grouping has its interaction effect for 1932 confidently related to that for at least one other grouping.

Groupings 1 and 5 are confidently different in all four elections; Groupings 1 and 6, in three of them; and Groupings 2 and 5, in three.

On the whole, these are fairly specific descriptions of the structure in Table 1-6e. Perhaps the most notable are the greater diversity in the 1932 election, the antithesis of Grouping 1 (Maine, New Hampshire, and Vermont) and Grouping 5 (Minnesota and the Dakotas), and the focus on early election versus late election. We must keep in mind, however, that this table—and hence all that we can infer from it—involves two-factor interactions, which remain after we have swept out *both* sets of main effects. To remind ourselves about what we found in the main effects for elections and the main effects for groupings, we would look back at Table 12-3 and Figure 12-3, respectively.

If we wanted to show confidence intervals for the conditional comparisons, we would use an interference plot for the 13 interaction effects (as in Figure 12-3) within each of the four elections and a list of the six intervals (as in Table 12-3b) within each of the 13 groupings.

12D. BICOMPARISONS (DOUBLE DIFFERENCES)

We analyze interactions in two-way tables because changing one factor from one version to a second version may produce different effects on the response for different versions of the supporting factor. If this is our main concern, we can address it directly by working with double differences. We first form differences in one direction (each corresponding to a change in one factor), and then we look at the differences (corresponding to a change in the second factor) of pairs of these differences that correspond to the same change in the first factor. That is, if we now use a and b as subscripts for rows and i and j as subscripts for columns, we may first form the differences $y_{ai} - y_{bi}$ between row entries within each column. Then for each pair of rows a and b we form the difference $(y_{ai} - y_{bi}) - (y_{aj} - y_{bj})$ for each pair of columns i and j. We refer to such a difference as a *bicomparison*.

Because of the differencing along the first factor (with the second factor constant), removing anything that is affected only by the second factor has no effect on the double differences: splitting out a main effect for the second factor does not alter any bicomparison. For each i, $(y_{ai} - g_i) - (y_{bi} - g_i) = y_{ai} - y_{bi}$. The same argument applies with the factors interchanged, so it does not matter how we split up or downsweep main effects. All the two-way tables that involve a given interaction will have the same set of bicomparisons.

We now describe an analysis for bicomparisons, using the grouping-by-election (GE) interactions as an illustration. For this 13×4 table we have $78 = \binom{13}{2}$ pairs of groupings and $6 = \binom{4}{2}$ pairs of elections, so that we face $78 \times 6 = 468$ bicomparisons. Clearly we want to avoid writing all of them out. We also need to think about what allowance to use with bicomparisons, each of which has the form

$$y_{ai} - y_{bi} - y_{aj} + y_{bj}$$

and hence has variance $4\sigma^2$, as compared to $2\sigma^2$ for a single difference such as $y_{ai} - y_{bi}$.

We approach bicomparisons in the same way as we did simple comparisons in Section 12B when we introduced the Studentized range. There we divided the range of k independent Gaussian variables by the square root of an independent estimate of their common variance. The range in that numerator is equivalent to the value of the largest simple comparison. Now we instead take as the numerator the largest bicomparison. As the denominator we continue to use the square root of an independent estimate of σ^2, the variance of an individual entry (such as y_{ai}). The theoretical distribution of this new ratio resembles that of a Studentized range, but it is not quite the same. We call it the *Studentized birange*.

The Studentized birange distribution does not coincide with a Studentized range distribution because of dependence within the numerator. When we write the bicomparisons as differences of simple (conditional) comparisons and then regard the numerator as the range of those simple comparisons, those comparisons are correlated. It turns out, however, that we can approximate a Studentized birange distribution in terms of a suitably chosen Studentized range distribution. Theoretical calculations (which we do not describe here), together with some simulation results, indicate that we can use a Studentized range distribution with

$$k = 1.1\,(\text{degrees of freedom for interaction table}) + 1.$$

The other parameter, ν, comes with the independent denominator and needs no modification. Finally, we multiply this Studentized range distribution by $\sqrt{2}$ when approximating the Studentized birange, because (as mentioned above) the variance of a bicomparison is twice the variance of a simple comparison.

This approximation of the Studentized birange distribution in terms of the Studentized range distribution is conservative, because the theoretical calcu-

Table 12-8. Shortcut Approach to Identifying Double Differences That Have Confident Sign in the Grouping-by-Election Interactions

(a) The single differences within grouping (calculated from Table 1-6e)

	Pair of Elections					
Grouping	'36–'32	'40–'36	'40–'32	'44–'40	'44–'36	'44–'32
1	−14	97	84	8	105	92
2	18	44	63	25	69	88
3	57	−7	49	15	8	64
4	22	−19	3	10	−9	13
5	−45	−112	−157	28	−84	−129
6	−55	−55	−110	5	−51	−105
7	0	43	43	−19	24	24
8	−27	48	21	−13	35	8
9	−12	48	36	−22	25	13
10	−57	17	−40	−19	−3	−60
11	39	−42	−4	−15	−58	−19
12	28	−11	17	−9	−20	7
13	46	−50	−5	9	−42	4

Table 12-8. *(Continued)*

(b) The single differences ordered within column (The grouping number appears in parentheses only where the entry thus labeled differs by more than $\sqrt{2}\, q^{(.05)}_{40.6,\,78}\mathrm{SE} = 127.9$ from another entry in the same column.)

Pair of Elections

'36–'32	'40–'36	'40–'32	'44–'40	'44–'36	'44–'32
57	97 (1)	84 (1)	28	105 (1)	92 (1)
46	48 (8)	63 (2)	25	69 (2)	88 (2)
39	48 (9)	49 (3)	15	35	64 (3)
28	44 (2)	43 (7)	10	25	24 (7)
22	43 (7)	36 (9)	9	24	13 (9)
18	17 (10)	21 (8)	8	8	13 (4)
0	−7	17 (12)	5	−3	8 (8)
−12	−11	3 (4)	−9	−9	7 (12)
−14	−19	−4 (11)	−13	−20	4 (13)
−27	−42 (11)	−5 (13)	−15	−42 (13)	−19
−45	−50 (13)	−40	−19	−51 (6)	−60 (10)
−55	−55 (6)	−110 (6)	−19	−58 (11)	−105 (6)
−57	−112 (5)	−157 (5)	−22	−84 (5)	−129 (5)

(c) Pairs of groupings whose differences for a pair of elections (i.e., within a column) are confidently positive in panel (b)

Pair of Elections

'36–'32	'40–'36	'40–'32	'44–'40	'44–'36	'44–'32
	1 − 11	1 − 6		1 − 13	1 − 10
	1 − 13	1 − 5		1 − 6	1 − 6
	1 − 6	2 − 6		1 − 11	1 − 5
	1 − 5	2 − 5		1 − 5	2 − 10
	8 − 5	3 − 6		2 − 5	2 − 6
	9 − 5	3 − 5			2 − 5
	2 − 5	7 − 6			3 − 6
	7 − 5	7 − 5			3 − 5
	10 − 5	9 − 6			7 − 6
		9 − 5			7 − 5
		8 − 6			9 − 5
		8 − 5			4 − 5
		12 − 5			8 − 5
		4 − 5			12 − 5
		11 − 5			13 − 5
		13 − 5			
Count: 0	9	16	0	5	15

lations show that we could replace the constant multiplier 1.1 by a multiplier that depends on the size of the two-way table and decreases toward 1 as the table becomes large. The simplification to the constant multiplier has only a modest impact, however, because the percentage points of the Studentized range increase rather slowly as k increases (and ν stays the same).

In the election example the grouping-by-election interactions have $12 \times 3 = 36$ degrees of freedom. Thus the approximation for the Studentized birange of this two-way table uses the Studentized range with $k = 1.1(36) + 1 = 40.6$. From Table 12-1 we continue to use $\nu = 78$ and SE = 15.78. By interpolating on both k and ν in the table for the Studentized range we get $q_{40.6,\,78}^{(.05)} = 5.73$. Thus our allowance is $\sqrt{2}\,q_{40.6,\,78}^{(.05)}\mathrm{SE} = 127.9$.

The GE interactions have 468 double differences. How would we have dealt with such a number of differences in the one-way case? Surely not by calculating all of them initially. Rather, as we did for the grouping effects in Table 12-4, we would have written down the individual values in order and asked, for each value in turn, "If we start from this one, which others are greater by at least qSE, so that we are confident about the sign of the corresponding difference?" If we take each of the six pairs of elections in turn and look at the 13 corresponding single differences, we can do just the same thing, picking out all pairs of single differences (for a fixed pair of elections) that differ by more than $\sqrt{2}\,q$SE, so that we are confident about the sign of the corresponding double difference.

Table 12-8 goes through this process step by step for the GE interactions. Panel (a) shows the single differences, from one election to another, in a G by $\binom{E}{2}$ pattern (13×6). Panel (b) orders these differences within each of the $\binom{E}{2} = 6$ columns. Panel (c) lists all pairs of groupings for which the differences (both involving the same pair of elections) differ by more than $\sqrt{2}\,q_{40.6,\,78}^{(.05)}\mathrm{SE}$, and it also counts the number of such occurrences.

Table 12-9, in which all election differences continue to have the form "later minus earlier," organizes the bicomparisons of confident sign for more careful study. Panel (a) counts the occurrences of each grouping as the first element of a pair—as having an observed difference that is positive enough for us to conclude (be confident at 5% simultaneous) that the sign of the bicomparison (of the underlying values) is positive. Panel (b) does the same for occurrences as the second member of a pair. Although all the groupings except 5 and 6 appear at least once as the first member of a confident bicomparison, Groupings 1 and 2 have noticeably more total occurrences than the others. Thus the shifts in Roosevelt's share of the vote from an election to a later one were more positive in Groupings 1 and 2 than elsewhere. The summary of second members in Table 12-9b points even more strongly to Groupings 5 and 6, where the shifts from one election to another (later) election were more negative.

From Table 12-8c and Table 12-9 we see that pairing 1936 with 1932 and pairing 1944 with 1940 produced no bicomparison large enough to have

Table 12-9. Further Examination of the Groupings That Have Large Double Differences in the Grouping-by-Election Interactions

(a) Number of occurrences as the first (i.e., +) member

	Pair of Elections						
Grouping	'36–'32	'40–'36	'40–'32	'44–'40	'44–'36	'44–'32	Total
1		4	2		4	3	13
2		1	2		1	3	7
3			2			2	4
4			1			1	2
5							0
6							0
7		1	2			2	5
8		1	2			1	4
9		1	2			1	4
10		1					1
11			1				1
12			1			1	2
13			1			1	2

(b) Number of occurrences as the second (i.e., −) member

	Pair of Elections						
Grouping	'36–'32	'40–'36	'40–'32	'44–'40	'44–'36	'44–'32	Total
1							0
2							0
3							0
4							0
5		6	10		2	9	27
6		1	6		1	4	12
7							0
8							0
9							0
10						2	2
11		1			1		2
12							0
13		1			1		2

confident direction. Indeed, these two pairings show great similarity. Thus the important shifts seem to have taken place between 1936 and 1940. Perhaps this pattern, which we also saw in the election effects in Table 12-3, reflects the opening of World War II.

Two subsets of groupings have emerged from this analysis of bicomparisons. The New England states (Groupings 1 and 2) and the West North Central states (Groupings 5 and 6) changed in quite different ways from 1932 or 1936 to 1940 or 1944. A look at the fitted interactions in Table 1-6e reveals that Groupings 1 and 2 went from substantial negative interactions in 1932 and 1936 to substantial positive interactions in 1940 and 1944, whereas Groupings 5 and 6 did the opposite. This analysis of change accords well with the conditional comparisons summarized at the end of Section 12C, and it reveals a feature of the data different from the groupings' typical level of response over the four elections. The main effects (Table 12-4) for all four of these groupings are negative, with Grouping 1 lowest of all. Thus among the groupings that tended to give Roosevelt a lower share of the vote were two subsets with different patterns of change over time, as well as other groupings whose pattern attracted no particular notice.

12E. COMPARING NESTED EFFECTS

Among the types of question that we listed early in Section 12A, one still needs an inference procedure: comparisons among states. Because we have introduced the additional structure of the groupings into this example, however, and have already handled comparisons among groupings, we now compare states that belong to the same grouping. For example, in Grouping 8 we might ask whether FDR received stronger support from South Carolina than from North Carolina (on average over the four elections). In the language of Chapter 4, the states are nested within the groupings. Thus this feature of the election example allows us to illustrate techniques that we would apply more generally to similar situations involving nesting.

When one factor is nested in another, as states are nested within groupings in the election example, we can present the corresponding bordering table (which is actually a stack of one-way arrays) either as a two-way array or as a one-way nest. Table 12-10 illustrates these alternatives, using the names of states in the first three groupings. If we choose the two-way arrangement (to save space), we must remember that the inner factor (e.g., S) has no consistent meaning from one version of the outer factor (e.g., G) to another. Thus we avoid acting as if anything that would depend on the inner classification (columns in our illustration) has a meaning across versions of the outer factor. We dare not use bicomparisons, which require that both rows and columns be meaningful.

Instead, we restrict the comparisons so that they obey the nested structure. Then simple differences, within a single version of the outer factor (in our example, within a single grouping, denoted by a), take the form (for states i and j in grouping a)

$$(y_{ai} - \bar{y}_a) - (y_{aj} - \bar{y}_a) = y_{ai} - y_{aj}.$$

Thus we can use the Studentized range to analyze these differences, just as we handled differences among main effects in Section 12B. We work with a

Table 12-10. Presentation of the Bordering Table for the Effects of an Inner Factor Nested Within an Outer Factor[a]

(a) As a two-way array (Each row corresponds to a grouping, but the columns have no consistent meaning.)

Maine	New Hampshire	Vermont
Connecticut	Massachusetts	Rhode Island
New Jersey	New York	Pennsylvania

(b) As a one-way nest

Maine New Hampshire Vermont
Connecticut Massachusetts Rhode Island
New Jersey New York Pennsylvania

[a]The names of the states in the first three groupings represent the corresponding effects.

5% simultaneous error rate overall and use Bonferroni's inequality to divide it among the versions of the outer factor (the groupings).

For the effects of states nested within grouping in the election example, we thus use an error rate of (5/13)% in each of the 13 groupings. From the number of states per grouping we get $k = 3$, and from Table 12-1 we have SE = 13.66 and $\nu = 78$. Interpolation between tabulated percentage points of the Studentized range gives $q_{3,78}^{(.05/13)} = 4.69$, and hence the allowance is $q_{3,78}^{(.05/13)}\text{SE} = 64.15$. The actual effects that we wish to examine appear in Table 1-6d. Putting in numerical order the effects in Grouping 1, we have

−35	Vermont
−11	Maine
46	New Hampshire

The comparison "New Hampshire minus Vermont" is confidently positive (81 > 64.15), but neither comparison involving Maine yields a confident direction.

Table 12-11 assembles the comparisons that have confident direction within all 13 groupings. In Groupings 8, 9, and 10 the three effects spread out enough that all three comparisons yield confident direction. The condensed

Table 12-11. Comparisons Among State Effects Within Grouping That Yield Confident Direction

Grouping	Direction
1	New Hampshire $\gg$ Vermont
5	Minnesota $\gg$ South Dakota
6	Missouri $\gg$ Kansas
7	Virginia $\gg$ Maryland
	Virginia $\gg$ Delaware
8	South Carolina $\gg$ Georgia $\gg$ North Carolina
9	Alabama $\gg$ Tennessee $\gg$ Kentucky
10	Mississippi $\gg$ Arkansas $\gg$ Oklahoma
11	Montana $\gg$ Wyoming
12	Arizona $\gg$ New Mexico

notation "South Carolina $\gg$ Georgia $\gg$ North Carolina," for example, implies also that South Carolina had confidently greater support for FDR than North Carolina. Generally, a state that appears at the upper end of a comparison in the list gave a sizable share of its vote to Roosevelt, and a state at the lower end favored the Republican candidates. Georgia, Tennessee, and Arkansas (in Groupings 8, 9, and 10, respectively) occupy a middle ground. We note, however, the relative nature of these comparisons, which involve only the other states in the same grouping. (The states do have familiar meanings apart from the groupings, so we could compare the average share of the vote that Roosevelt received in any two states; but we would start from an analysis that regarded state and election as crossed in a two-way layout.) On the whole, the states within most groupings did not show homogeneous voting behavior. Groupings 8, 9, and 10 seem especially heterogeneous.

If we wanted to present confidence intervals for the three comparisons within each grouping, we would use the shifted-list format of Table 12-3b.

12F. SUMMARY

As a basis for making confidence statements about differences that arise in one-way and two-way bordering tables, we use two main tools: Bonferroni's inequality and the Studentized range. The election example offers an opportunity to apply these to one-way tables of main effects, to a two-way table of interaction effects, and to an assemblage of one-way tables that arises from a nested factor.

The analysis of the one-way tables is straightforward; we make directional or interval statements for differences between main effects.

The two-way table presents more possibilities. We may look at conditional comparisons or at bicomparisons. The conditional comparisons, which involve the effects of one factor at a given version of the other factor, form two families, according to which factor we fix. We allocate the error rate equally between the two families and then use Bonferroni's inequality again to divide it equally among the versions of the conditioning factor. These steps reduce the inferences to simple comparisons, once we adjust the allowance for the reduced variability of the conditional comparisons.

For the bicomparisons we introduced the Studentized birange, whose distribution we approximate by a suitably chosen Studentized range distribution.

When one factor is nested in another, only the main effects for the outer factor can meaningfully be swept out. In this case we work with the individual nested effects, which take the form $y_{ai} - \bar{y}_a$. Because differences between nested effects take the form $y_{ai} - y_{aj}$, we apply the Studentized range to them, separately for each version of the outer factor.

When both one-way margins of the two-way table can be swept out but only one has been (often because the other margin has been recombined with the two-way table), we can again analyze conditional comparisons or bicomparisons. The procedures for conditional comparisons depend on the details of the downsweeping in ways that lie beyond the scope of the present chapter. For the bicomparisons, however, the downsweeping can affect only the error term. Thus, although we have not illustrated the details, we would again apply the Studentized birange.

Some of these approaches can be applied to three-way and higher-way bordering tables, but we do not do so in the present chapter.

RELATED LITERATURE

Hochberg, Y. and Tamhane, A. C. (1987). *Multiple Comparison Procedures*. New York: Wiley.

McGill, R., Tukey, J. W., and Larsen, W. A. (1978). "Variations of box plots," *The American Statistician*, 32, 12–16.

Miller, R. G., Jr. (1981). *Simultaneous Statistical Inference*, 2nd ed. New York: Springer-Verlag.

Tukey, J. W. (1991). "The philosophy of multiple comparisons," *Statistical Science*, 6, 100–116.

EXERCISES

1. Determine the probability that at least one of 10 independent 95% confidence intervals will fail to cover the corresponding parameter. How does this simultaneous error rate compare with the result of applying Bonferroni's inequality? Make the same calculations for 99.5% confidence intervals.

2. An experiment produced the five treatment means 0.76, 1.05, 1.61, 2.18, and 2.21, each with standard error .00864 (on 140 d.f.). Give 95% simultaneous confidence intervals, based on the Studentized range, for the pairwise differences among the main effects.

3. For a factor with five versions, suppose the treatment means have an error term with 20 d.f. Compare the lengths of the confidence intervals for all pairwise differences that arise from the following three procedures: individual t, Bonferroni t, and Studentized range. Repeat with 10 d.f. for error and with ∞ d.f.

4. In an interference plot (such as Figure 12-3), verify that the difference between two bow ties (pairing the upper end of each with the lower end of the other) gives the confidence interval for the difference between the corresponding underlying effects.

5. Make an interference plot for the five treatment means given in Exercise 2.

6. Consider the main effects for groupings (Table 12-4).

- **(a)** Calculate the half-length for a set of confidence intervals (based on Student's t and Bonferroni's inequality) that give 95% simultaneous confidence for the 13 effects.
- **(b)** Calculate the half-length (i.e., allowance) for a similar set of confidence intervals that give 95% simultaneous confidence for all pairwise comparisons among the 13 effects.
- **(c)** Compare the two half-lengths calculated above to the value $\frac{1}{2}q^{(.05)}_{13,33.4}\text{SE}$ used in the interference plot for the grouping effects (Figure 12-3).

7. Ignore the groupings shown in Table 1-5 and thus regard the election data as a 39 × 4 table. Analyze the data in this structure and make simultaneous comparisons among the state effects.

CHAPTER 13

Introduction to Transformation

John D. Emerson
Middlebury College

Tables of measurements—particularly those involving more than two factors—often lead to rather complicated descriptions that include interaction effects. Sometimes a single operation on the data offers a way to simplify subsequent analysis and description. Transformation of the data may lead to:

A better description of the data using main effects (a simple point system);
A description with fewer interaction effects or relatively smaller interactions;
Simpler dependence of a response on one or more factors; or
Closer conformity to the conventional assumptions, such as equal variance or Gaussian error structure, that often accompany classical ANOVA.

This chapter introduces transformation of a response variable for one-way and two-way tables. Transformation aims first at promoting simplicity of description. It may also lead to fits whose residuals have more nearly constant variance and a more nearly symmetric, mound-shaped distribution. The first section introduces several transformations that are familiar from everyday experience. Section 13B outlines some goals of transformation in data analysis, including the analysis of variance.

Section 13C introduces basic mathematical concepts that arise when using transformations. These concepts include linearity, monotonicity, direction of curvature, power transformations, and other ideas that are useful in this chapter and elsewhere.

Section 13D describes and illustrates transformation in one-way ANOVA. It introduces a spread-versus-level plot for seeking a power transformation that may make variability more stable across groups. The next section moves to two-way ANOVA. Here an important goal is to get nearly additive structure—to reduce or eliminate interaction. A diagnostic plot often helps to find a power transformation that produces additive effects.

Section 13F presents and illustrates specialized transformations that are well suited to transforming proportions or percentages.

13A. TRANSFORMATION IN EVERYDAY EXPERIENCE

Transforming data (often by a logarithm or a reciprocal) lets us think about, and analyze, the data in a scale different from that of the recorded measurements. Hoaglin (1988) emphasizes that results of transformations are surprisingly commonplace in everyday experience; his examples include the Richter scale for earthquakes, gasoline consumption in miles per gallon from mileage tests on automobiles, and the system of gages for copper wire used in homes and in electrical equipment.

On sunny summer mornings (all too rare in Vermont), I enjoy starting the day with a 3-mile run. I pace myself to complete the run in exactly 24 minutes, giving an average of 8 minutes per mile. I can easily calculate my average speed by using a reciprocal transformation: 8 minutes per mile is equivalent to $1/8$ mile per minute, or $60 \times 1/8 = 7.5$ miles per hour. The transformation from the elapsed time (for a course of known length) to the average speed is so common that such data are routinely reported in the transformed scale.

Two musical tones are an octave apart if the higher note has a frequency that is twice that of the lower note. A standard piano has 88 keys, with adjacent keys one "half-step" apart. The piano tones range from a low A at a frequency of 27.5 Hz to a high C at 4186 Hz. There are $7\frac{1}{3}$ ($= \frac{88}{12}$) octaves on a piano. But whereas the difference in frequency between the two lowest As is 27.5, the difference between the two highest As is 1760. Neither octaves nor adjacent keys are equally spaced in frequency. By taking the logarithms (conveniently using base 2) of the frequencies, we transform the frequency data to a scale in which adjacent notes are equally spaced. In the log-base-2 scale the frequencies of any two notes an octave apart differ by one unit, and in an "equitempered" scale the difference between any two adjacent notes is $\frac{1}{12}$. See Hoaglin (1988) for further discussion.

Perhaps the simplest and most familiar types of transformation involve a linear change of scale, a shift of the origin, or both together. Converting distance from miles to kilometers and converting time from minutes to hours illustrate linear changes of scale. Measuring distance from the North Pole instead of measuring from the equator is a shift of the origin. Changing temperature to degrees Celsius, °C, from degrees Fahrenheit, °F, combines both:

$$°\text{C} = \tfrac{5}{9}(°\text{F} - 32).$$

As used in statistics and data analysis, transformations do not always have a ready physical interpretation, nor does an analysis of a physical situation that gives rise to data necessarily suggest a transformation that is somehow "natural" and likely to be effective in achieving a stated purpose. Often, understanding the context that produced a data set can lead to a usable and interpretable new scale for analysis. Thus, many transformations used in statistics are much like those encountered in everyday experience, but this is not always true.

Linear transformations alone cannot bring substantive improvement to the usual analyses of a data set. Converting from miles to kilometers can hardly improve an analysis of a collection of distances. However, linear transformations used before proceeding with nonlinear (though often still simple) transformations can help to bring improvement to the analysis of a data set. In one instance $y = x + c$ is followed by $z = \sqrt{y}$ to give $z = \sqrt{x + c}$.

13B. GOALS OF TRANSFORMATION

Data analysts and statisticians search for attractive properties in a data set by using transformations. Transforming a single variable may reduce skewness, thereby promoting symmetry and even Gaussian shape in the transformed data. For data in several groups at different levels, a suitably chosen transformation may help to stabilize variability—that is, make the spread more nearly equal from group to group. With y-versus-x data, transformation of x or y or both may lead to a nearly linear relationship in the new scale(s). For factorial data, transformation of the response or of the levels of one or more factors may lead to a simpler fit—a fit with smaller or fewer interactions, perhaps one that uses only main effects. When factors are measured, transformation of a factor's levels may increase the effectiveness of a linear dependence in describing the corresponding main effects.

Simplicity of description is a primary goal in analysis of variance. Transformation of nonnegative amounts or counts by powers or logarithms often gives this simplicity. Other responses, such as percentages, may benefit from other types of transformation.

In a two-way table with replication, a fit that allows for arbitrary interaction can be written

$$\hat{y}_{ij} = m + a_i + b_j + (ab)_{ij}.$$

A power transformation of the form $z_{ij} = y_{ij}^p$ (if $p = 0$, we use $z_{ij} = \log y_{ij}$) may lead to a fit having main effects only,

$$\hat{z}_{ij} = m^* + a_i^* + b_j^*,$$

which adequately describes systematic variation in the data table. (The values of the effects in the two fits will differ, as the $*$ on each term reminds us.)

If we want to think in terms of the y_{ij}, we are fitting

$$\hat{\hat{y}}_{ij} = \left(\hat{z}_{ij}\right)^{1/p} = \left(m^* + a_i^* + b_j^*\right)^{1/p}.$$

It is convenient to work with the z_{ij}, where the calculations are simpler and more conventional (two good reasons for thinking in terms of the z's also). One view of what we are doing is that we are trying to replace the $(I - 1)(J - 1)$ constants needed for a general $(ab)_{ij}$ by the single constant p.

Similarly, transformation of the response to a new scale may simplify the structure in a three-way table, from

$$m + a_i + b_j + c_k + (ab)_{ij} + (ac)_{ik} + (bc)_{jk}$$

in the original scale to, say, the form

$$m + a_i + b_j + c_k + (ab)_{ij},$$

or, most fortunately, to the form

$$m + a_i + b_j + c_k.$$

The values of the effects in the fits again differ, even though our notation does not try to suggest this.

Again, we might think of fitting

$$\hat{y}_{ijk} = (m + a_i + b_j + c_k)^{1/p}$$

to the y's. A good choice of the power p may make working with the z's simpler, cleaner, and more effective.

Although other goals of transformation in analysis of variance have secondary importance, we should not ignore them. Transformation may lead to residuals that are more nearly symmetrically distributed. It may also lead to more stable variance throughout an overlay of residuals. We recall that in classical analysis of variance the "errors" are often assumed to have independent Gaussian distributions with common variance. The assumption of independence of measurements is important; correlation within clusters or within other groupings, or serial correlation, can invalidate inferences that neglect such correlations, and addressing this issue is difficult (Miller 1986, Chap. 4). Although statistical research indicates that most departures from other assumptions do not seriously invalidate any positive findings of a classical analysis of variance unless the design is grossly unbalanced, unequal variances can make the classical analysis inefficient. Miller (1986, Sects. 4.2, 4.3, 4.6, and 4.8) discusses these issues and gives references to the research literature. Because we usually consider balanced designs in this volume, departures from the typical assumptions have only secondary importance in our selection of transformations, as compared with removing or reducing nonadditivity.

In sum, an important purpose in considering the transformation of a response, y, is to move toward a scale needing few, if any, interaction terms

—or at least much smaller interaction terms—to describe the structure of a table of transformed data. In any event, we regard *reduction of the relative size* of interactions as desirable. Ideally, we might seek a power, p, for which y^p is well described using only main effects in the fit. Because this uses only one constant in place of $(I - 1)(J - 1)$ constants for a two-factor interaction, we cannot hope for universal success. For three-way and higher-way tables, reducing the number of interaction effects needed to account for the structure of a table is almost always important progress; reducing their relative size can be worthwhile.

13C. BASIC CONCEPTS FOR TRANSFORMATION

We might choose among several statistical transformations to change the structure of a table of measurements, perhaps to summarize it with fewer overlays. What sort of transformations are most likely to be useful? We first discuss some ideas about transformations. These ideas will lead us to uses of transformations for one-way and two-way ANOVA and then for proportions, in Sections 13D through 13F.

Factorial data may require a transformation to re-express the data in a scale that simplifies a description of variability. To accomplish the change, we transform the data by applying a single mathematical function, f, to all data values. To change (and simplify) the structure of a table of measurements, it is no help to rely solely on linear transformations, which change only the origin and the units of measurement. Instead, we need transformations that can fundamentally alter the shape of a set of data. For example, $f(y_{ij}) = y_{ij}^{1/2}$ —in place of y_{ij}—may be able to bring benefits to a data table, but $f(y_{ij}) = 10y_{ij} - 1$ cannot do so.

Linear Transformations

A *linear transformation* of y is a transformation that may be expressed as $T(y) = a + by$ for some constants a and b. Linear transformations can assist us in data analysis, for example, when multiplying by a power of 10 to get rid of decimal points. They can also provide a more convenient location of the origin and/or unit of measurement.

Monotonic Transformations

A monotonic transformation of y is a function that either preserves the order of all y values or reverses their order. For example, $f(y) = \sqrt{y}$ preserves the ordering of positive measurements, and $1/y$ reverses their order. When y can be either positive or negative, $\sqrt{y}$ is not always defined. If y can assume both positive and negative values, neither $1/y$ nor y^2 is a monotonic transformation. In practice, we ordinarily transform only nonnegative data. By requiring monotonicity we avoid transforming two different measurements to the same value.

Accelerating and Decelerating Transformations

In statistical applications, most transformations bend in only one direction. Thus they are either accelerating (curved up) or decelerating (curved down). Sometimes the language cupped or capped is used, and these help people remember the shape; they may be misleading when we discuss monotone transformations because the bottom of the cup and the top of the cap do not ordinarily occur. Two one-bend curves are

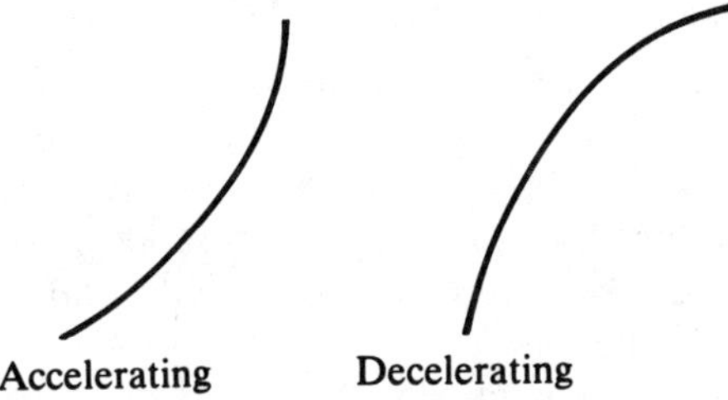

Examples include $f(y) = y^2$ for $y \geq 0$, accelerating, and $f(y) = \sqrt{y}$ for $y \geq 0$, decelerating. The function $y^{1/3}$ is accelerating for $y < 0$ and decelerating for $y > 0$, with a trouble spot at $y = 0$, where the slope is infinite. Thus it bends in one direction and then in the other, like this two-bend curve.

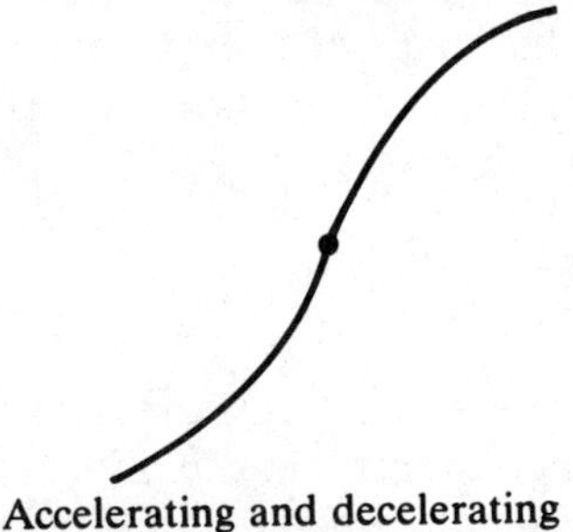

Accelerating and decelerating

Strength of Transformations

We compare transformations by noting how much or how little they change the shape of a data set; this idea leads to the vague concept of *strength*. A transformation that does not curve at all is a straight line and has strength 0. A transformation that curves a lot is stronger than one that curves a little. For example, in Figure 13-1, $\sqrt{y}$ appears stronger for y between 0 and 1 than it is for y between 4 and 5.

In practice when dealing with positive measurements, we are most likely to use transformations that are monotonic increasing and decelerating. The following are among the power transformations that have these properties

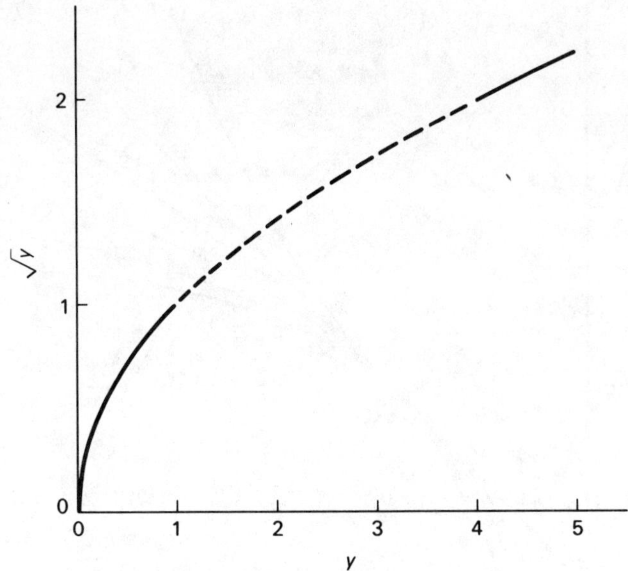

Figure 13-1. The curve for $\sqrt{y}$ versus y suggests that the transformation is stronger between 0 and 1 than it is between 4 and 5.

(except that y^1 is straight); we list them in increasing order of strength:

$$y^1, \quad \sqrt{y}, \quad \log y, \quad -\frac{1}{\sqrt{y}}, \quad -\frac{1}{y}.$$

For an effective comparison, we translate these functions and change their scales, so that their strengths can be more fairly compared and the appearance of the analyses made more similar. We now introduce the ideas needed for this comparison.

A Family of Power Transformations

A power transformation of y is any transformation that may be written

$$z = ay^p + b,$$

where a, b, and p are arbitrary constants. When $p = 0$, we replace y^0 by $\log y$ in order to have the family of curves change smoothly as p approaches zero. Figure 13-2 sketches members of a family of power transformations in the same coordinate system. For convenience in comparing their strengths at a common point, we have chosen values of a and b so that all curves pass through $(y, z) = (1, 0)$ and have slope equal to 1 there. (This requires

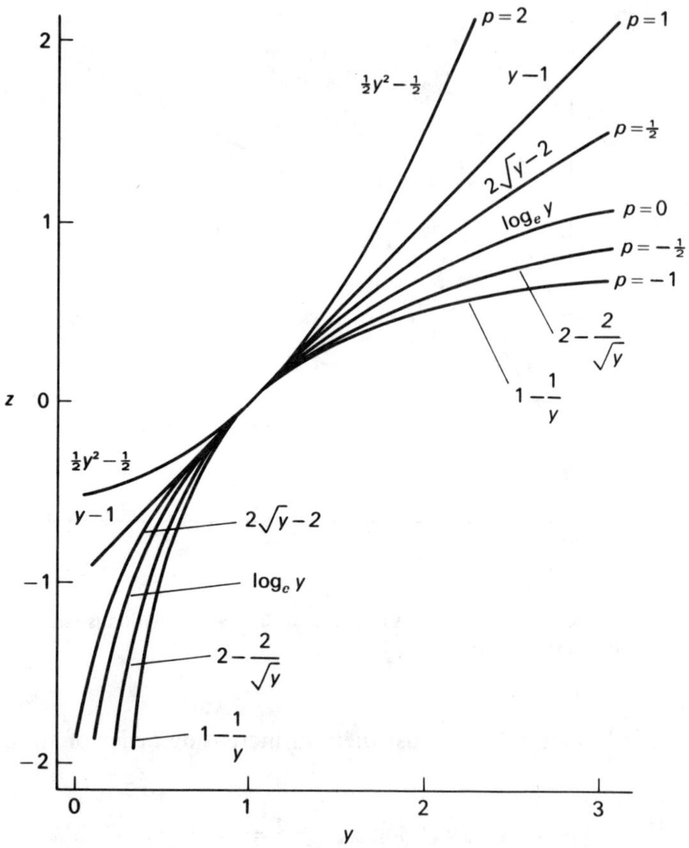

Figure 13-2. Graphs of some members of a family of power transformations. *Source:* David C. Hoaglin, Frederick Mosteller, and John W. Tukey (Eds.) (1983). *Understanding Robust and Exploratory Data Analysis*. New York: Wiley (Figure 8-1, p. 250). Copyright © 1983 John Wiley & Sons, Inc. Reprinted by permission of John Wiley & Sons, Inc.

$a = 1/p$ and $b = -1/p$.) The curves farthest from the line $z = y$ have greatest strength. Thus, $1 - 1/y$ is stronger than $2 - 2/\sqrt{y}$, which in turn is stronger than $\log_e y$. Similarly, $\frac{1}{2}y^2 - \frac{1}{2}$ is stronger than $y - 1$, which has strength 0; but the curvature of $\frac{1}{2}y^2 - \frac{1}{2}$ is opposite to that of all curves whose powers are less than 1. UREDA (Chaps. 4 and 8) gives background for these curves and offers a unified treatment of power transformations.

Other Transformations

Transformations other than power transformations are sometimes used. The "simple family" of transformations has the form $(y + k)^p$; and when $y \geq -k$, it is a one-bend transformation (Tukey 1957). A special case is $\log(y + k)$, the logarithmic family; Berry (1987) examines the logarithmic family in detail, with special focus on regression applications.

Two-bend transformations may be well suited to examining proportions (or percentages) when some proportions are close to 0 and others are close to 1. A two-bend transformation can remove "compression" at both ends of the scale from 0 to 1. The logistic transformation has the form

$$f(y) = \tfrac{1}{2}\log_e\left(\frac{y}{1-y}\right), \quad \text{for } 0 < y < 1.$$

Section 13F examines two-bend transformations and their application to data consisting of proportions or percentages.

Transformations and Means

Because transformations, especially strong transformations, can change relationships among data values substantially, they do not preserve averages. For example, the mean of 1, 4, and 25 is 10, and the square root of 10 is about 3.2. But the mean of the square roots of the three numbers is $(1 + 2 + 5)/3 = 2.7$ and not 3.2.

Similarly, transformations do not preserve relationships among variances. In fact, one benefit of transformation is that it can take several groups of data with unequal variances and give groups of transformed data with variances that are nearly equal. We illustrate this idea in the next section.

13D. TRANSFORMATION IN ONE-WAY ANOVA

For I groups and J observations in each group, one classical ANOVA model represents the jth observation in group i as

$$y_{ij} = \mu_i + \varepsilon_{ij}, \qquad i = 1, 2, \ldots, I; \quad j = 1, 2, \ldots, J,$$

where the independent ε_{ij} are sometimes assumed to have the distributions Gau$(0, \sigma_i^2)$. Practical experience with positive measurements reveals that larger variance often goes with larger means. For example, maple trees of various ages have different mean heights, and the older ones have both larger mean height and larger variance of height. Furthermore, within a particular group of individuals the measurements often display right-skewness. A well-chosen transformation can sometimes make the within-group variances more nearly equal and reduce or eliminate skewness. An example illustrates that a log transformation can bring benefits to a data set with groups defined by a single factor.

EXAMPLE: SMOOTHNESS OF PAPER

Lashof and Mandel (1960) gave measurements on the smoothness of paper made in different laboratories. The full study involved 14 materials—kinds of paper—and 14 laboratories. One aim of the study was to compare the Bekk method for measuring smoothness of paper with two other methods. A second aim was to determine contributions of the measuring instrument, observer, and laboratory to the variability for papers whose smoothness differed substantially.

The Bekk instrumentation method, as well as the other methods considered, measures the smoothness of paper by determining air leakage between the paper and a smooth glass surface. At the time of this study in the 1950s, the Bekk method had been adopted as a standard for measuring smoothness in the United States, The Netherlands, Czechoslovakia, and the U.S.S.R., and it had been recommended as a standard in France (Lashof and Mandel 1960, p. 386). Other methods for measuring smoothness were faster and cheaper than the Bekk method, but they were slightly less reliable. In this chapter, we use only data from the Bekk instrument, as reported by Mandel (1964, p. 325). Table 13-1 presents 32 measurements on each of 5 different materials, arranged from left to right in increasing order of the measured quantity.

The group means and standard deviations presented in Table 13-1 show a very strong relationship: a larger standard deviation goes with a larger mean. Figure 13-3 shows this tendency graphically using a parallel boxplot display. More detailed boxplots of the data for the five materials separately show that each data set is fairly symmetric. Thus the chief feature of the data that leads us to explore for a transformation is the strong tendency for increasing spread to accompany increasing mean. Changes in variability among groups are sometimes described by the technical term *heteroscedasticity*, but "differing variance" or "differing variability" will be more generally understood.

We now ask for a transformation of the measurements that produces standard deviations that are relatively stable across the five groups. This transformation must compress the scale for the largest measurements—those based on material 5, which range from 113.5 to 247.3—substantially more than it compresses the scale for the smallest measurements—those based on material 1, which range from 4.3 to 8.1. Many power transformations (e.g., those with $p = \frac{1}{2}, \frac{1}{3}, \frac{1}{4}$, as well as the log transformation ($p = 0$)) have just this sort of impact. The spread-versus-level plot (see below) offers one effective approach to choosing the value of p. A roughly linear plot with slope b suggests $p = 1 - b$ as the (approximate) power.

The Spread-Versus-Level Plot

A spread-versus-level plot is an ordinary plot of the logarithm of a measure of spread in each group against the logarithm of a measure of location (level) in the same group. Of course, many statistics measure spread (e.g., standard

Table 13-1. Measurements on the Smoothness of Paper Using the Bekk Method

	Material (Group)				
	1	2	3	4	5
	6.5	12.9	38.7	166.3	125.9
	6.5	12.5	47.7	151.8	113.5
	6.6	12.8	44.5	141.0	123.4
	8.1	11.0	45.5	149.4	174.0
	6.5	9.3	41.5	151.7	130.1
	6.5	13.7	46.0	166.1	158.2
	6.8	13.1	43.8	148.6	144.5
	6.5	14.0	58.0	158.3	180.0
	6.0	13.2	42.9	178.5	196.7
	6.0	14.2	39.3	183.3	230.8
	6.1	12.2	39.7	150.8	146.0
	5.8	13.1	45.8	145.0	247.3
	5.7	9.9	43.3	162.9	183.7
	5.7	9.6	41.4	184.1	237.2
	5.7	12.6	41.8	160.5	229.3
	6.0	9.0	39.2	170.8	185.9
	6.0	11.0	39.0	150.0	150.0
	5.0	10.0	44.0	150.0	165.0
	6.0	12.0	43.0	150.0	172.0
	6.0	13.0	43.0	160.0	162.0
	6.0	11.0	40.0	150.0	170.0
	6.0	10.0	35.0	160.0	150.0
	6.0	11.0	34.0	182.0	158.0
	6.0	12.0	45.0	140.0	198.0
	5.0	11.4	35.4	121.8	123.8
	5.0	10.0	37.2	127.4	162.0
	4.9	8.8	34.8	145.0	128.4
	4.8	8.2	41.2	162.4	153.0
	4.6	10.0	42.6	122.2	164.4
	4.5	8.4	37.8	124.0	140.0
	4.8	10.0	34.8	110.2	130.2
	4.3	12.6	34.0	141.2	198.8
Mean	5.81	11.33	41.25	152.04	166.63
Standard deviation	0.80	1.73	4.92	18.22	35.13
Standard deviation of mean (standard error)	0.14	0.31	0.87	3.22	6.21

Source: John Mandel (1964). *The Statistical Analysis of Experimental Data*. New York: Wiley. (Data from Table 13.3, p. 325). Reproduced with permission of John Mandel.

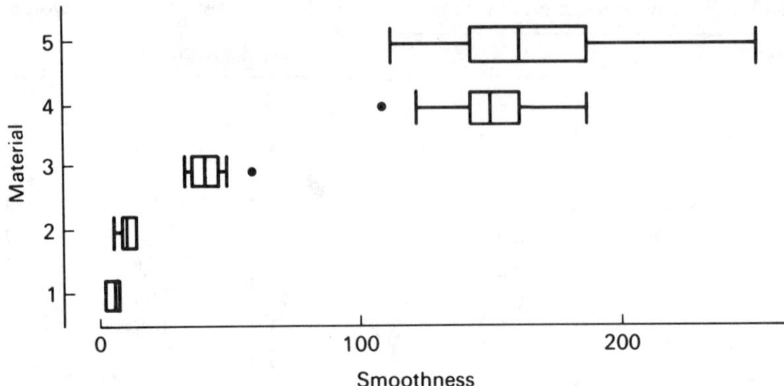

Figure 13-3. Side-by-side boxplots of smoothness of paper for five materials.

deviation, interquartile range, range, variance), and many statistics measure location (e.g., mean, median, midrange). To gain resistance to the potential impact of outliers, one common method employs the interquartile range (or fourth-spread) and the median; the spread-versus-level plot then displays log(interquartile range) on the y axis and log(median) on the x axis. UREDA (Chap. 3) gives a detailed development of this plot and illustrates its use. The treatment there justifies using logarithms of both spread and level in the plot.

In keeping with traditional practices in analysis of variance, we shall, in this instance, make use of spread-versus-level plots that display

$$\log_{10}(s_i) \quad \text{against} \quad \log_{10}(\bar{y}_i);$$

s_i is the sample standard deviation, and $\bar{y}_i$ is the sample mean, for the ith group. We use these measures because they are often readily available, and because our other calculations focus on means and sample variances.

Example: smoothness of paper

Figure 13-4 plots the logarithm of the group standard deviation against the logarithm of the group mean. The five points show a strong linear trend, and the slope of a line fitted to these points is very nearly equal to 1 (a least-squares line has slope 1.03). Using the diagnostic relationship

$$\text{power} = 1 - \text{slope},$$

we conclude that this plot suggests the power $p = 0$, and thus a log transformation.

Table 13-2 presents the data on paper smoothness in the log (base-10) scale. The group standard deviations are now relatively stable across the different groups (materials). The parallel boxplots in the log scale (Figure

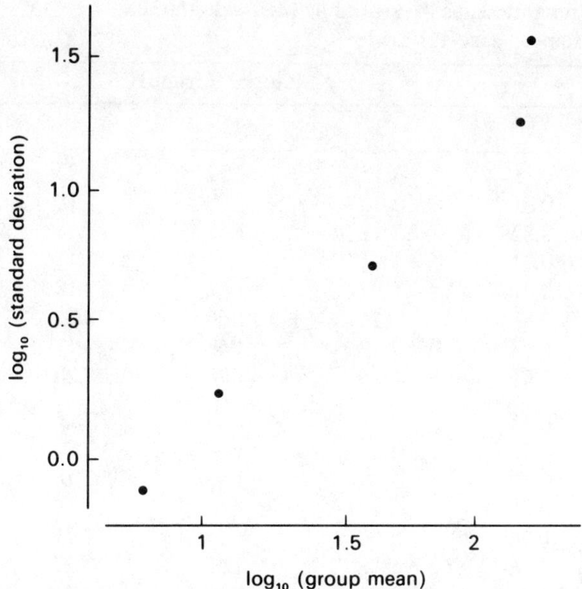

Figure 13-4. Spread-versus-level plot for smoothness of paper. Plot of $\log_{10}(s_i)$ against $\log_{10}(\bar{y}_i)$ from Table 13-1. The regression line has slope 1.03.

13-5), when compared with those in Figure 13-3, reveal graphically just how much more stable are the spreads of the five groups.

Analysis of Variance in the Transformed Scale

The transformed data on smoothness of paper more nearly exhibit the equal-variance property assumed in classical analysis of variance than do the data in the raw scale. One may thus ask about the choice of scale for subsequent analysis. Is an analysis of variance for log(smoothness) more appropriate than one for smoothness? How much difference does the choice of scale for the analysis make? Do the benefits of this transformation offset its additional complication?

Example: smoothness of paper

Table 13-3 summarizes the analysis of variance in the original scale and in the log scale. Because the magnitudes of the measurements reported in the two scales are different, the sums of squares and mean squares across the scales are not comparable. The values of F, however, are unit-free (dimensionless), so that comparison of the two F ratios is meaningful. The value of F for the analysis in the log scale is larger by a factor of 5 than that for the analysis of the raw data. One-way analysis of the raw measurements reveals

Table 13-2. Smoothness, as Measured by the Bekk Method, on the Logarithmic (Base-10) Scale

	Material (Group)				
	1	2	3	4	5
	0.81	1.11	1.59	2.22	2.10
	0.81	1.10	1.68	2.18	2.06
	0.82	1.11	1.65	2.15	2.09
	0.91	1.04	1.66	2.17	2.24
	0.81	0.97	1.62	2.18	2.11
	0.81	1.14	1.66	2.22	2.20
	0.83	1.12	1.64	2.17	2.16
	0.81	1.15	1.76	2.20	2.26
	0.78	1.12	1.63	2.25	2.29
	0.78	1.15	1.59	2.26	2.36
	0.79	1.09	1.60	2.18	2.16
	0.76	1.12	1.66	2.16	2.39
	0.76	1.00	1.64	2.21	2.26
	0.76	0.98	1.62	2.27	2.38
	0.76	1.10	1.62	2.21	2.36
	0.78	0.95	1.59	2.23	2.27
	0.78	1.04	1.59	2.18	2.18
	0.70	1.00	1.64	2.18	2.21
	0.78	1.08	1.63	2.18	2.24
	0.78	1.11	1.63	2.20	2.21
	0.78	1.04	1.60	2.18	2.23
	0.78	1.00	1.54	2.20	2.18
	0.78	1.04	1.53	2.26	2.20
	0.78	1.08	1.65	2.15	2.30
	0.70	1.06	1.55	2.09	2.09
	0.70	1.00	1.57	2.11	2.21
	0.69	0.94	1.54	2.16	2.11
	0.68	0.91	1.61	2.21	2.18
	0.66	1.00	1.63	2.09	2.22
	0.65	0.92	1.58	2.09	2.15
	0.68	1.00	1.54	2.04	2.11
	0.63	1.10	1.53	2.15	2.30
Mean	0.76	1.05	1.61	2.18	2.21
Standard deviation	0.06	0.07	0.05	0.05	0.09

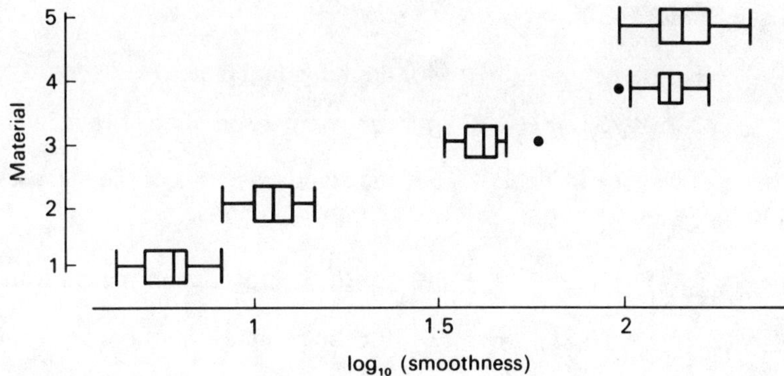

Figure 13-5. Side-by-side boxplots for smoothness of paper in log (base-10) scale.

that differences among means for the five materials account for 94% of the total sum-of-squares variation, whereas a similar analysis in the log scale accounts for almost 99% of the variation. Thus transformation brings obvious benefits to the analysis.

Given the large differences in smoothness among the five papers, we should not have been doing our analysis of variance just to see that the five smoothnesses were not all the same. The ratio of the MS for materials to its error term is so large that significance of the differences among materials is not in question for either analysis. One of the more likely reasons would be to compare various papers with one another, taking account of the errors of measurement. If we did only an analysis of variance for the raw scale, and stopped there, we would have used (beginning with an error mean square of

Table 13-3. One-Way Analysis of Variance[a] for the Data on the Bekk Smoothness of Paper Made of Five Materials

Source of Variation	d.f.	SS	MS	F
		(a) Raw Scale (R^2 = 94.0%)		
Material	4	777,947	194,487	610
Error	155	49,412	319	
Total	159	827,359		
		(b) Log (Base-10) Scale (R^2 = 98.8%)		
Material	4	54.81	13.7	3171
Error	155	0.670	0.004	
Total	159	55.48		

[a]Analyses are (*a*) in the original scale and units and (*b*) in the log scale with base-10 units.

319, see Table 13-3)

$$\pm\sqrt{319/32} = \pm 3.16 \quad \text{for an individual mean,}$$
$$\pm\sqrt{2}\sqrt{319/32} = \pm 4.47 \quad \text{for the comparison of any two means.}$$

However, if we look in Table 13-1 at the standard errors of the means, we are told to use

$$\pm\sqrt{(.14)^2 + (.31)^2} = \pm .34 \quad \text{for the comparison of papers 1 and 2,}$$
$$\pm\sqrt{(3.22)^2 + (6.21)^2} = \pm 7.0 \quad \text{for the comparison of papers 4 and 5.}$$

It is clear that we should not use the same standard error for comparing one of these pairs of *raw* means as we use for the other pair. Going on from a one-way analysis of variance of raw smoothness will lead us astray.

In contrast, the standard errors of the different means in the log scale are reasonably similar. Going on from an analysis of variance of the logarithmic data will help us, not hurt us. This is a much more important benefit from using logarithms in this example.

A plot of residuals against fitted values can reveal systematic structure in the data not extracted by the fit. We examined and compared these plots for the two fits described above; the residuals from the fit in the raw scale exhibit a strong wedge-shaped pattern around the horizontal axis, whereas those from the fit in the log scale have roughly similar scatter for the five materials. An exercise invites the reader to construct and examine these plots.

If all the values have to be positive (as amounts or counts) and the ratio of maximum to minimum is 10 or more, the data are almost certain to benefit from a logarithmic transformation. Had we not wanted to illustrate the spread-versus-level plot, we might have started our analysis of the data on a logarithmic scale. Mandel (1964, pp. 326–9) makes a similar point.

Section 13E continues the analysis of the smoothness-of-paper data. There we explore the impact of a second factor on the variability.

13E. TRANSFORMATION IN TWO-WAY ANOVA

When two factors can affect a response, with I levels of the first factor and J levels of the second, a general model for a balanced design with K replications per cell is

$$y_{ijk} = \mu + \alpha_i + \beta_j + (\alpha\beta)_{ij} + \varepsilon_{ijk}$$

($i = 1, 2, \ldots, I$; $j = 1, 2, \ldots, J$; $k = 1, 2, \ldots, K$). The ε_{ijk} are often assumed independent with a common normal distribution, Gau$(0, \sigma^2)$.

Transformation now has the important aim of eliminating the interaction term—the $(\alpha\beta)_{ij}$—or at least of reducing the relative magnitude of interaction between the row factor and the column factor. Experience suggests that a transformation that reduces interaction often brings other benefits—for example, by leading to residuals whose variances are more stable across different cells and whose distributions are more nearly symmetrical. But in a two-way table of measurements, the reduction of interaction is typically a primary consideration. We turn to a plot for choosing a power transformation that may reduce such nonadditivity.

A Diagnostic Plot for Two-Way Tables

Our strategy for determining which transformation may be useful takes two forms, depending on whether the number of replications per cell (K) is greater than 1. If $K > 1$, the full-effects fit

$$\hat{y}_{ijk} = m + a_i + b_j + (ab)_{ij}$$

includes an overlay, $(ab)_{ij}$, that directly summarizes the interactions. These effects may make a substantial but unsystematic contribution throughout the two-way layout, they may be concentrated primarily in a few cells, or they may follow a systematic pattern that allows a suitably chosen transformation to reduce the nonadditivity. To reveal such systematic behavior, we make a *diagnostic plot* of the interaction effect $(ab)_{ij}$ against the corresponding *comparison value*, $a_i b_j / m$.

A power transformation, $z = y^p$ for some suitable value of p, may remove the nonadditivity. When the diagnostic plot is roughly linear, one-minus-slope is the approximate value of p to try. UREDA (Sect. 8F) and Mosteller and Tukey (1977, Sects. 9E and 9F) discuss the origins of the diagnostic plot for the two-way table. In practice, we often choose powers that are simple integer multiples of $\frac{1}{2}$, like $p = 1$ (no transformation), $p = \frac{1}{2}$, $p = 0$ (logarithm), $p = -\frac{1}{2}$, and $p = -1$.

When $K = 1$, the fit cannot include $(ab)_{ij}$. Thus we work with the main-effects-only fit,

$$\hat{y}_{ij} = m + a_i + b_j$$

(it is customary to omit the subscript k), whose residuals, e_{ij}, combine interaction and fluctuation. Now the diagnostic plot uses e_{ij} versus $a_i b_j / m$, but we still interpret one-minus-slope as the suggested value of p.

Example: smoothness of paper

The arrangement of the smoothness data in Table 13-1 as four sets of eight rows reflects the (concealed) presence of a second factor. In fact, the four groupings of rows correspond to four different laboratories. Thus the data in Table 13-1 have a structure with two factors, one with four levels and the second with five levels, arising in a balanced factorial design with 8 replications per cell.

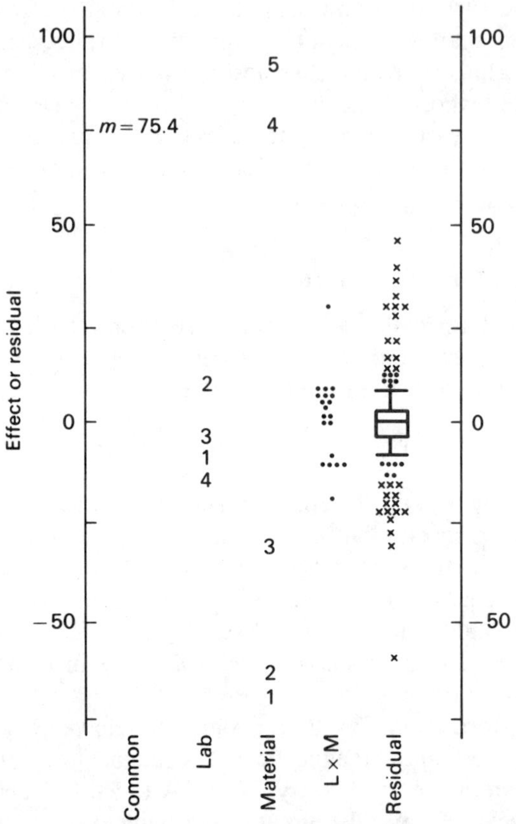

Figure 13-6. Effects and residuals for the full-effects analysis of the data on smoothness of paper. For the two factors, lab and material, the plotting symbols identify the four labs and, respectively, the five materials. The vertical arrangement of these symbols above the factor labels illustrates the approximate values of the effects. For the residuals, a dot denotes an outlier, and a cross denotes a far outlier.

A value splitting on the smoothness data produced overlays for the common value, row effects (laboratory), column effects (material), row-by-column effects (laboratory-by-material interaction), and the residuals. Figure 13-6 displays side by side the four types of effects and the residuals.

This display reveals that the magnitudes of the effects for material are roughly the same as the common value (75.4). The effects for laboratories are much smaller and, in fact, are smaller in size than many of the effects for laboratory-by-material interaction and many residuals. The residuals from the full-effects fit include 42 outliers, of which 29 are "far outside" (see UREDA, pp. 60–65).

In summary, the fit is somewhat disappointing because:

1. A plot of these residuals against fitted values (Figure 13-7) reveals, through a strong wedge-shaped pattern, that the variability of the residuals increases considerably with the fitted values.

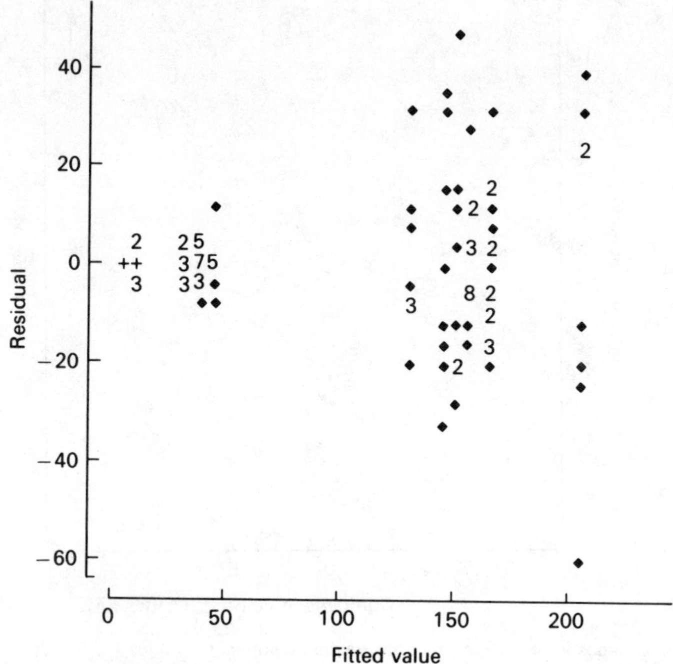

Figure 13-7. Plot of residuals against fitted values for the full-effects fit to the data on smoothness of paper. Plotting symbols: ◆, single point; n, multiple points very close together ($n = 2, 3, \ldots, 9$); +, more than nine points very close together.

2. The residuals have 25% outliers.
3. Some interaction effects are larger than the laboratory main effects.
4. Many residuals are larger than laboratory main effects and larger than some interaction effects.

(All these statements are unaffected by $y \to ay + b$, whose effect we have agreed to treat as trivial. Since comparing common term with main effects would be changed by $y \to ay + b$, this is not likely to be a helpful comparison.) The last two statements are not as important as they may seem; we almost always prefer smaller interactions, and, in this special study, we like to see small differences between laboratories. The first point shows clearly and graphically what we saw at the close of Section 13D—that if we use the same standard error for all materials, for different ranges of fitted values, we will be in trouble.

Diagnosing a Power for Transformation

In search of a power transformation to reduce or eliminate nonadditivity in the smoothness data, we make a diagnostic plot of the interactions $(ab)_{ij}$ against the comparison values $a_i b_j / m$. This plot (Figure 13-8) shows some

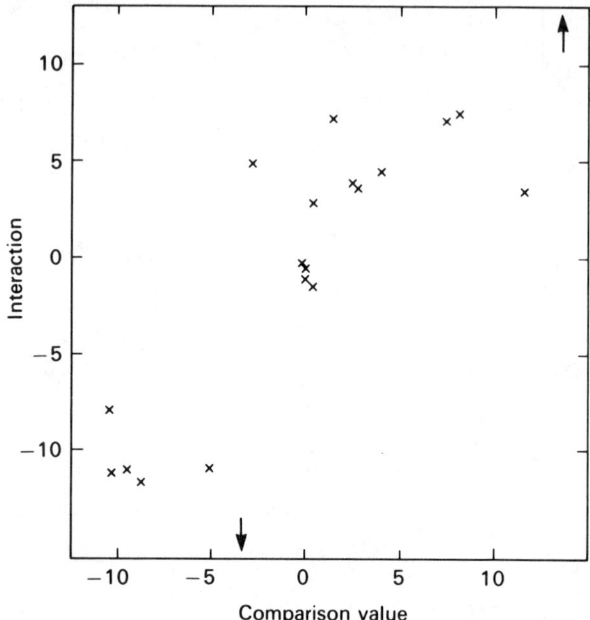

Figure 13-8. Diagnostic plot of interactions against comparison values for the data on smoothness of paper. The two points off-scale (indicated by arrows) have coordinates (−3.47, −20.06) and (13.63, 29.22). The least-squares line through the origin (not shown) has slope 1.24.

linear trend, and the least-squares regression line through the origin has a slope of 1.24. Thus one-minus-slope gives −0.24, suggesting the reciprocal fourth root. The amount of scatter in the diagnostic plot, however, indicates that this suggestion is only approximate. In addition, the one large positive and one large negative interaction, represented by arrows in Figure 13-8, have raised the slope. In practice, we begin with a nearby power that is a multiple of 0.5. Thus for these data we choose $p = 0$, the log transformation. We note that a slope of 1 seems a better fit to the rest of the points in the diagnostic plot than a slope of 1.24.

Figure 13-9 gives a side-by-side display of the main effects, interactions, and residuals from a full-effects fit in the log (base-10) scale. The changes from the fit in the raw scale, summarized in Figure 13-6, are substantial. The interaction terms have relatively smaller magnitudes when compared to the main effects for material and for laboratory. The residuals also have relatively more modest magnitudes than do the other effects. Only 5 of 160 residuals are outliers, and none is "far outside"; this is about what we should expect from chance (Hoaglin et al. 1986).

A plot of the residuals against the fitted values (Figure 13-10) shows nearly stable variance. The groupings of the fitted values from left to right are largely determined by the materials. Because the materials are numbered

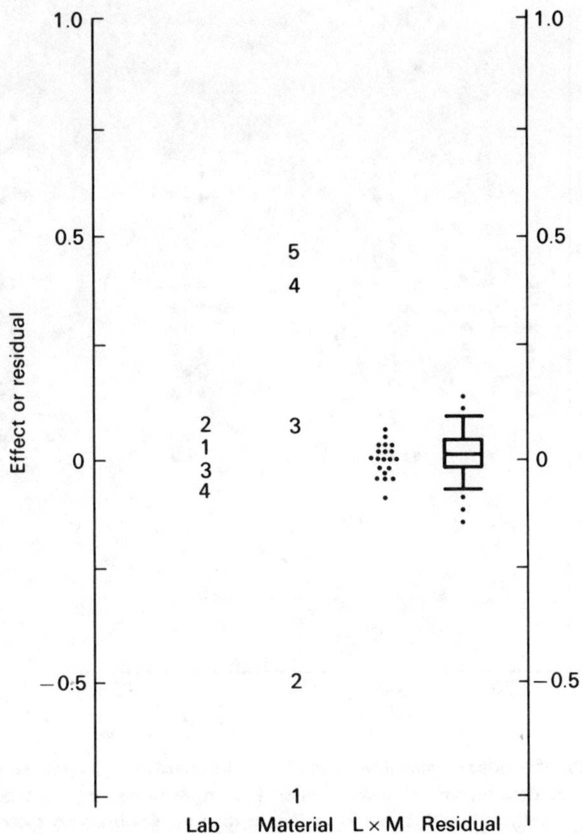

Figure 13-9. Effects and residuals from the full-effects analysis of the data on smoothness of paper in the log (base-10) scale. For the residuals, a dot denotes an outlier. The plotting symbols for main effects are the identifying numbers for the labs and the materials.

from 1 to 5, in order of increasing effect, the fitted values for material 1 are farthest to the left. Next come the points for material 2, and then, at the center and very isolated, the points for material 3. Only the points plotted for materials 4 and 5 at the right do not separate clearly.

Comparing ANOVA Tables

Table 13-4 gives in four panels the summary ANOVA tables for four two-way analyses of variance—in the original scale and in the log scale and, for each scale, with and without an overlay for interaction. These ANOVA tables can be usefully compared, especially in terms of F, with one another and with those arising from the one-way ANOVA given in Table 13-3.

We sometimes want to see how much of the fit is accounted for by main effects. We first examine the fits in the original scale. The two-way analysis

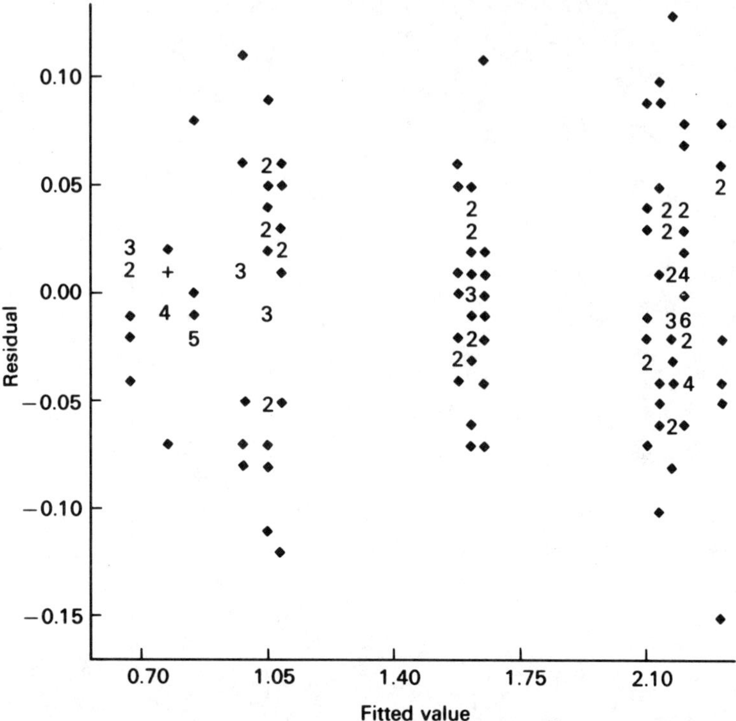

Figure 13-10. Plot of residuals against fitted values in the transformed log (base-10) scale, for the full-effects fit to the data on smoothness of paper. Plotting symbols: ♦, single point; n, multiple points very close together ($n = 2, 3, \ldots, 9$); +, more than nine points very close together.

using only main effects (Table 13-4a) leads to summary statistics—values of F and of R^2—that are little larger than those from the one-way ANOVA, whose summary appears in Table 13-3a. Even when interaction is swept out, using 12 additional degrees of freedom, the summary measures of the fit improve only modestly. In particular, the full-effects analysis using 19 degrees of freedom, Table 13-4b, accounts for 97.1% of the variability in the squared scale, whereas a one-way ANOVA of smoothness, Table 13-3a, uses only 4 degrees of freedom yet explains 94.0% of the squared variation.

The summary ANOVA tables for the analyses in the log scale tell a different story. We have seen that transformation to the log scale enables a one-way analysis using materials to account for most of the variation in the data. In the log scale, a two-way main-effects-only analysis, Table 13-4c, accounts for over 99% of squared-scale variability. The full-effects analysis gives larger F values (compared to those from analysis in the original scale) for material effects (5737 versus 1118) and laboratory effects (27.4 versus 16.1), and it gives smaller values for interaction (4.86 versus 7.97). Each of

Table 13-4. Two-Way Analysis of Variance[a] for the Data on the Bekk Smoothness of Paper, Measured in Four Labs for Five Materials

Source of Variation	d.f.	SS	MS	*F*
(a) Raw Scale—Main Effects Only (R^2 = 95.0%, 7 d.f.)				
Material	4	777,947	194,487	720
Lab	3	8,409	2,803	10
Error	152	41,003	270	
Total	159	827,359		
(b) Raw Scale—with Interaction (R^2 = 97.1%, 19 d.f.)				
Material	4	777,947	194,487	1,118
Lab	3	8,409	2,803	16
Interaction	12	16,639	1,387	8.0
Error	140	24,364	174	
Total	159	827,359		
(c) Log Scale—Main Effects Only (R^2 = 99.2%, 7 d.f.)				
Material	4	54.81	13.7	4431
Lab	3	0.196	0.065	21
Error	152	0.473	0.0031	
Total	159	55.48		
(d) Log Scale—with Interaction (R^2 = 99.4%, 19 d.f.)				
Material	4	54.81	13.70	5737
Lab	3	0.196	0.065	27
Interaction	12	0.139	0.012	4.9
Error	140	0.334	0.0024	
Total	159	55.48		

[a]Analyses (*a*) and (*b*) are in the original scale, and (*c*) and (*d*) are in the log scale.
Note: For material only, R^2 = 94.0% in the raw scale and 98.8% in the log scale.

these results supports our previous findings that the log scale brings benefit to the analysis.

Discounting Material

We already knew, as soon as we inspected Table 13-1, that variation among the five materials dominated the data. Laboratory-to-laboratory variation, averaged over the five materials, has now been clearly detected even though it is much smaller; the ratios (lab MS)/(error MS) are 10, 16, 21, and 27 in the four analyses. Both changing to logarithms and including an interaction overlay increased this ratio, clarifying considerably our assessment of differences among laboratories. The typical difference went from about 3 times its standard error ($\sqrt{10}$ = 3.16) to about 5 times its standard error ($\sqrt{27}$ = 5.20), a gain equivalent to that of making the experiment 2.7 times as large ($\frac{27}{10} = 2.7$). Of this gain, a factor of 2.1 was due to the re-expression.

Looking at an R^2 that includes differences between materials is of somewhat questionable relevance, because the differences between materials were planned to be large—probably to cover the whole range of the measuring instrument. The fit should not get credit for a lot of what the original investigators deliberately put in! A more appropriate measure of the effectiveness of the analysis is an R^2 that omits the SSs for the materials overlay entirely. This leads to $[8409/(827{,}359 - 777{,}947)] \times 100\% = 17\%$ for the main effects in the raw scale, and to $[.196/(55.48 - 54.81)] \times 100\% = 30\%$ for the main effects in the log scale—a considerable improvement. Similar calculations for main effects and interactions give an R^2 of 51% in each of the scales. This result is less encouraging about logs, but it still shows an overall gain for re-expression. In sum, the analyses, even with materials discounted, clearly favor taking out the interaction, and they do not contradict the indication by the overall R^2 that the log scale is better.

Benefits of Transformation

Ample evidence supports the log scale as a better scale for analyzing smoothness of paper than the original scale. The analysis of the sources of variability in this data table benefits from the log transformation in several ways:

1. The relative sizes of the interaction effects are reduced when compared to the spread of the two sets of main effects.
2. The residuals are made smaller relative to the other effects.
3. The residuals have far fewer outlying values.
4. The residuals display within-cell variability that is much more nearly stable.

Disadvantages of Transformation

The benefits of transformation, when they appear, do not come free. Transforming the original data to a new scale leaves the analyst one step removed from the original data. (This may be either good or bad.) It is harder to think about log(smoothness) than it is to think about smoothness, at least for some of those accustomed to working with the latter. (Some may find it natural to think about effects acting multiplicatively.)

The reader has certainly noticed that the task of comparing fits in the original scale with fits in the log scale need not be trivial. As in the paper-smoothness example, we are likely to be urged to compare ratios of mean squares (F's) and to use the word "relatively."

When we have to look more carefully at smaller differences, a technique called *matching* can often facilitate comparisons across scales. Section 13C introduced this technique but did not apply it to real data. Matching uses an additional linear transformation, both to retain the gains of the transformed scale, and to bring us much closer to complete numerical comparability of effects, overlays, and mean squares. UREDA (Sec. 4E) discusses matched transformations in more detail.

13F. TRANSFORMATION OF PERCENTAGES AND PROPORTIONS

Sometimes the values in a data table are expressed as proportions or percentages. For example, Table 5-9 gives the percentage of people who are smokers. The two-way data table has factors for age category and for income category. When data values are proportions or percentages, transformation can also bring benefits, especially if some of the percentages are close to 0% or 100%. Because percentages have two potential barriers—at 0% and 100%—power transformations are often inappropriate because they cannot produce similar changes near both barriers. (Power transformation of the form y^p could work well if all percentages y are near 0%, and power transformation of the form $(100\% - y)^p$ may work well if all percentages are near 100%.)

How should we compare changes in percentages? A change in the percentage of smokers in a certain subpopulation from 30% to 29% might not be difficult. But a 1-point reduction in another subpopulation, from 2% to 1%, seems relatively substantial: it represents cutting in half the number of smokers. Similarly, increasing the percentage of households that have television from 98% to 99% has to seem much more impressive (and difficult) than increasing it from 70% to 71%. The message here is that it seems very reasonable not to think about percentages additively, especially when they are near 0% or 100%. For analytic purposes we may want a transformation that stretches the two ends relative to the middle.

We do not, however, believe that a multiplicative scale is *always* appropriate for thinking about percentages. In biostatistics, people sometimes make the mistake of thinking that a reduction from 2% to 1% is more important than one from 10% to 8%. When the differences in percentages represent lives saved by a new medical treatment, and when the size of the population is the same for both situations, we have to prefer the second difference, while recognizing the first as more drastic.

In transforming percentages, it seems clear that we ought to work symmetrically on the two ends of the scale. How we transform 71% should somehow reflect what we do for 29%. What we do with 98% should reflect the transformation of 2%. The reflection here is, of course, across 50%. Indeed, it might be attractive and convenient if a transformation moved 50% to zero and worked symmetrically on either side. The amount of stretching could then increase from smallest in the middle to largest at the two ends, symmetrically.

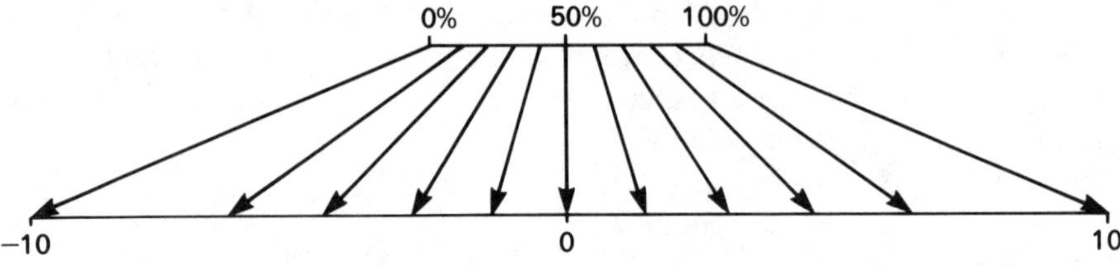

Folded Power Transformations

A simple transformation that takes 50% to 0 and stretches the scale only in a uniform way is

$$y - (100\% - y).$$

This transformation treats the two ends symmetrically; for example, it takes 98% to 96% and 2% to −96%, and it maps 71% to 42% and 29% to −42%. If y is the percentage of people who have a characteristic, then $100\% - y$ is the percentage who do not, and $y - (100\% - y)$ is the difference in percentages of those who do and those who don't. Tukey (1977, p. 498) refers to this differencing operation as "folding."

When combined with the concept of power transformation, the ideas just outlined lead naturally to a family of transformations for *proportions*:

$$y^p - (1 - y)^p \qquad \text{for } 0 < y < 1.$$

We refer to these as the *folded power transformations*. (When working with percentages, we would use $y^p - (100\% - y)^p$.) This family includes the folded log transformation ($p = 0$), when y is a proportion:

$$\log y - \log(1 - y).$$

Practical Application

Sometimes we may wish to try more than one folded power transformation on a data set consisting of proportions. In comparing the results, it is convenient if all transformations behave similarly near $\frac{1}{2}$ and differ only in the amount of stretching as we move toward the two ends. A slight change in the way the folded power transformations are defined, using a well-chosen multiplicative constant to change the units, can bring this advantage. We define the family of folded power transformations as follows:

$$T_p(y) = \frac{(2y)^p - (2 - 2y)^p}{2p}, \quad p \neq 0,$$

and

$$T_0(y) = \tfrac{1}{2}\{\log_e y - \log_e(1 - y)\}, \quad p = 0$$

(the logistic transformation). An exercise invites the reader to compare the members of this family, both near $y = \frac{1}{2}$, where they are almost identical, and near 0 and 1, where they behave symmetrically but differ in the amount of stretching.

For the most part, we use values of p that are 1 or smaller. The linear transformation $T_1(y)$ does not change the analysis. It is a peculiarity of the folded transformations that $p = 1$ and $p = 2$ yield identical transformations. However, we rarely use $p > 1$.

The Angular Transformation

Other transformations can stretch the ends of the unit interval but preserve the scale near $\frac{1}{2}$ and move that value to 0. We consider one of these other transformations as illustration—the angular transformation—because it is used often and because it has special properties for binomial data.

The *angular transformation* makes use of the inverse sine function; $\arcsin(x)$ means "the angle whose sine is x." The transformation is often defined as

$$g(y) = 2\arcsin(\sqrt{y}) - \frac{\pi}{2},$$

with $g(y)$ in units of radians ($\pi/2$ radians = 90°). The expression $\arcsin(\sqrt{y})$ maps $y = \frac{1}{2}$ to $\pi/4$, and it halves the scale near $\frac{1}{2}$; multiplying by 2 and subtracting $\pi/2$ lead to a version of the angular transformation whose properties at the center match those of the folded power transformations. In fact, Tukey (1970, Chap. 28) shows that this version of the angular transformation closely resembles a folded power transformation with $p = .41$, and so is not far from the folded square root.

The angular transformation has a mathematical property that holds special interest. When x, the number of successes in n independent trials, comes from a binomial distribution with success probability q, the variance of x is $nq(1 - q)$. If we work with $y = x/n$, the fraction of successes, then y estimates q, and its variance is $q(1 - q)/n$, making the standard deviation equal to $\sqrt{q(1 - q)/n}$. The angular transformation makes the variance of

$g(y)$ nearly constant, except very near the ends, where no transformation can do this.

When the angular transformation is given in radians, and when we are dealing with binomial distributions, so that the proportion y is computed from n independent binomial trials, the approximate variance of $g(y)$ is

$$\sigma_g^2 \approx \frac{1}{n + \frac{1}{2}}.$$

Of course, the transformation cannot totally stabilize the variance, it just does a good job for the middle values of q. When $q = 0$ or $q = 1$, the variance of $g(y)$ is still 0. When the variance of g is plotted against q for a given large n, the shape has ears like a cat, descending to zero at the ends, but with blips near $q = 0$ and $q = 1$. The larger the value of n, the more the ears are pushed toward the extremes. But they cannot go away. They do get narrower. The stabilization of the variance then refers to the relatively flat part of the function between the ears, and this has an average ordinate of about $1/(n + \frac{1}{2})$ when g is measured in radians.

EXAMPLE: QUANTIFYING PROBABILITY EXPRESSIONS

Mosteller and Youtz (1990) studied the numerical interpretations that science writers gave to 52 probability expressions—for example, *certain*, *very high probability*, *often*, and *unlikely*. The findings summarized the results of a survey of 637 members of the Council for the Advancement of Science Writing. Survey participants were science writers in the United States and Canada, from whom 238 usable questionnaires were received. Each respondent gave an estimate between 0% and 100% for the relative frequency represented by each probability expression. Table 13-5 summarizes the median and the quartiles of the estimates given by these respondents for the 52 expressions.

Expressions with associated percentages near 0% or near 100% tended to have much shorter interquartile ranges than expressions with percentages between 40% and 60%. Notable exceptions to this rule arose, however; for example, *more often than not*, *as often as not*, *less often than not*, *better than even chance*, *even chance*, and *less than an even chance* had relatively smaller interquartile ranges than one might expect, given that their probabilities place them in the middle. Mosteller and Youtz also identified expressions—*possible*, *not unreasonable*, and *might happen*—with unusually large interquartile ranges.

We analyze some subsets of these expressions to quantify the effect of modifiers. We use average values rather than the medians of Table 13-5.

One kind of question is what power, p, will be best at producing an additive model for expressions as they stand versus expressions introduced by *very*; for example, *infrequent* versus *very infrequent* and *unlikely* versus *very*

Table 13-5. Quartiles, Median, and Interquartile Range for the Science Writers' Estimates for 52 Probability Expressions[a]

Expression	25%	Median	75%	IQR
Always	99.6	99.7	99.8	.3
Almost always	89.7	91.7	95.2	5.5
Certain	98.7	99.6	99.8	1.1
Almost certain	87.5	90.2	95.0	7.5
Very frequent	75.3	82.6	89.7	14.5
Frequent	60.0	72.2	75.3	15.2
Not infrequent	32.7	49.6	57.3	24.6
Infrequent	10.1	17.3	22.6	12.5
Very infrequent	3.6	5.2	10.0	6.4
Very high probability	89.8	92.5	95.2	5.4
High probability	77.1	82.3	87.2	10.1
Moderate probability	40.1	52.4	58.7	18.5
Low probability	7.8	15.0	22.3	14.5
Very low probability	1.9	4.9	7.6	5.7
Very likely	80.1	87.5	90.2	10.1
Likely	62.6	71.1	77.6	15.0
Unlikely	9.8	17.2	22.7	13.0
Very unlikely	2.7	5.0	9.8	7.1
Very probable	81.5	89.7	90.4	8.9
Probable	64.7	70.2	77.7	13.0
Improbable	7.6	12.5	22.3	14.7
Very improbable	1.5	4.8	7.5	5.9
Very often	77.5	82.8	89.9	12.4
Often	65.0	72.5	75.4	10.4
More often than not	57.1	59.8	60.4	3.3
As often as not	49.8	50.0	50.3	.6
Less often than not	34.8	40.0	42.7	7.9

Expression	25%	Median	75%	IQR
Not often	10.3	19.7	24.8	14.5
Not very often	5.3	10.1	19.6	14.3
Possible	7.5	38.5	50.2	42.7
Impossible	.2	.3	.5	.3
High chance	77.5	80.4	89.1	11.7
Better than even chance	53.3	57.6	60.2	6.9
Even chance	49.7	50.0	50.2	.5
Less than an even chance	39.6	40.2	45.0	5.4
Poor chance	8.4	10.3	19.7	11.3
Low chance	5.0	9.8	12.8	7.8
Liable to happen	59.8	68.2	77.7	17.9
Might happen	19.9	37.6	50.1	30.2
Usually	65.6	75.1	82.2	16.7
Unusually	9.9	17.4	26.1	16.3
Sometimes	17.5	25.0	35.0	17.5
Once in a while	9.9	15.3	22.4	12.5
Not unreasonable	23.5	37.6	52.6	29.1
Occasionally	12.5	20.0	27.7	15.2
Now and then	9.8	15.1	25.0	15.1
Seldom	7.4	10.2	17.5	10.1
Very seldom	3.2	4.9	7.7	4.5
Rarely	3.6	7.2	10.0	6.5
Very rarely	1.2	3.0	5.0	3.8
Almost never	1.2	2.9	4.6	3.4
Never	.1	.3	.4	.3

[a] Writers gave a percentage, from 0% to 100%, that they attached to each probability expression.

Source: Frederick Mosteller and Cleo Youtz (1990). "Quantifying probabilistic expressions," *Statistical Science*, 5, 2–34 (with discussion) (Table 2, p. 6). Reproduced by permission of the Institute of Mathematical Statistics.

Table 13-6. Seven Low-Probability Expressions with Their Average Probability Estimated by Science Writers and Analyses of Variance for Various Folded Transformations Indexed by Values of p

(a) Raw Data

	No Modifier	Modifier *very*
Infrequent	.1735	.0714
Low probability	.1580	.0570
Unlikely	.1714	.0637
Improbable	.1558	.0569
Not often	.1869	.1256
Seldom	.1255	.0592
Rarely	.0746	.0363

(b) SS for ANOVA Tables

	d.f.	$p = 1$	$p = \frac{1}{2}$	$p = \frac{1}{4}$	$p = 0$	$p = -\frac{1}{4}$	$p = -\frac{1}{2}$
Rows	6	.0451	.1014	.1878	.3876	.8693	2.0758
Columns	1	.0947	.2067	.3716	.7366	1.5710	3.5342
Error	6	.0086	.0154	.0246	.0456	.0999	.2570
Total		.1483	.3235	.5840	1.1699	2.5402	5.8670
$\frac{100\text{ (Error SS)}}{\text{Total SS}}$		5.79%	4.76%	4.22%	3.90%	3.93%	4.38%

unlikely. Mosteller and Youtz multiplied the original probability by a constant to generate an estimate of the probability when *very* was attached. This choice is like using a logarithmic transformation.

Table 13-6 shows the raw data and the associated analyses of variance for seven expressions with low probabilities where *very* reduces the probability. We give the analysis of variance for several values of p using the corresponding matched folded transformation.

One way of assessing the preference for the power p in the transformation uses the percentage of the sums of squares allotted to error and interaction (labeled "Error" in Table 13-6). We wish this to be small. In the example, $p = 0$, corresponding to the logarithm, approximately minimizes the error plus interaction sum of squares.

This is the transformation we would have chosen for this problem had we not been choosing the value of p by examining outcomes for alternative possible values. Our reasoning would have been that logarithms change multiplication into addition. We are not so used to folded logs, though. With relatively small sets of data, the estimate of p may be rather unstable, and so finding an answer so close to $p = 0$ is partly luck. In some related problems we got different results, and an exercise at the end of the chapter illustrates this.

Roughly, what happens is that, for these expressions of low probability, the adjoining of *very* multiplies the proportion associated with the unadorned expression by about 0.45.

Binomial Variability

As suggested earlier, the angular transformation holds special interest because it stabilizes binomial variance.

EXAMPLE: EASTERN DIVISION FOOTBALL

As an example that uses the angular transformation, we consider some records from professional football. We chose eight years with the same numbers of games played each year (16), omitting 1982 because of the players' strike. We ask how much variation there is among the teams on average and how much from year to year within teams, compared with binomial variation.

We chose to study the Eastern Division in the American Conference of the National Football League. It has five teams, and we replace each team's won–lost record each year by the angular-transformed values before carrying out a one-way analysis of variance. If the teams were randomly paired, we might expect approximately binomial variation. For the most part the same teams play each other twice. This should produce a stratifying effect, so that the error variation from "measurement error" or "sampling variation" might be a bit less than binomial.

Table 13-7 gives the raw data, and Table 13-8 gives the analysis of variance of the transformed rates of winning. The two main components of this one-way analysis of variance are teams and year-to-year variation within teams.

The variation between teams has $F = .7934/.1239 = 6.40$, with 4 and 35 degrees of freedom. Thus we have reason to believe that the teams vary in their average strength over this period of time, Indianapolis (formerly, Baltimore) being weak and Miami very strong. In addition, the teams have varied from year to year. If each team had had a constant probability of success each year, the residual mean square would be about .0606 ($= 1/16.5$). Instead, we find .124 with 35 degrees of freedom. The ratio actual/ideal is thus a little more than 2, $.124/.0606 = 2.05$, and so over half of the residual mean square is estimated to come from year-to-year variation in a team's probability of winning, over and above the binomial variation. We see signs of this variation in the extreme years 1980, 1981 for Buffalo, 1981 for New England, and 1980 for the New York Jets. Thus in this problem we are able to extract a little more information than merely team and residual mean squares, because we have a special way of assessing the contribution of the

Table 13-7. Number of Games Won[a] (Out of 16) by Each Team in the Eastern Division of the American Conference of the National Football League

Year	Buffalo	Indianapolis[b]	Miami	New England	New York Jets
1978	5	5	11	11	8
1979	7	5	10	9	8
1980	11	7	8	10	4
1981	10	2	11.5	2	10.5
1983	8	7	12	8	7
1984	2	4	14	9	7
1985	2	5	12	11	11
1986	4	3	8	11	10

[a]Ties are counted as one-half of a game won.
[b]Moved from Baltimore to Indianapolis beginning 1984.
Source: Zander Hollander (Ed.) (1988). *The Complete Handbook of Pro Football*, 14th ed. New York: New American Library. (Data from pp. 338–342.)

binomial error, which enters this table in a way analogous to a measurement error.

From our variance-components approach, we can follow the procedure described in Section 9A to estimate the variance component between teams as (BMS − WMS)/8 = (.7934 − .1239)/8 = .08369, in squared radians. Thus, in the angular scale, the random effect for team has mean 0 and (estimated) variance .08369. We get mean ±2 standard-deviation limits in proportions by calculating those limits in the angular scale and then undoing the angular transformation, $2 \arcsin \sqrt{p} - \pi/2$. For the upper limit we take,

Table 13-8. One-Way ANOVA of Transformed Football Data

	d.f.	SS	MS
Between teams	4	3.1738	.7934
Residual	35	4.3363	.1239
Total	39	7.5101	

Common = 1.532 − 1.571 = −.039

Effects:	Buffalo	Indianapolis	Miami	New England	New York Jets
(radians)	−.20	−.39	.39	.14	.06

Data in Table 13-7 have been re-expressed using the angular transformation,

$$T = 2 \arcsin \sqrt{\frac{x}{16}} - \frac{\pi}{2}$$

expressed in radians, where $x/16$ is the proportion of games won. $\pi/2$ is approximately 1.571.

adding back $\pi/2$,

$$\pi/2 + 2\sqrt{.08369} = 2.1494.$$

Undoing the transformation, we have

$$\begin{aligned} 2\arcsin\sqrt{p} &= 2.1494 \text{ (in radians)} \\ \arcsin\sqrt{p} &= 1.0747 \\ \sin\left(\arcsin\sqrt{p}\right) = \sqrt{p} &= .8794 \\ p &= .7734. \end{aligned}$$

Thus the upper limit is about .77, and the lower limit is about .23, by symmetry around $p = .5$. The observed proportions run from .30 for Indianapolis to .68 for Miami.

We have only five teams, and so about 1/3 of the probability associated with their distribution of p's would fall outside the limits given by the best and poorest records, because 5 observations divide a distribution into 6 parts, averaging a probability of 1/6 each. Thus the true p's might fall at about one rather than two standard deviations from .5, or limits from .36 to .64. When some binomial variation is added on, the observed range of .30 to .68 from Indianapolis to Miami seems reasonable.

13G. SUMMARY

Transformations can sometimes enable us to view data in a scale that simplifies the description of variability and its components. In analysis of variance, transformation aims primarily to eliminate interaction or reduce the relative magnitude of interaction terms. Power transformations, with the log transformation as a special case, often serve this purpose; they may also help to stabilize the variability across different groups. Folded power transformations, including the logistic transformation, are well-suited to transforming proportions and percentages.

REFERENCES

Berry, D. A. (1987). "Logarithmic transformations in ANOVA," *Biometrics*, 43, 439–456.

Hoaglin, D. C. (1988). "Transformations in everyday experience," *Chance*, 1(4), 40–45.

Hoaglin, D. C., Iglewicz, B., and Tukey, J. W. (1986). "Performance of some resistant rules for outlier labeling," *Journal of the American Statistical Association*, 81, 991–999.

Lashof, T. W. and Mandel, J. (1960). "Measurement of the smoothness of paper," *Tappi*, 43, 385–399.

Mandel, J. (1964). *The Statistical Analysis of Experimental Data*. New York: Wiley.

Miller, R. G., Jr. (1986). *Beyond ANOVA, Basics of Applied Statistics*. New York: Wiley.

Mosteller, F. and Youtz, C. (1990). "Quantifying probablistic expressions," *Statistical Science*, 5, 2–34 (with discussion).

Mosteller, F. and Tukey, J. W. (1977). *Data Analysis and Regression*. Reading, MA: Addison-Wesley.

Tukey, J. W. (1957). "On the comparative anatomy of transformations," *Annals of Mathematical Statistics*, 28, 602–632.

Tukey, J. W. (1970). *Exploratory Data Analysis—Limited Preliminary Edition*. Reading, MA: Addison-Wesley.

Tukey, J. W. (1977). *Exploratory Data Analysis*. Reading, MA: Addison-Wesley.

EXERCISES

1. Power transformations with negative values of p are often useful. Explain why $T(y) = -y^{-1}$ and $T(y) = -y^{-2}$ are more convenient than $T(y) = y^{-1}$ and $T(y) = y^{-2}$.

2. Re-examine the smoothness data of Table 13-1 using a fourth-root transformation and one-way analysis of variance. Compare the ANOVA tables and the residuals with those described in Section 13C using the log scale. Which transformation is "stronger," the log or the fourth root?

3. Mathematicians sometimes prefer the natural log (ln or $\log_e$) to base-10 logs. Computer scientists often use base-2 logs, because they relate well to binary representations of numbers used in computer hardware. Explain why the analysis of a data set in the log scale depends only superficially on the choice of the base for the logarithm.

4. Give the plots of residuals against fitted values for six fits to the data on smoothness of paper:

(a) One-way ANOVA in original scale.

(b) Two-way ANOVA in original scale (main effects only).

(c) Two-way ANOVA in original scale (with interaction).

(d) One-way ANOVA in log scale.

(e) Two-way ANOVA in log scale (main effects only).

(f) Two-way ANOVA in log scale (with interaction).

5. Transform the smoothness of paper data using the reciprocal fourth-root transformation. Analyze the sources of variability and give all overlays associated with a two-way analysis of variance. Compare your findings with those of the analysis in the log scale.

6. Table 5-9 presents data on the percentage of smokers by age category and by income category. Using the results of a simple additive fit, construct a diagnostic plot for this table. Does it suggest a power transformation? If so, re-analyze the data table in an appropriate transformed scale.

7. One version of the folded power transformations for proportions y $(0 < y < 1)$ is

$$T_p(y) = \frac{(2y)^p - (2 - 2y)^p}{2p}, \qquad p \neq 0$$

and

$$T_0(y) = \frac{\log_e(y) - \log_e(1 - y)}{2}, \qquad p = 0.$$

Sketch graphs of these transformations for $p = 1, \frac{1}{2}, \frac{1}{4}, 0, -\frac{1}{2}$, and -1. Then add points for $y = .49$ and $.51$. Discuss the similarities and differences among these graphs.

8. In the science writers data, some expressions increase their estimated probability when *very* is adjoined. Use the following data to see what folded transformation would explain the largest fraction of the total sum of squares by main effects in the manner of the example in the text. (Mosteller and Youtz found what multiplier to apply to reduce the distance of the unmodified proportion from 1 to agree with the finding for *very*.)

	Unmodified	Modified by *very*
Frequent	.6700	.8094
High Probability	.7988	.9103
Likely	.6997	.8535
Probable	.7044	.8662
Often	.6998	.8237

9. Table 8-1 gives the percentage of Americans never married (by gender, race, and age group) at each of five selected years from 1950 through 1983.

(a) Re-express the data using the folded square-root transformation

$$T_{1/2}(y) = (2y)^{1/2} - (2 - 2y)^{1/2}.$$

(b) Carry out a full-effects analysis in the new scale. Compare the ANOVA table to that given in Table 8-3, and interpret any differences.

(c) Construct a side-by-side plot attuned to mean squares for your new analysis, and compare it to the plot shown in Figure 8-9. Does the re-expression bring dramatic changes in the relative magnitudes of the effects overlays?

(d) Discuss any gains and losses that come with analyzing the data in the re-expressed scale.

10. Repeat Exercise 9 using the logistic transformation

$$T_0(y) = \tfrac{1}{2}\{\log_e y - \log_e(1 - y)\}.$$

In addition to answering questions (a) through (d) for the logistic transformation, compare your findings here with those for $T_{1/2}(y)$.

11. Explain why we might expect a folded power transformation to be better suited to re-expressing the data on percentage never married than a power transformation. Choose one power transformation and compare its full-effects fit to that for a folded power transformation having the same value of p.

12. If $t = \frac{1}{2}\{\log_e y - \log_e(1 - y)\}$, solve this equation for y to show that

$$y = \frac{e^{2t}}{1 + e^{2t}}.$$

Explain how you might use this result, following an analysis based on the logistic re-expression, to reinterpret the results of the analysis.

13. The analysis of the data on marriage in Section 8D and the re-analyses in Exercises 9–11 invite a charge of "data-dredging." Discuss this issue as you reach some (perhaps tentative) conclusions about what you believe is a preferred analysis for these data. What cautions would you give about the interpretation of your findings? It will be useful to distinguish the findings of exploratory analyses from those, like F ratios and P values, that are inferential in nature.

Appendix: Tables

Table A-1. Critical Values of Student's t Distributions, Including the Gaussian Distribution (at $\nu = \infty$).

Table A-2. Percentage Points of the Chi-Squared Distributions for Selected Tail Areas.

Table A-3. Percentage Points of the F Distributions for Upper Tail Areas of .05 and .01.

Table A-4. Percentage Points of the Studentized Range Distributions for Selected Tail Areas.

Student's *t*

Many tables give critical values of Student's t distributions for a variety of tail areas, but few of these tables include tail areas smaller than .005. Table A-1, from Kafadar and Tukey (1988), extends to smaller tail areas and facilitates interpolation. This table uses a finer than usual mesh of tail areas, constructed to take advantage of the roughly linear relation between the critical value and the logarithm of the tail area. The column headed "Rough dg" gives the single tail area in "decigalts" (a special logarithmic transformation of tail area), rounded to the nearest integer.

Applications of Bonferroni's inequality may require interpolation in the tail area. As the inclusion of the rough dg in the table suggests, this interpolation is linear in the logarithm of the tail area. For example, to calculate the (5/3)% point of $|t_{24}|$, we begin with the one-sided tail area, $(5/6)\% = .833\%$, which lies between 1% and .8%. For $\nu = 24$, Table A-1c gives

$$\begin{aligned} t_{24}(1\%) &= 2.492 \\ t_{24}(.8\%) &= \underline{2.592} \\ \text{difference} &= 0.100 \end{aligned}$$

and the logarithms of the tail areas are

$$\begin{aligned}\log_{10}(1\%) &= -2.0000\\ \log_{10}((5/6)\%) &= -2.0792\\ \log_{10}(.8\%) &= -2.0969.\end{aligned}$$

Thus the interpolated value is

$$2.492 + (0.100)\frac{-2.0792 - (-2.0000)}{-2.0969 - (-2.0000)} = 2.574.$$

For values of ν that are not included in Tables A-1c and A-1d, it is desirable to interpolate linearly in $1/\nu$. Beyond $\nu = 20$ the tables include values of ν that correspond to integer values of $120/\nu$, in order to make such interpolation easy. For example, to find the (one-sided) .8% point of t_{28}, we locate in Table A-1c

$$\begin{aligned}t_{24}(.8\%) &= 2.592\\ t_{30}(.8\%) &= \underline{2.553}\\ \text{difference} &= -0.039.\end{aligned}$$

Then $120/28 = 4.2587$, and the interpolated value is

$$2.592 + (-0.039)\frac{4.2587 - 5}{4 - 5} = 2.564.$$

Studentized Range

In its simplest form the Studentized range $q_{k,\nu}$ arises as the ratio of the range of a standard Gaussian sample of k to the square root of an independent χ_ν^2/ν variable on ν degrees of freedom. In this book the Gaussian observations are usually means of independent samples of the same size, and the denominator is an independent estimate of their common standard error.

Perhaps the most extensive table of percentage points of the Studentized range distribution is that of Harter (1969, Table B2), which has

$$\begin{aligned}k &= 2(1)20(2)40(10)100,\\ \nu &= 1(1)20, 24, 30, 40, 60, 120, \infty,\end{aligned}$$

and upper tail area

$$\alpha = .001, .005, .01, .025, .05, .1(.1).9, .95, .975, .99, .995, .999.$$

Table A-4 reproduces the percentage points for $\alpha = .05$, .025, .01, .005, and .001, $k = 2(1)16(2)20(4)40$, and the above values of ν.

For combinations of α, ν, and k that lie within the range of Table A-4 but do not appear in the table, we approximate the value of $q_{k,\nu}^{(\alpha)}$ by interpolation. We use logarithmic interpolation in α and harmonic interpolation in ν, just as described above for the t distributions. Logarithmic interpolation is also appropriate for k, especially among the larger values. For example, to approximate the 1% point when $k = 19$ and $\nu = 40$, we would interpolate between $q_{18,40}^{(.01)} = 6.119$ and $q_{20,40}^{(.01)} = 6.209$:

$$q_{19,40}^{(.01)} \approx 6.119 + \frac{\log_{10} 19 - \log_{10} 18}{\log_{10} 20 - \log_{10} 18}(6.209 - 6.119) = 6.165,$$

which agrees with the value given by Harter (1969).

Approximations for the percentage points of the Studentized range have received less study than, say, the t or F distributions. The computer program of Lund and Lund (1983) approximates $q_{k,\nu}^{(\alpha)}$ for $.01 \le \alpha \le .1$ (see also Royston 1987). Kurtz et al. (1965) give approximation formulas for the 5% point and 1% point of $q_{k,\nu}$ with $k \ge 3$.

Interpolation for Chi-Squared and F

Use of the tables of percentage points of the chi-squared distributions (Table A-2) and the F distributions (Table A-3) also may sometimes require interpolation.

For numbers of degrees of freedom (ν) not given in Table A-2, interpolation can take advantage of the nearly linear relation between the square root of the χ_ν^2 percentage point and the square root of ν. Alternatively, a cube-root transformation of a χ_ν^2/ν variable yields a variable whose distribution is well approximated by a Gaussian distribution with mean $1 - 2/(9\nu)$ and variance $2/(9\nu)$—the Wilson–Hilferty approximation. Thus, if z_p is the $100p$th percentile of the standard Gaussian distribution (i.e., $P\{Z \le z_p\} = p$), available from Table A-1, then the $100p$th percentile of a chi-squared distribution on ν degrees of freedom is approximately

$$\nu\left[1 - \frac{2}{9\nu} + z_p\sqrt{\frac{2}{9\nu}}\right]^3.$$

This approach largely avoids the need to interpolate with respect to either ν or p.

For the F distributions Table A-3 gives percentage points only for upper tail areas of .05 and .01. For a value of either the numerator degrees of freedom (ν_1) or the denominator degrees of freedom (ν_2) that is not included in the table, it is desirable to interpolate linearly in $1/\nu_1$ or $1/\nu_2$, respectively.

REFERENCES

Harter, H. L. (1969). *Order Statistics and Their Use in Testing and Estimation, Volume 1: Tests Based on Range and Studentized Range of Samples from a Normal Population.* Washington, DC: U.S. Government Printing Office.

Kafadar, K. and Tukey, J. W. (1988). "A bidec *t* table," *Journal of the American Statistical Association*, 83, 532–539.

Kurtz, T. E., Link, R. F., Tukey, J. W., and Wallace, D. L. (1965). "Short-cut multiple comparisons for balanced single and double classifications: Part 2. Derivations and approximations," *Biometrika*, 52, 485–498.

Lund, R. E. and Lund, J. R. (1983). "Algorithm AS190: Probabilities and upper quantiles for the Studentized range," *Applied Statistics*, 32, 204–210. Correction: 34, 104.

Royston, J. P. (1987). "A remark on algorithm AS190: Probabilities and upper quantiles for the Studentized range," *Applied Statistics*, 36, 119.

Table A-1a. Critical Values of Student's t for ν Degrees of Freedom, $1 \leq \nu \leq 10$

Rough dg	Single tail area	v=1	v=2	v=3	v=4	v=5	v=6	v=7	v=8	v=9	v=10
0	100%	-∞									
1	80	-1.376	-1.061	-0.978	-0.941	-.920	-0.906	-0.896	-0.889	-0.883	-0.879
2	62.5	-0.414	-0.365	-0.349	-0.341	-.337	-0.334	-0.331	-0.330	-0.329	-0.328
3	50%	0	0	0	0	0	0	0	0	0	0
4	40	0.325	0.289	0.277	0.271	.267	0.265	0.263	0.262	0.261	0.260
5	32	0.635	0.546	0.518	0.505	.497	0.492	0.489	0.486	0.484	0.482
6	25%	1.000	0.816	0.765	0.741	.727	0.716	0.711	0.706	0.703	0.700
7	20	1.376	1.061	0.978	0.941	.920	0.906	0.896	0.889	0.883	0.879
8	16	1.819	1.312	1.189	1.134	1.104	1.084	1.070	1.060	1.053	1.046
9	12.5	2.414	1.604	1.423	1.344	1.301	1.273	1.254	1.240	1.230	1.221
10	10%	3.078	1.886	1.638	1.533	1.476	1.440	1.415	1.397	1.383	1.372
11	8	3.895	2.189	1.859	1.723	1.649	1.603	1.572	1.549	1.532	1.518
12	6.25	5.027	2.556	2.113	1.936	1.841	1.782	1.742	1.713	1.691	1.674
13	5%	6.314	2.920	2.353	2.132	2.015	1.943	1.895	1.860	1.833	1.812
14	4	7.916	3.320	2.605	2.333	2.191	2.104	2.046	2.004	1.973	1.948
15	3.2	9.914	3.761	2.871	2.540	2.369	2.266	2.197	2.148	2.111	2.082
16	2.5%	12.706	4.303	3.182	2.776	2.571	2.447	2.365	2.306	2.262	2.228
17	2	15.895	4.849	3.482	2.999	2.757	2.612	2.517	2.449	2.398	2.359
18	1.6	19.878	5.455	3.800	3.229	2.947	2.780	2.670	2.592	2.534	2.490
19	1.25	25.452	6.205	4.177	3.495	3.163	2.969	2.841	2.752	2.685	2.634
20	1%	31.821	6.965	4.541	3.747	3.365	3.143	2.998	2.896	2.821	2.764
21	.8	39.780	7.811	4.930	4.010	3.573	3.320	3.157	3.043	2.958	2.894
22	.625	50.923	8.860	5.392	4.315	3.810	3.521	3.335	3.206	3.111	3.038
23	.5%	63.657	9.925	5.841	4.604	4.032	3.707	3.499	3.355	3.250	3.169
24	.4	79.573	11.113	6.322	4.908	4.262	3.898	3.667	3.507	3.390	3.301
25	.32	99.468	12.440	6.838	5.226	4.500	4.094	3.837	3.660	3.531	3.434
26	.25%	127.321	14.089	7.453	5.598	4.773	4.317	4.029	3.833	3.690	3.581
27	.2	159.153	15.764	8.053	5.951	5.030	4.524	4.207	3.991	3.835	3.716
28	.16	198.942	17.635	8.696	6.324	5.297	4.737	4.389	4.152	3.981	3.852
29	.125	254.647	19.962	9.465	6.758	5.604	4.981	4.595	4.334	4.146	4.005
30	.1%	318.309	22.327	10.215	7.173	5.893	5.208	4.785	4.501	4.297	4.144
31	.08	397.887	24.970	11.021	7.610	6.194	5.442	4.981	4.671	4.450	4.285
32	.0625	509.295	28.258	11.984	8.122	6.541	5.709	5.202	4.864	4.622	4.442
33	.05%	636.619	31.599	12.924	8.610	6.869	5.959	5.408	5.041	4.781	4.587
34	.04	795.774	35.334	13.936	9.126	7.210	6.217	5.619	5.223	4.942	4.733
35	.032	994.718	39.509	15.025	9.670	7.565	6.483	5.835	5.408	5.106	4.882
36	.025%	1273.239	44.705	16.326	10.306	7.976	6.788	6.082	5.617	5.291	5.049
37	.02	1591.549	49.985	17.598	10.915	8.363	7.074	6.311	5.811	5.461	5.202
38	.016	1989.437	55.888	18.967	11.559	8.768	7.369	6.546	6.009	5.634	5.358
39	.0125	2546.479	63.234	20.604	12.312	9.235	7.708	6.814	6.234	5.830	5.533
40	.01%	3183.099	70.700	22.204	13.034	9.678	8.025	7.063	6.442	6.010	5.694

Source: K. Kafadar and J. W. Tukey (1988). "A bidec *t* table," *Journal of the American Statistical Association*, 83, 532–539.

Table A-1b. Critical Values of Student's t for ν Degrees of Freedom, $10 \leq \nu \leq 20$

Rough dg	Single tail area	v=10	v=11	v=12	v=13	v=14	v=15	v=16	v=17	v=18	v=19	v=20
0	100%	$-\infty$										
1	80	-0.879	-0.876	-0.873	-0.870	-0.868	-0.866	-0.865	-0.863	-0.862	-0.861	-0.860
2	62.5	-0.328	-0.327	-0.326	-0.325	-0.325	-0.324	-0.324	-0.324	-0.324	-0.323	-0.323
3	50	0	0	0	0	0	0	0	0	0	0	0
4	40	0.260	0.260	0.259	0.259	0.258	0.258	0.258	0.257	0.257	0.257	0.257
5	32	0.482	0.481	0.480	0.479	0.478	0.477	0.477	0.476	0.476	0.475	0.475
6	25%	0.700	0.697	0.695	0.694	0.692	0.691	0.690	0.689	0.688	0.688	0.687
7	20	0.879	0.876	0.873	0.870	0.868	0.866	0.865	0.863	0.862	0.861	0.860
8	16	1.046	1.041	1.037	1.034	1.031	1.029	1.026	1.024	1.023	1.021	1.020
9	12.5	1.221	1.214	1.209	1.204	1.200	1.197	1.194	1.191	1.189	1.187	1.185
10	10%	1.372	1.363	1.356	1.350	1.345	1.341	1.337	1.333	1.330	1.328	1.325
11	8	1.518	1.507	1.498	1.490	1.484	1.478	1.474	1.469	1.466	1.462	1.459
12	6.25	1.674	1.661	1.649	1.640	1.632	1.625	1.619	1.614	1.609	1.605	1.601
13	5%	1.812	1.796	1.782	1.771	1.761	1.753	1.746	1.740	1.734	1.729	1.725
14	4	1.948	1.928	1.912	1.899	1.887	1.878	1.869	1.862	1.855	1.850	1.844
15	3.2	2.082	2.059	2.040	2.024	2.011	2.000	1.990	1.981	1.973	1.967	1.961
16	2.5%	2.228	2.201	2.179	2.160	2.145	2.131	2.120	2.110	2.101	2.093	2.086
17	2	2.359	2.328	2.303	2.282	2.264	2.249	2.235	2.224	2.214	2.205	2.197
18	1.6	2.490	2.454	2.425	2.401	2.381	2.364	2.349	2.336	2.324	2.314	2.305
19	1.25	2.634	2.593	2.560	2.533	2.510	2.490	2.473	2.458	2.445	2.433	2.423
20	1%	2.764	2.718	2.681	2.650	2.624	2.602	2.583	2.567	2.552	2.539	2.528
21	.8	2.894	2.843	2.801	2.767	2.739	2.714	2.693	2.675	2.658	2.644	2.631
22	.625	3.038	2.981	2.934	2.896	2.864	2.837	2.813	2.793	2.775	2.759	2.744
23	.5%	3.169	3.106	3.055	3.012	2.977	2.947	2.921	2.898	2.878	2.861	2.845
24	.4	3.301	3.231	3.175	3.128	3.089	3.056	3.028	3.003	2.982	2.962	2.945
25	.32	3.434	3.357	3.295	3.244	3.202	3.166	3.134	3.108	3.084	3.063	3.045
26	.25%	3.581	3.497	3.428	3.372	3.326	3.286	3.252	3.222	3.197	3.174	3.153
27	.2	3.716	3.624	3.550	3.489	3.438	3.395	3.358	3.326	3.298	3.273	3.251
28	.16	3.852	3.752	3.671	3.605	3.550	3.504	3.464	3.429	3.399	3.372	3.348
29	.125	4.005	3.895	3.807	3.735	3.675	3.624	3.581	3.543	3.510	3.481	3.455
30	.1%	4.144	4.025	3.930	3.852	3.787	3.733	3.686	3.646	3.610	3.579	3.552
31	.08	4.285	4.156	4.054	3.970	3.901	3.842	3.792	3.749	3.711	3.677	3.648
32	.0625	4.442	4.303	4.192	4.101	4.026	3.963	3.909	3.862	3.822	3.786	3.754
33	.05%	4.587	4.437	4.318	4.221	4.140	4.073	4.015	3.965	3.922	3.883	3.850
34	.04	4.733	4.573	4.445	4.341	4.255	4.183	4.121	4.068	4.022	3.981	3.945
35	.032	4.882	4.710	4.573	4.462	4.371	4.294	4.228	4.171	4.122	4.079	4.040
36	.025%	5.049	4.863	4.716	4.597	4.499	4.417	4.346	4.286	4.233	4.187	4.146
37	.02	5.202	5.004	4.847	4.721	4.616	4.528	4.454	4.390	4.334	4.285	4.241
38	.016	5.358	5.146	4.980	4.845	4.734	4.641	4.562	4.494	4.435	4.383	4.337
39	.0125	5.533	5.306	5.128	4.984	4.865	4.766	4.682	4.609	4.547	4.491	4.443
40	.01%	5.694	5.453	5.263	5.111	4.985	4.880	4.791	4.714	4.648	4.590	4.539

Source: K. Kafadar and J. W. Tukey (1988). "A bidec t table," *Journal of the American Statistical Association*, 83, 532–539. Copyright 1987 by the American Mathematical Society. Reproduced by permission.

Table A-1c. Critical Values of Student's t for ν Degrees of Freedom, $20 \leq \nu \leq \infty$ $(6 \geq 120 / \nu \geq 0)$

Rough dg	Single tail area	ν=20	ν=24	ν=30	ν=40	ν=60	ν=120	∞
0	100%							
1	80	-0.860	-.857	-.854	-.851	-.848	-.845	-.842
2	62.5	-0.323	-.322	-.322	-.321	-.320	-.320	-.319
3	50%	0	0	0	0	0	0	0
4	40	0.257	.256	.256	.255	.254	.254	.253
5	32	0.475	.474	.472	.471	.470	.469	.468
6	25%	0.687	.685	.683	.681	.679	.677	.674
7	20	0.860	.857	.854	.851	.848	.845	.842
8	16	1.020	1.015	1.011	1.007	1.003	.999	.994
9	12.5	1.185	1.179	1.173	1.167	1.162	1.156	1.150
10	10%	1.325	1.318	1.310	1.303	1.296	1.289	1.282
11	8	1.459	1.450	1.441	1.432	1.423	1.414	1.405
12	6.25	1.601	1.590	1.578	1.567	1.556	1.545	1.534
13	5%	1.725	1.711	1.697	1.684	1.671	1.658	1.645
14	4	1.844	1.828	1.812	1.796	1.781	1.766	1.751
15	3.2	1.961	1.942	1.923	1.905	1.887	1.869	1.852
16	2.5%	2.086	2.064	2.042	2.021	2.000	1.980	1.960
17	2	2.197	2.172	2.147	2.123	2.099	2.076	2.054
18	1.6	2.305	2.277	2.249	2.222	2.196	2.170	2.144
19	1.25	2.423	2.391	2.360	2.329	2.299	2.270	2.241
20	1%	2.528	2.492	2.457	2.423	2.390	2.358	2.326
21	.8	2.631	2.592	2.553	2.516	2.479	2.444	2.409
22	.625	2.744	2.700	2.657	2.616	2.575	2.536	2.498
23	.5%	2.845	2.797	2.750	2.704	2.660	2.617	2.576
24	.4	2.945	2.892	2.841	2.792	2.744	2.697	2.652
25	.32	3.045	2.987	2.931	2.878	2.826	2.775	2.727
26	.25%	3.153	3.091	3.030	2.971	2.915	2.860	2.807
27	.2	3.251	3.183	3.118	3.055	2.994	2.935	2.878
28	.16	3.348	3.275	3.205	3.137	3.071	3.008	2.948
29	.125	3.455	3.376	3.300	3.227	3.156	3.088	3.023
30	.1%	3.552	3.467	3.385	3.307	3.232	3.160	3.090
31	.08	3.648	3.557	3.470	3.386	3.306	3.229	3.156
32	.0625	3.754	3.656	3.563	3.473	3.388	3.306	3.227
33	.05%	3.850	3.745	3.646	3.551	3.460	3.373	3.291
34	.04	3.945	3.834	3.729	3.628	3.532	3.440	3.353
35	.032	4.040	3.923	3.811	3.705	3.603	3.506	3.414
36	.025%	4.146	4.021	3.902	3.788	3.681	3.578	3.481
37	.02	4.241	4.109	3.983	3.864	3.750	3.642	3.540
38	.016	4.337	4.197	4.064	3.938	3.819	3.706	3.599
39	.0125	4.443	4.294	4.154	4.020	3.895	3.775	3.662
40	.01%	4.539	4.382	4.234	4.094	3.962	3.837	3.719

Source: K. Kafadar and J. W. Tukey (1988). "A bidec t table," *Journal of the American Statistical Association*, 83, 532–539.

Table A-1d. Critical Values of Student's t: Extension of Tables A-1b and A-1c to $40 \le$ Rough dg ≤ 60 for $11 \le \nu \le \infty$

Rough dg	Single tail area	v=11	v=12	v=13	v=14	v=15	v=16	v=17	v=18	v=19	v=20
40	.01%	5.453	5.263	5.111	4.985	4.880	4.791	4.714	4.648	4.590	4.539
41	.008	5.601	5.400	5.239	5.106	4.995	4.901	4.820	4.750	4.689	4.634
42	.00625	5.768	5.554	5.382	5.240	5.123	5.023	4.937	4.863	4.798	4.741
43	.005%	5.921	5.694	5.513	5.363	5.239	5.134	5.044	4.966	4.897	4.837
44	.004	6.077	5.837	5.645	5.487	5.357	5.246	5.151	5.069	4.997	4.934
45	.0032	6.234	5.981	5.778	5.613	5.475	5.358	5.259	5.173	5.097	5.031
46	.0025%	6.412	6.143	5.928	5.753	5.607	5.484	5.379	5.288	5.209	5.139
47	.002	6.574	6.291	6.065	5.881	5.727	5.598	5.488	5.393	5.310	5.237
48	.0016	6.740	6.442	6.204	6.010	5.849	5.714	5.598	5.498	5.411	5.335
49	.00125	6.926	6.610	6.359	6.154	5.985	5.842	5.721	5.616	5.524	5.444
50	.001%	7.097	6.765	6.501	6.287	6.109	5.959	5.832	5.722	5.627	5.543
51	.0008	7.271	6.922	6.645	6.420	6.234	6.078	5.945	5.830	5.730	5.642
52	.000625	7.467	7.099	6.807	6.570	6.374	6.210	6.070	5.950	5.845	5.753
53	.0005%	7.647	7.261	6.954	6.706	6.502	6.330	6.184	6.058	5.949	5.854
54	.0004	7.831	7.425	7.104	6.845	6.631	6.452	6.299	6.168	6.054	5.955
55	.00032	8.017	7.592	7.256	6.985	6.761	6.574	6.415	6.279	6.160	6.057
56	.00025%	8.227	7.779	7.426	7.142	6.907	6.711	6.545	6.402	6.278	6.170
57	.0002	8.421	7.952	7.583	7.285	7.041	6.836	6.663	6.514	6.385	6.273
58	.00016	8.617	8.127	7.741	7.431	7.175	6.962	6.782	6.627	6.493	6.376
59	.000125	8.839	8.323	7.919	7.593	7.326	7.104	6.915	6.754	6.614	6.492
60	.0001%	9.043	8.504	8.082	7.743	7.464	7.233	7.036	6.869	6.723	6.597

Rough dg	Single tail area	v=20	v=24	v=30	v=40	v=60	v=120	∞
40	.01%	4.539	4.382	4.234	4.094	3.962	3.837	3.719
41	.008	4.634	4.470	4.314	4.168	4.029	3.898	3.775
42	.00625	4.741	4.567	4.403	4.248	4.103	3.965	3.836
43	.005%	4.837	4.654	4.482	4.321	4.169	4.025	3.891
44	.004	4.934	4.742	4.562	4.393	4.234	4.085	3.944
45	.0032	5.031	4.830	4.641	4.465	4.299	4.144	3.998
46	.0025%	5.139	4.927	4.729	4.544	4.370	4.208	4.056
47	.002	5.237	5.015	4.808	4.615	4.435	4.266	4.107
48	.0016	5.335	5.103	4.888	4.686	4.498	4.323	4.159
49	.00125	5.444	5.201	4.975	4.764	4.568	4.385	4.215
50	.001%	5.543	5.290	5.054	4.835	4.631	4.442	4.265
51	.0008	5.642	5.378	5.133	4.905	4.694	4.497	4.314
52	.000625	5.753	5.477	5.220	4.983	4.763	4.558	4.369
53	.0005%	5.854	5.566	5.299	5.053	4.825	4.613	4.417
54	.0004	5.955	5.655	5.378	5.123	4.886	4.668	4.465
55	.00032	6.057	5.745	5.458	5.192	4.948	4.722	4.513
56	.00025%	6.170	5.845	5.545	5.269	5.015	4.781	4.566
57	.0002	6.273	5.935	5.624	5.339	5.076	4.834	4.611
58	.00016	6.376	6.026	5.704	5.408	5.137	4.887	4.658
59	.000125	6.492	6.126	5.792	5.485	5.203	4.945	4.710
60	.0001%	6.597	6.218	5.871	5.554	5.264	4.997	4.753

Source: K. Kafadar and J. W. Tukey (1988). "A bidec t table," *Journal of the American Statistical Association*, 83, 532–539.

Table A-2. Percentage Points of the Chi-Squared Distributions (ν Is the Degrees of Freedom, and α Is the Upper Tail Area)

ν \ α	.995	.990	.975	.950	.900	.500	.100	.050	.025	.010	.005
1	.00 +	.00 +	.00 +	.00 +	.02	.45	2.71	3.84	5.02	6.63	7.88
2	.01	.02	.05	.10	.21	1.39	4.61	5.99	7.38	9.21	10.60
3	.07	.11	.22	.35	.58	2.37	6.25	7.81	9.35	11.34	12.84
4	.21	.30	.48	.71	1.06	3.36	7.78	9.49	11.14	13.28	14.86
5	.41	.55	.83	1.15	1.61	4.35	9.24	11.07	12.83	15.09	16.75
6	.68	.87	1.24	1.64	2.20	5.35	10.64	12.59	14.45	16.81	18.55
7	.99	1.24	1.69	2.17	2.83	6.35	12.02	14.07	16.01	18.48	20.28
8	1.34	1.65	2.18	2.73	3.49	7.34	13.36	15.51	17.53	20.09	21.96
9	1.73	2.09	2.70	3.33	4.17	8.34	14.68	16.92	19.02	21.67	23.59
10	2.16	2.56	3.25	3.94	4.87	9.34	15.99	18.31	20.48	23.21	25.19
11	2.60	3.05	3.82	4.57	5.58	10.34	17.28	19.68	21.92	24.72	26.76
12	3.07	3.57	4.40	5.23	6.30	11.34	18.55	21.03	23.34	26.22	28.30
13	3.57	4.11	5.01	5.89	7.04	12.34	19.81	22.36	24.74	27.69	29.82
14	4.07	4.66	5.63	6.57	7.79	13.34	21.06	23.68	26.12	29.14	31.32
15	4.60	5.23	6.27	7.26	8.55	14.34	22.31	25.00	27.49	30.58	32.80
16	5.14	5.81	6.91	7.96	9.31	15.34	23.54	26.30	28.85	32.00	34.27
17	5.70	6.41	7.56	8.67	10.09	16.34	24.77	27.59	30.19	33.41	35.72
18	6.26	7.01	8.23	9.39	10.87	17.34	25.99	28.87	31.53	34.81	37.16
19	6.84	7.63	8.91	10.12	11.65	18.34	27.20	30.14	32.85	36.19	38.58
20	7.43	8.26	9.59	10.85	12.44	19.34	28.41	31.41	34.17	37.57	40.00
21	8.03	8.90	10.28	11.59	13.24	20.34	29.62	32.67	35.48	38.93	41.40
22	8.64	9.54	10.98	12.34	14.04	21.34	30.81	33.92	36.78	40.29	42.80
23	9.26	10.20	11.69	13.09	14.85	22.34	32.01	35.17	38.08	41.64	44.18
24	9.89	10.86	12.40	13.85	15.66	23.34	33.20	36.42	39.36	42.98	45.56
25	10.52	11.52	13.12	14.61	16.47	24.34	34.38	37.65	40.65	44.31	46.93
26	11.16	12.20	13.84	15.38	17.29	25.34	35.56	38.89	41.92	45.64	48.29
27	11.81	12.88	14.57	16.15	18.11	26.34	36.74	40.11	43.19	46.96	49.64
28	12.46	13.56	15.31	16.93	18.94	27.34	37.92	41.34	44.46	48.28	50.99
29	13.12	14.26	16.05	17.71	19.77	28.34	39.09	42.56	45.72	49.59	52.34
30	13.79	14.95	16.79	18.49	20.60	29.34	40.26	43.77	46.98	50.89	53.67
40	20.71	22.16	24.43	26.51	29.05	39.34	51.81	55.76	59.34	63.69	66.77
50	27.99	29.71	32.36	34.76	37.69	49.33	63.17	67.50	71.42	76.15	79.49
60	35.53	37.48	40.48	43.19	46.46	59.33	74.40	79.08	83.30	88.38	91.95
70	43.28	45.44	48.76	51.74	55.33	69.33	85.53	90.53	95.02	100.42	104.22
80	51.17	53.54	57.15	60.39	64.28	79.33	96.58	101.88	106.63	112.33	116.32
90	59.20	61.75	65.65	69.13	73.29	89.33	107.57	113.14	118.14	124.12	128.30
100	67.33	70.06	74.22	77.93	82.36	99.33	118.50	124.34	129.56	135.81	140.17

Source: W. W. Hines and D. C. Montgomery (1980). *Probability and Statistics in Engineering and Management Science*, 2nd ed. New York: Wiley. Copyright © 1980 John Wiley & Sons, Inc. Reprinted by permission of John Wiley & Sons, Inc.

Table A-3a. Percentage Points of the *F* Distributions for Upper Tail Area .05

ν_2 \ ν_1	1	2	3	4	5	6	7	8	9
	Degrees of Freedom for the Numerator (ν_1)								
1	161.4	199.5	215.7	224.6	230.2	234.0	236.8	238.9	240.5
2	18.51	19.00	19.16	19.25	19.30	19.33	19.35	19.37	19.38
3	10.13	9.55	9.28	9.12	9.01	8.94	8.89	8.85	8.81
4	7.71	6.94	6.59	6.39	6.26	6.16	6.09	6.04	6.00
5	6.61	5.79	5.41	5.19	5.05	4.95	4.88	4.82	4.77
6	5.99	5.14	4.76	4.53	4.39	4.28	4.21	4.15	4.10
7	5.59	4.74	4.35	4.12	3.97	3.87	3.79	3.73	3.68
8	5.32	4.46	4.07	3.84	3.69	3.58	3.50	3.44	3.39
9	5.12	4.26	3.86	3.63	3.48	3.37	3.29	3.23	3.18
10	4.96	4.10	3.71	3.48	3.33	3.22	3.14	3.07	3.02
11	4.84	3.98	3.59	3.36	3.20	3.09	3.01	2.95	2.90
12	4.75	3.89	3.49	3.26	3.11	3.00	2.91	2.85	2.80
13	4.67	3.81	3.41	3.18	3.03	2.92	2.83	2.77	2.71
14	4.60	3.74	3.34	3.11	2.96	2.85	2.76	2.70	2.65
15	4.54	3.68	3.29	3.06	2.90	2.79	2.71	2.64	2.59
16	4.49	3.63	3.24	3.01	2.85	2.74	2.66	2.59	2.54
17	4.45	3.59	3.20	2.96	2.81	2.70	2.61	2.55	2.49
18	4.41	3.55	3.16	2.93	2.77	2.66	2.58	2.51	2.46
19	4.38	3.52	3.13	2.90	2.74	2.63	2.54	2.48	2.42
20	4.35	3.49	3.10	2.87	2.71	2.60	2.51	2.45	2.39
21	4.32	3.47	3.07	2.84	2.68	2.57	2.49	2.42	2.37
22	4.30	3.44	3.05	2.82	2.66	2.55	2.46	2.40	2.34
23	4.28	3.42	3.03	2.80	2.64	2.53	2.44	2.37	2.32
24	4.26	3.40	3.01	2.78	2.62	2.51	2.42	2.36	2.30
25	4.24	3.39	2.99	2.76	2.60	2.49	2.40	2.34	2.28
26	4.23	3.37	2.98	2.74	2.59	2.47	2.39	2.32	2.27
27	4.21	3.35	2.96	2.73	2.57	2.46	2.37	2.31	2.25
28	4.20	3.34	2.95	2.71	2.56	2.45	2.36	2.29	2.24
29	4.18	3.33	2.93	2.70	2.55	2.43	2.35	2.28	2.22
30	4.17	3.32	2.92	2.69	2.53	2.42	2.33	2.27	2.21
40	4.08	3.23	2.84	2.61	2.45	2.34	2.25	2.18	2.12
60	4.00	3.15	2.76	2.53	2.37	2.25	2.17	2.10	2.04
120	3.92	3.07	2.68	2.45	2.29	2.17	2.09	2.02	1.96
∞	3.84	3.00	2.60	2.37	2.21	2.10	2.01	1.94	1.88

Degrees of Freedom for the Denominator (ν_2)

Table A-3a. *(Continued)*

10	12	15	20	24	30	40	60	120	∞
241.9	243.9	245.9	248.0	249.1	250.1	251.1	252.2	253.3	254.3
19.40	19.41	19.43	19.45	19.45	19.46	19.47	19.48	19.49	19.50
8.79	8.74	8.70	8.66	8.64	8.62	8.59	8.57	8.55	8.53
5.96	5.91	5.86	5.80	5.77	5.75	5.72	5.69	5.66	5.63
4.74	4.68	4.62	4.56	4.53	4.50	4.46	4.43	4.40	4.36
4.06	4.00	3.94	3.87	3.84	3.81	3.77	3.74	3.70	3.67
3.64	3.57	3.51	3.44	3.41	3.38	3.34	3.30	3.27	3.23
3.35	3.28	3.22	3.15	3.12	3.08	3.04	3.01	2.97	2.93
3.14	3.07	3.01	2.94	2.90	2.86	2.83	2.79	2.75	2.71
2.98	2.91	2.85	2.77	2.74	2.70	2.66	2.62	2.58	2.54
2.85	2.79	2.72	2.65	2.61	2.57	2.53	2.49	2.45	2.40
2.75	2.69	2.62	2.54	2.51	2.47	2.43	2.38	2.34	2.30
2.67	2.60	2.53	2.46	2.42	2.38	2.34	2.30	2.25	2.21
2.60	2.53	2.46	2.39	2.35	2.31	2.27	2.22	2.18	2.13
2.54	2.48	2.40	2.33	2.29	2.25	2.20	2.16	2.11	2.07
2.49	2.42	2.35	2.28	2.24	2.19	2.15	2.11	2.06	2.01
2.45	2.38	2.31	2.23	2.19	2.15	2.10	2.06	2.01	1.96
2.41	2.34	2.27	2.19	2.15	2.11	2.06	2.02	1.97	1.92
2.38	2.31	2.23	2.16	2.11	2.07	2.03	1.98	1.93	1.88
2.35	2.28	2.20	2.12	2.08	2.04	1.99	1.95	1.90	1.84
2.32	2.25	2.18	2.10	2.05	2.01	1.96	1.92	1.87	1.81
2.30	2.23	2.15	2.07	2.03	1.98	1.94	1.89	1.84	1.78
2.27	2.20	2.13	2.05	2.01	1.96	1.91	1.86	1.81	1.76
2.25	2.18	2.11	2.03	1.98	1.94	1.89	1.84	1.79	1.73
2.24	2.16	2.09	2.01	1.96	1.92	1.87	1.82	1.77	1.71
2.22	2.15	2.07	1.99	1.95	1.90	1.85	1.80	1.75	1.69
2.20	2.13	2.06	1.97	1.93	1.88	1.84	1.79	1.73	1.67
2.19	2.12	2.04	1.96	1.91	1.87	1.82	1.77	1.71	1.65
2.18	2.10	2.03	1.94	1.90	1.85	1.81	1.75	1.70	1.64
2.16	2.09	2.01	1.93	1.89	1.84	1.79	1.74	1.68	1.62
2.08	2.00	1.92	1.84	1.79	1.74	1.69	1.64	1.58	1.51
1.99	1.92	1.84	1.75	1.70	1.65	1.59	1.53	1.47	1.39
1.91	1.83	1.75	1.66	1.61	1.55	1.50	1.43	1.35	1.25
1.83	1.75	1.67	1.57	1.52	1.46	1.39	1.32	1.22	1.00

Table A-3b. Percentage Points of the F distributions for Upper Tail Area .01

ν_2 \ ν_1	Degrees of Freedom for the Numerator (ν_1) 1	2	3	4	5	6	7	8	9
1	4052	4999.5	5403	5625	5764	5859	5928	5982	6022
2	98.50	99.00	99.17	99.25	99.30	99.33	99.36	99.37	99.39
3	34.12	30.82	29.46	28.71	28.24	27.91	27.67	27.49	27.35
4	21.20	18.00	16.69	15.98	15.52	15.21	14.98	14.80	14.66
5	16.26	13.27	12.06	11.39	10.97	10.67	10.46	10.29	10.16
6	13.75	10.92	9.78	9.15	8.75	8.47	8.26	8.10	7.98
7	12.25	9.55	8.45	7.85	7.46	7.19	6.99	6.84	6.72
8	11.26	8.65	7.59	7.01	6.63	6.37	6.18	6.03	5.91
9	10.56	8.02	6.99	6.42	6.06	5.80	5.61	5.47	5.35
10	10.04	7.56	6.55	5.99	5.64	5.39	5.20	5.06	4.94
11	9.65	7.21	6.22	5.67	5.32	5.07	4.89	4.74	4.63
12	9.33	6.93	5.95	5.41	5.06	4.82	4.64	4.50	4.39
13	9.07	6.70	5.74	5.21	4.86	4.62	4.44	4.30	4.19
14	8.86	6.51	5.56	5.04	4.69	4.46	4.28	4.14	4.03
15	8.68	6.36	5.42	4.89	4.56	4.32	4.14	4.00	3.89
16	8.53	6.23	5.29	4.77	4.44	4.20	4.03	3.89	3.78
17	8.40	6.11	5.18	4.67	4.34	4.10	3.93	3.79	3.68
18	8.29	6.01	5.09	4.58	4.25	4.01	3.84	3.71	3.60
19	8.18	5.93	5.01	4.50	4.17	3.94	3.77	3.63	3.52
20	8.10	5.85	4.94	4.43	4.10	3.87	3.70	3.56	3.46
21	8.02	5.78	4.87	4.37	4.04	3.81	3.64	3.51	3.40
22	7.95	5.72	4.82	4.31	3.99	3.76	3.59	3.45	3.35
23	7.88	5.66	4.76	4.26	3.94	3.71	3.54	3.41	3.30
24	7.82	5.61	4.72	4.22	3.90	3.67	3.50	3.36	3.26
25	7.77	5.57	4.68	4.18	3.85	3.63	3.46	3.32	3.22
26	7.72	5.53	4.64	4.14	3.82	3.59	3.42	3.29	3.18
27	7.68	5.49	4.60	4.11	3.78	3.56	3.39	3.26	3.15
28	7.64	5.45	4.57	4.07	3.75	3.53	3.36	3.23	3.12
29	7.60	5.42	4.54	4.04	3.73	3.50	3.33	3.20	3.09
30	7.56	5.39	4.51	4.02	3.70	3.47	3.30	3.17	3.07
40	7.31	5.18	4.31	3.83	3.51	3.29	3.12	2.99	2.89
60	7.08	4.98	4.13	3.65	3.34	3.12	2.95	2.82	2.72
120	6.85	4.79	3.95	3.48	3.17	2.96	2.79	2.66	2.56
∞	6.63	4.61	3.78	3.32	3.02	2.80	2.64	2.51	2.41

Degrees of Freedom for the Denominator (ν_2)

Table A-3b. *(Continued)*

10	12	15	20	24	30	40	60	120	∞
6056	6106	6157	6209	6235	6261	6287	6313	6339	6366
99.40	99.42	99.43	99.45	99.46	99.47	99.47	99.48	99.49	99.50
27.23	27.05	26.87	26.69	26.60	26.50	26.41	26.32	26.22	26.13
14.55	14.37	14.20	14.02	13.93	13.84	13.75	13.65	13.56	13.46
10.05	9.89	9.72	9.55	9.47	9.38	9.29	9.20	9.11	9.02
7.87	7.72	7.56	7.40	7.31	7.23	7.14	7.06	6.97	6.88
6.62	6.47	6.31	6.16	6.07	5.99	5.91	5.82	5.74	5.65
5.81	5.67	5.52	5.36	5.28	5.20	5.12	5.03	4.95	4.86
5.26	5.11	4.96	4.81	4.73	4.65	4.57	4.48	4.40	4.31
4.85	4.71	4.56	4.41	4.33	4.25	4.17	4.08	4.00	3.91
4.54	4.40	4.25	4.10	4.02	3.94	3.86	3.78	3.69	3.60
4.30	4.16	4.01	3.86	3.78	3.70	3.62	3.54	3.45	3.36
4.10	3.96	3.82	3.66	3.59	3.51	3.43	3.34	3.25	3.17
3.94	3.80	3.66	3.51	3.43	3.35	3.27	3.18	3.09	3.00
3.80	3.67	3.52	3.37	3.29	3.21	3.13	3.05	2.96	2.87
3.69	3.55	3.41	3.26	3.18	3.10	3.02	2.93	2.84	2.75
3.59	3.46	3.31	3.16	3.08	3.00	2.92	2.83	2.75	2.65
3.51	3.37	3.23	3.08	3.00	2.92	2.84	2.75	2.66	2.57
3.43	3.30	3.15	3.00	2.92	2.84	2.76	2.67	2.58	2.49
3.37	3.23	3.09	2.94	2.86	2.78	2.69	2.61	2.52	2.42
3.31	3.17	3.03	2.88	2.80	2.72	2.64	2.55	2.46	2.36
3.26	3.12	2.98	2.83	2.75	2.67	2.58	2.50	2.40	2.31
3.21	3.07	2.93	2.78	2.70	2.62	2.54	2.45	2.35	2.26
3.17	3.03	2.89	2.74	2.66	2.58	2.49	2.40	2.31	2.21
3.13	2.99	2.85	2.70	2.62	2.54	2.45	2.36	2.27	2.17
3.09	2.96	2.81	2.66	2.58	2.50	2.42	2.33	2.23	2.13
3.06	2.93	2.78	2.63	2.55	2.47	2.38	2.29	2.20	2.10
3.03	2.90	2.75	2.60	2.52	2.44	2.35	2.26	2.17	2.06
3.00	2.87	2.73	2.57	2.49	2.41	2.33	2.23	2.14	2.03
2.98	2.84	2.70	2.55	2.47	2.39	2.30	2.21	2.11	2.01
2.80	2.66	2.52	2.37	2.29	2.20	2.11	2.02	1.92	1.80
2.63	2.50	2.35	2.20	2.12	2.03	1.94	1.84	1.73	1.60
2.47	2.34	2.19	2.03	1.95	1.86	1.76	1.66	1.53	1.38
2.32	2.18	2.04	1.88	1.79	1.70	1.59	1.47	1.32	1.00

Source: Adapted from E. S. Pearson and H. O. Hartley (1966). *Biometrika Tables for Statisticians*, Vol. 1, 3rd ed. Cambridge: Cambridge University Press. Reproduced by permission of the Biometrika Trustees.

Table A-4a. Upper 5% Point of the Studentized Range Distribution, $q_{k,\nu}^{(.05)}$

$\nu \backslash k$	2	3	4	5	6	7	8	9	10	11	12
1	17.97	26.98	32.82	37.08	40.41	43.12	45.40	47.36	49.07	50.59	51.96
2	6.085	8.331	9.798	10.88	11.74	12.44	13.03	13.54	13.99	14.39	14.75
3	4.501	5.910	6.825	7.502	8.037	8.478	8.853	9.177	9.462	9.717	9.946
4	3.927	5.040	5.757	6.287	6.707	7.053	7.347	7.602	7.826	8.027	8.208
5	3.635	4.602	5.218	5.673	6.033	6.330	6.582	6.802	6.995	7.168	7.324
6	3.461	4.339	4.896	5.305	5.628	5.895	6.122	6.319	6.493	6.649	6.789
7	3.344	4.165	4.681	5.060	5.359	5.606	5.815	5.998	6.158	6.302	6.431
8	3.261	4.041	4.529	4.886	5.167	5.399	5.597	5.767	5.918	6.054	6.175
9	3.199	3.949	4.415	4.756	5.024	5.244	5.432	5.595	5.739	5.867	5.983
10	3.151	3.877	4.327	4.654	4.912	5.124	5.305	5.461	5.599	5.722	5.833
11	3.113	3.820	4.256	4.574	4.823	5.028	5.202	5.353	5.487	5.605	5.713
12	3.082	3.773	4.199	4.508	4.751	4.950	5.119	5.265	5.395	5.511	5.615
13	3.055	3.735	4.151	4.453	4.690	4.885	5.049	5.192	5.318	5.431	5.533
14	3.033	3.702	4.111	4.407	4.639	4.829	4.990	5.131	5.254	5.364	5.463
15	3.014	3.674	4.076	4.367	4.595	4.782	4.940	5.077	5.198	5.306	5.404
16	2.998	3.649	4.046	4.333	4.557	4.741	4.897	5.031	5.150	5.256	5.352
17	2.984	3.628	4.020	4.303	4.524	4.705	4.858	4.991	5.108	5.212	5.307
18	2.971	3.609	3.997	4.277	4.495	4.673	4.824	4.956	5.071	5.174	5.267
19	2.960	3.593	3.977	4.253	4.469	4.645	4.794	4.924	5.038	5.140	5.231
20	2.950	3.578	3.958	4.232	4.445	4.620	4.768	4.896	5.008	5.108	5.199
24	2.919	3.532	3.901	4.166	4.373	4.541	4.684	4.807	4.915	5.012	5.099
30	2.888	3.486	3.845	4.102	4.302	4.464	4.602	4.720	4.824	4.917	5.001
40	2.858	3.442	3.791	4.039	4.232	4.389	4.521	4.635	4.735	4.824	4.904
60	2.829	3.399	3.737	3.977	4.163	4.314	4.441	4.550	4.646	4.732	4.808
120	2.800	3.356	3.685	3.917	4.096	4.241	4.363	4.468	4.560	4.641	4.714
∞	2.772	3.314	3.633	3.858	4.030	4.170	4.286	4.387	4.474	4.552	4.622

Table A-4a. *(Continued)*

$\nu \backslash k$	13	14	15	16	18	20	24	28	32	36	40
1	53.20	54.33	55.36	56.32	58.04	59.56	62.12	64.23	66.01	67.56	68.92
2	15.08	15.38	15.65	15.91	16.37	16.77	17.45	18.02	18.50	18.92	19.28
3	10.15	10.35	10.53	10.69	10.98	11.24	11.68	12.05	12.36	12.63	12.87
4	8.373	8.525	8.664	8.794	9.028	9.233	9.584	9.875	10.12	10.34	10.53
5	7.466	7.596	7.717	7.828	8.030	8.208	8.512	8.764	8.979	9.165	9.330
6	6.917	7.034	7.143	7.244	7.426	7.587	7.861	8.088	8.283	8.452	8.601
7	6.550	6.658	6.759	6.852	7.020	7.170	7.423	7.634	7.814	7.972	8.110
8	6.287	6.389	6.483	6.571	6.729	6.870	7.109	7.307	7.477	7.625	7.756
9	6.089	6.186	6.276	6.359	6.510	6.644	6.871	7.061	7.222	7.363	7.488
10	5.935	6.028	6.114	6.194	6.339	6.467	6.686	6.868	7.023	7.159	7.279
11	5.811	5.901	5.984	6.062	6.202	6.326	6.536	6.712	6.863	6.994	7.110
12	5.710	5.798	5.878	5.953	6.089	6.209	6.414	6.585	6.731	6.858	6.970
13	5.625	5.711	5.789	5.862	5.995	6.112	6.312	6.478	6.620	6.744	6.854
14	5.554	5.637	5.714	5.786	5.915	6.029	6.224	6.387	6.526	6.647	6.754
15	5.493	5.574	5.649	5.720	5.846	5.958	6.149	6.309	6.445	6.564	6.669
16	5.439	5.520	5.593	5.662	5.786	5.897	6.084	6.241	6.374	6.491	6.594
17	5.392	5.471	5.544	5.612	5.734	5.842	6.027	6.181	6.313	6.427	6.529
18	5.352	5.429	5.501	5.568	5.688	5.794	5.977	6.128	6.258	6.371	6.471
19	5.315	5.391	5.462	5.528	5.647	5.752	5.932	6.081	6.209	6.321	6.419
20	5.282	5.357	5.427	5.493	5.610	5.714	5.891	6.039	6.165	6.275	6.373
24	5.179	5.251	5.319	5.381	5.494	5.594	5.764	5.906	6.027	6.132	6.226
30	5.077	5.147	5.211	5.271	5.379	5.475	5.638	5.774	5.889	5.990	6.080
40	4.977	5.044	5.106	5.163	5.266	5.358	5.513	5.642	5.753	5.849	5.934
60	4.878	4.942	5.001	5.056	5.154	5.241	5.389	5.512	5.617	5.708	5.789
120	4.781	4.842	4.898	4.950	5.044	5.126	5.266	5.382	5.481	5.568	5.644
∞	4.685	4.743	4.796	4.845	4.934	5.012	5.144	5.253	5.346	5.427	5.498

Table A-4b. Upper 2.5% Point of the Studentized Range Distribution, $q_{k,\nu}^{(.025)}$

$\nu \backslash k$	2	3	4	5	6	7	8	9	10	11	12
1	35.99	54.00	65.69	74.22	80.87	86.29	90.85	94.77	98.20	101.3	104.0
2	8.776	11.94	14.01	15.54	16.75	17.74	18.58	19.31	19.95	20.52	21.03
3	5.907	7.661	8.808	9.660	10.34	10.89	11.37	11.78	12.14	12.46	12.75
4	4.943	6.244	7.088	7.716	8.213	8.625	8.976	9.279	9.548	9.788	10.01
5	4.474	5.558	6.257	6.775	7.186	7.527	7.816	8.068	8.291	8.490	8.670
6	4.199	5.158	5.772	6.226	6.586	6.884	7.138	7.359	7.554	7.729	7.887
7	4.018	4.897	5.455	5.868	6.194	6.464	6.695	6.895	7.072	7.230	7.373
8	3.892	4.714	5.233	5.616	5.919	6.169	6.382	6.568	6.732	6.879	7.011
9	3.797	4.578	5.069	5.430	5.715	5.950	6.151	6.325	6.479	6.617	6.742
10	3.725	4.474	4.943	5.287	5.558	5.782	5.972	6.138	6.285	6.416	6.534
11	3.667	4.391	4.843	5.173	5.433	5.648	5.831	5.989	6.130	6.256	6.369
12	3.620	4.325	4.762	5.081	5.332	5.540	5.716	5.869	6.004	6.125	6.235
13	3.582	4.269	4.694	5.004	5.248	5.449	5.620	5.769	5.900	6.017	6.123
14	3.550	4.222	4.638	4.940	5.178	5.374	5.540	5.684	5.811	5.926	6.029
15	3.522	4.182	4.589	4.885	5.118	5.309	5.471	5.612	5.737	5.848	5.949
16	3.498	4.148	4.548	4.838	5.066	5.253	5.412	5.550	5.672	5.781	5.879
17	3.477	4.118	4.512	4.797	5.020	5.204	5.361	5.496	5.615	5.722	5.818
18	3.458	4.092	4.480	4.761	4.981	5.162	5.315	5.448	5.565	5.670	5.765
19	3.442	4.068	4.451	4.728	4.945	5.123	5.275	5.405	5.521	5.624	5.718
20	3.427	4.047	4.426	4.700	4.914	5.089	5.238	5.368	5.481	5.583	5.675
24	3.381	3.983	4.347	4.610	4.816	4.984	5.216	5.250	5.358	5.455	5.543
30	3.337	3.919	4.271	4.523	4.720	4.881	5.017	5.134	5.238	5.330	5.414
40	3.294	3.858	4.197	4.439	4.627	4.780	4.910	5.022	5.120	5.208	5.288
60	3.251	3.798	4.124	4.356	4.536	4.682	4.806	4.912	5.006	5.089	5.164
120	3.210	3.739	4.053	4.276	4.447	4.587	4.704	4.805	4.894	4.972	5.043
∞	3.170	3.682	3.984	4.197	4.361	4.494	4.605	4.700	4.784	4.858	4.925

Table A-4b. *(Continued)*

$\nu \backslash k$	13	14	15	16	18	20	24	28	32	36	40
1	106.5	108.8	110.8	112.7	116.2	119.2	124.3	128.6	132.1	135.2	137.9
2	21.49	21.91	22.30	22.67	23.32	23.89	24.87	25.67	26.35	26.95	27.47
3	13.01	13.26	13.48	13.69	14.06	14.39	14.95	15.41	15.81	16.15	16.46
4	10.20	10.39	10.55	10.71	10.99	11.23	11.66	12.00	12.30	12.56	12.79
5	8.834	8.984	9.124	9.253	9.486	9.693	10.04	10.34	10.59	10.80	11.00
6	8.031	8.163	8.286	8.399	8.605	8.787	9.097	9.355	9.575	9.767	9.938
7	7.504	7.624	7.735	7.839	8.025	8.191	8.473	8.708	8.909	9.084	9.239
8	7.132	7.244	7.347	7.443	7.616	7.769	8.031	8.250	8.436	8.599	8.743
9	6.856	6.961	7.058	7.148	7.311	7.455	7.702	7.908	8.084	8.237	8.373
10	6.643	6.742	6.834	6.920	7.075	7.212	7.447	7.643	7.810	7.956	8.086
11	6.473	6.568	6.657	6.739	6.887	7.019	7.244	7.431	7.592	7.732	7.856
12	6.335	6.427	6.512	6.591	6.734	6.861	7.078	7.258	7.413	7.548	7.668
13	6.220	6.309	6.392	6.468	6.607	6.730	6.939	7.115	7.265	7.396	7.512
14	6.123	6.210	6.290	6.364	6.499	6.619	6.823	6.993	7.139	7.266	7.379
15	6.041	6.125	6.203	6.276	6.407	6.523	6.723	6.889	7.031	7.155	7.265
16	5.969	6.052	6.128	6.199	6.328	6.441	6.636	6.799	6.938	7.059	7.167
17	5.907	5.987	6.062	6.132	6.258	6.370	6.560	6.720	6.856	6.975	7.081
18	5.852	5.931	6.004	6.073	6.197	6.306	6.493	6.650	6.784	6.900	7.005
19	5.803	5.881	5.954	6.020	6.142	6.250	6.434	6.588	6.719	6.835	6.936
20	5.759	5.836	5.907	5.974	6.093	6.200	6.381	6.532	6.662	6.775	6.876
24	5.623	5.697	5.764	5.827	5.941	6.043	6.215	6.359	6.482	6.589	6.685
30	5.490	5.560	5.624	5.684	5.792	5.888	6.052	6.188	6.305	6.406	6.497
40	5.360	5.426	5.487	5.544	5.646	5.737	5.891	6.020	6.130	6.226	6.311
60	5.232	5.295	5.352	5.406	5.503	5.588	5.733	5.854	5.958	6.048	6.127
120	5.107	5.166	5.221	5.271	5.362	5.442	5.578	5.691	5.788	5.872	5.946
∞	4.985	5.041	5.092	5.139	5.224	5.299	5.425	5.530	5.620	5.698	5.766

Table A-4c. Upper 1% Point of the Studentized Range Distribution, $q_{k,\nu}^{(.01)}$

$\nu \backslash k$	2	3	4	5	6	7	8	9	10	11	12
1	90.03	135.0	164.3	185.6	202.2	215.8	227.2	237.0	245.6	253.2	260.0
2	14.04	19.02	22.29	24.72	26.63	28.20	29.53	30.68	31.69	32.59	33.40
3	8.261	10.62	12.17	13.33	14.24	15.00	15.64	16.20	16.69	17.13	17.53
4	6.512	8.120	9.173	9.958	10.58	11.10	11.55	11.93	12.27	12.57	12.84
5	5.702	6.976	7.804	8.421	8.913	9.321	9.669	9.972	10.24	10.48	10.70
6	5.243	6.331	7.033	7.556	7.973	8.318	8.613	8.869	9.097	9.301	9.485
7	4.949	5.919	6.543	7.005	7.373	7.679	7.939	8.166	8.368	8.548	8.711
8	4.746	5.635	6.204	6.625	6.960	7.237	7.474	7.681	7.863	8.027	8.176
9	4.596	5.428	5.957	6.348	6.658	6.915	7.134	7.325	7.495	7.647	7.784
10	4.482	5.270	5.769	6.136	6.428	6.669	6.875	7.055	7.213	7.356	7.485
11	4.392	5.146	5.621	5.970	6.247	6.476	6.672	6.842	6.992	7.128	7.250
12	4.320	5.046	5.502	5.836	6.101	6.321	6.507	6.670	6.814	6.943	7.060
13	4.260	4.964	5.404	5.727	5.981	6.192	6.372	6.528	6.667	6.791	6.903
14	4.210	4.895	5.322	5.634	5.881	6.085	6.258	6.409	6.543	6.664	6.772
15	4.168	4.836	5.252	5.556	5.796	5.994	6.162	6.309	6.439	6.555	6.660
16	4.131	4.786	5.192	5.489	5.722	5.915	6.079	6.222	6.349	6.462	6.564
17	4.099	4.742	5.140	5.430	5.659	5.847	6.007	6.147	6.270	6.381	6.480
18	4.071	4.703	5.094	5.379	5.603	5.788	5.944	6.081	6.201	6.310	6.407
19	4.046	4.670	5.054	5.334	5.554	5.735	5.889	6.022	6.141	6.247	6.342
20	4.024	4.639	5.018	5.294	5.510	5.688	5.839	5.970	6.087	6.191	6.285
24	3.956	4.546	4.907	5.168	5.374	5.542	5.685	5.809	5.919	6.017	6.106
30	3.889	4.455	4.799	5.048	5.242	5.401	5.536	5.653	5.756	5.849	5.932
40	3.825	4.367	4.696	4.931	5.114	5.265	5.392	5.502	5.599	5.686	5.764
60	3.762	4.282	4.595	4.818	4.991	5.133	5.253	5.356	5.447	5.528	5.601
120	3.702	4.200	4.497	4.709	4.872	5.005	5.118	5.214	5.299	5.375	5.443
∞	3.643	4.120	4.403	4.603	4.757	4.882	4.987	5.078	5.157	5.227	5.290

Table A-4c. *(Continued)*

$\nu \setminus k$	13	14	15	16	18	20	24	28	32	36	40
1	266.2	271.8	277.0	281.8	290.4	298.0	310.8	321.3	330.3	338.0	344.8
2	34.13	34.81	35.43	36.00	37.03	37.95	39.49	40.76	41.84	42.78	43.61
3	17.89	18.22	18.52	18.81	19.32	19.77	20.53	21.16	21.70	22.17	22.59
4	13.09	13.32	13.53	13.73	14.08	14.40	14.93	15.37	15.75	16.08	16.37
5	10.89	11.08	11.24	11.40	11.68	11.93	12.36	12.71	13.02	13.28	13.52
6	9.653	9.808	9.951	10.08	10.32	10.54	10.91	11.21	11.47	11.69	11.90
7	8.860	8.997	9.124	9.242	9.456	9.646	9.970	10.24	10.47	10.67	10.85
8	8.312	8.436	8.552	8.659	8.854	9.027	9.322	9.569	9.779	9.964	10.13
9	7.910	8.025	8.132	8.232	8.412	8.573	8.847	9.075	9.271	9.443	9.594
10	7.603	7.712	7.812	7.906	8.076	8.226	8.483	8.698	8.883	9.044	9.187
11	7.362	7.465	7.560	7.649	7.809	7.952	8.196	8.400	8.575	8.728	8.864
12	7.167	7.265	7.356	7.441	7.594	7.731	7.964	8.159	8.327	8.473	8.603
13	7.006	7.101	7.188	7.269	7.417	7.548	7.772	7.960	8.121	8.262	8.387
14	6.871	6.962	7.047	7.126	7.268	7.395	7.611	7.792	7.948	8.084	8.204
15	6.757	6.845	6.927	7.003	7.142	7.264	7.474	7.650	7.800	7.932	8.049
16	6.658	6.744	6.823	6.898	7.032	7.152	7.356	7.527	7.673	7.802	7.916
17	6.572	6.656	6.734	6.806	6.937	7.053	7.253	7.420	7.563	7.687	7.799
18	6.497	6.579	6.655	6.725	6.854	6.968	7.163	7.325	7.465	7.587	7.696
19	6.430	6.510	6.585	6.654	6.780	6.891	7.082	7.242	7.379	7.498	7.605
20	6.371	6.450	6.523	6.591	6.714	6.823	7.011	7.168	7.302	7.419	7.523
24	6.186	6.261	6.330	6.394	6.510	6.612	6.789	6.936	7.062	7.173	7.270
30	6.008	6.078	6.143	6.203	6.311	6.407	6.572	6.710	6.828	6.932	7.023
40	5.835	5.900	5.961	6.017	6.119	6.209	6.362	6.490	6.600	6.697	6.782
60	5.667	5.728	5.785	5.837	5.931	6.015	6.158	6.277	6.378	6.467	6.546
120	5.505	5.562	5.614	5.662	5.750	5.827	5.959	6.069	6.162	6.244	6.316
∞	5.348	5.400	5.448	5.493	5.574	5.645	5.766	5.866	5.952	6.026	6.092

Table A-4d. Upper 0.5% Point of the Studentized Range Distribution, $q_{k,\nu}^{(.005)}$

$\nu \backslash k$	2	3	4	5	6	7	8	9	10	11	12
1	180.1	270.1	328.5	371.2	404.4	431.6	454.4	474.0	491.1	506.3	520.0
2	19.93	26.97	31.60	35.02	37.73	39.95	41.83	43.46	44.89	46.16	47.31
3	10.55	13.50	15.45	16.91	18.06	19.01	19.83	20.53	21.15	21.70	22.20
4	7.916	9.814	11.06	11.99	12.74	13.35	13.88	14.33	14.74	15.10	15.42
5	6.751	8.196	9.141	9.847	10.41	10.88	11.28	11.63	11.93	12.21	12.46
6	6.105	7.306	8.088	8.670	9.135	9.522	9.852	10.14	10.40	10.63	10.83
7	5.699	6.750	7.429	7.935	8.339	8.674	8.961	9.211	9.433	9.632	9.812
8	5.420	6.370	6.981	7.435	7.797	8.097	8.354	8.578	8.777	8.955	9.117
9	5.218	6.096	6.657	7.074	7.405	7.680	7.915	8.120	8.303	8.466	8.614
10	5.065	5.888	6.412	6.800	7.109	7.365	7.584	7.775	7.944	8.096	8.234
11	4.945	5.727	6.222	6.588	6.878	7.119	7.325	7.505	7.664	7.807	7.937
12	4.849	5.597	6.068	6.416	6.693	6.922	7.118	7.288	7.439	7.575	7.697
13	4.770	5.490	5.943	6.277	6.541	6.760	6.947	7.111	7.255	7.384	7.502
14	4.704	5.401	5.838	6.160	6.414	6.626	6.805	6.962	7.101	7.225	7.338
15	4.647	5.325	5.750	6.061	6.308	6.511	6.685	6.837	6.971	7.091	7.200
16	4.599	5.261	5.674	5.977	6.216	6.413	6.582	6.729	6.859	6.976	7.081
17	4.557	5.205	5.608	5.903	6.136	6.329	6.493	6.636	6.763	6.876	6.979
18	4.521	5.156	5.550	5.839	6.067	6.255	6.415	6.554	6.678	6.788	6.888
19	4.488	5.113	5.500	5.783	6.005	6.189	6.346	6.482	6.603	6.711	6.809
20	4.460	5.074	5.455	5.732	5.951	6.131	6.285	6.418	6.537	6.642	6.738
24	4.371	4.955	5.315	5.577	5.783	5.952	6.096	6.221	6.332	6.431	6.520
30	4.285	4.841	5.181	5.428	5.621	5.780	5.914	6.031	6.135	6.227	6.310
40	4.202	4.731	5.053	5.284	5.465	5.614	5.739	5.848	5.944	6.030	6.108
60	4.122	4.625	4.928	5.146	5.316	5.454	5.571	5.673	5.762	5.841	5.913
120	4.045	4.523	4.809	5.013	5.172	5.301	5.410	5.504	5.586	5.660	5.726
∞	3.970	4.424	4.694	4.886	5.033	5.154	5.255	5.341	5.418	5.485	5.546

Table A-4d. *(Continued)*

$\nu \backslash k$	13	14	15	16	18	20	24	28	32	36	40
1	532.4	543.6	554.0	563.6	580.9	596.0	621.7	642.7	660.6	676.0	689.6
2	48.35	49.30	50.17	50.99	52.45	53.74	55.92	57.73	59.26	60.59	61.76
3	22.66	23.08	23.46	23.82	24.46	25.03	26.00	26.80	27.48	28.07	28.60
4	15.72	15.99	16.24	16.48	16.90	17.28	17.91	18.44	18.89	19.28	19.63
5	12.69	12.90	13.09	13.27	13.60	13.89	14.38	14.79	15.13	15.44	15.71
6	11.02	11.20	11.36	11.51	11.78	12.02	12.43	12.77	13.06	13.32	13.54
7	9.977	10.13	10.27	10.40	10.64	10.85	11.21	11.50	11.76	11.99	12.18
8	9.265	9.401	9.527	9.644	9.857	10.04	10.37	10.64	10.87	11.07	11.25
9	8.749	8.874	8.990	9.097	9.292	9.465	9.761	10.01	10.22	10.41	10.58
10	8.360	8.476	8.583	8.683	8.865	9.026	9.302	9.532	9.730	9.904	10.06
11	8.055	8.164	8.265	8.359	8.530	8.682	8.941	9.159	9.345	9.509	9.654
12	7.810	7.914	8.009	8.099	8.261	8.405	8.652	8.858	9.036	9.191	9.328
13	7.609	7.708	7.800	7.886	8.040	8.178	8.414	8.611	8.781	8.929	9.061
14	7.442	7.537	7.625	7.707	7.856	7.988	8.215	8.404	8.568	8.710	8.837
15	7.300	7.392	7.477	7.556	7.699	7.827	8.046	8.229	8.387	8.524	8.647
16	7.178	7.267	7.349	7.426	7.566	7.689	7.901	8.078	8.231	8.365	8.483
17	7.072	7.159	7.239	7.314	7.449	7.569	7.775	7.948	8.096	8.226	8.341
18	6.980	7.064	7.142	7.215	7.347	7.464	7.665	7.833	7.978	8.104	8.217
19	6.898	6.981	7.057	7.128	7.257	7.372	7.568	7.732	7.873	7.996	8.106
20	6.826	6.907	6.981	7.051	7.177	7.289	7.481	7.642	7.780	7.901	8.008
24	6.602	6.677	6.747	6.812	6.930	7.034	7.213	7.362	7.491	7.603	7.704
30	6.387	6.456	6.521	6.581	6.691	6.788	6.954	7.093	7.212	7.316	7.409
40	6.179	6.244	6.304	6.360	6.461	6.550	6.704	6.832	6.942	7.038	7.123
60	5.979	6.039	6.094	6.146	6.239	6.321	6.462	6.580	6.681	6.769	6.846
120	5.786	5.842	5.893	5.940	6.025	6.101	6.230	6.337	6.428	6.508	6.580
∞	5.602	5.652	5.699	5.742	5.820	5.889	6.006	6.103	6.186	6.258	6.322

Table A-4e. Upper 0.1% Point of the Studentized Range Distribution, $q_{k,\nu}^{(.001)}$

$\nu \backslash k$	2	3	4	5	6	7	8	9	10	11	12
1	900.3	1351.	1643.	1856.	2022.	2158.	2272.	2370.	2455.	2532.	2600.
2	44.69	60.42	70.77	78.43	84.49	89.46	93.67	97.30	100.5	103.3	105.9
3	18.28	23.32	26.65	29.13	31.11	32.74	34.12	35.33	36.39	37.34	38.20
4	12.18	14.99	16.84	18.23	19.34	20.26	21.04	21.73	22.33	22.87	23.36
5	9.714	11.67	12.96	13.93	14.71	15.35	15.90	16.38	16.81	17.18	17.53
6	8.427	9.960	10.97	11.72	12.32	12.83	13.26	13.63	13.97	14.27	14.54
7	7.648	8.930	9.768	10.40	10.90	11.32	11.68	11.99	12.27	12.52	12.74
8	7.130	8.250	8.978	9.522	9.958	10.32	10.64	10.91	11.15	11.36	11.56
9	6.762	7.768	8.419	8.906	9.295	9.619	9.897	10.14	10.36	10.55	10.73
10	6.487	7.411	8.006	8.450	8.804	9.099	9.352	9.573	9.769	9.946	10.11
11	6.275	7.136	7.687	8.098	8.426	8.699	8.933	9.138	9.319	9.482	9.630
12	6.106	6.917	7.436	7.821	8.127	8.383	8.601	8.793	8.962	9.115	9.254
13	5.970	6.740	7.231	7.595	7.885	8.126	8.333	8.513	8.673	8.817	8.948
14	5.856	6.594	7.062	7.409	7.685	7.915	8.110	8.282	8.434	8.571	8.696
15	5.760	6.470	6.920	7.252	7.517	7.736	7.925	8.088	8.234	8.365	8.483
16	5.678	6.365	6.799	7.119	7.374	7.585	7.766	7.923	8.063	8.189	8.303
17	5.608	6.275	6.695	7.005	7.250	7.454	7.629	7.781	7.916	8.037	8.148
18	5.546	6.196	6.604	6.905	7.143	7.341	7.510	7.657	7.788	7.906	8.012
19	5.492	6.127	6.525	6.817	7.049	7.242	7.405	7.549	7.676	7.790	7.893
20	5.444	6.065	6.454	6.740	6.966	7.154	7.313	7.453	7.577	7.688	7.788
24	5.297	5.877	6.238	6.503	6.712	6.884	7.031	7.159	7.272	7.374	7.467
30	5.156	5.698	6.033	6.278	6.470	6.628	6.763	6.880	6.984	7.077	7.162
40	5.022	5.528	5.838	6.063	6.240	6.386	6.509	6.616	6.711	6.796	6.872
60	4.894	5.365	5.653	5.860	6.022	6.155	6.268	6.366	6.451	6.528	6.598
120	4.771	5.211	5.476	5.667	5.815	5.937	6.039	6.128	6.206	6.276	6.339
∞	4.654	5.063	5.309	5.484	5.619	5.730	5.823	5.903	5.973	6.036	6.092

Table A-4e. *(Continued)*

$\nu \backslash k$	13	14	15	16	18	20	24	28	32	36	40
1	2662.	2718.	2770.	2818.	2904.	2980.	3108.	3213.	3303.	3380.	3448.
2	108.2	110.4	112.3	114.2	117.4	120.3	125.2	129.3	132.7	135.7	138.3
3	38.98	39.69	40.35	40.97	42.07	43.05	44.70	46.07	47.24	48.26	49.16
4	23.81	24.21	24.59	24.94	25.58	26.14	27.10	27.89	28.57	29.16	29.68
5	17.85	18.13	18.41	18.66	19.10	19.51	20.19	20.75	21.24	21.66	22.03
6	14.79	15.01	15.22	15.42	15.78	16.09	16.64	17.08	17.47	17.81	18.10
7	12.95	13.14	13.32	13.48	13.78	14.04	14.50	14.88	15.20	15.49	15.74
8	11.74	11.91	12.06	12.21	12.47	12.70	13.09	13.42	13.71	13.96	14.18
9	10.89	11.03	11.18	11.30	11.54	11.75	12.10	12.39	12.65	12.87	13.07
10	10.25	10.39	10.52	10.64	10.85	11.03	11.36	11.63	11.87	12.07	12.25
11	9.766	9.892	10.01	10.12	10.31	10.49	10.79	11.04	11.26	11.45	11.62
12	9.381	9.498	9.606	9.707	9.891	10.06	10.34	10.57	10.78	10.96	11.11
13	9.068	9.178	9.281	9.376	9.550	9.704	9.969	10.19	10.39	10.55	10.70
14	8.809	8.914	9.012	9.103	9.267	9.414	9.666	9.878	10.06	10.22	10.37
15	8.592	8.693	8.786	8.872	9.030	9.170	9.411	9.613	9.788	9.940	10.08
16	8.407	8.504	8.593	8.676	8.828	8.963	9.194	9.388	9.556	9.702	9.833
17	8.248	8.342	8.427	8.508	8.654	8.784	9.007	9.194	9.355	9.497	9.623
18	8.110	8.199	8.283	8.361	8.502	8.628	8.844	9.025	9.181	9.318	9.440
19	7.988	8.075	8.156	8.232	8.369	8.491	8.701	8.876	9.028	9.161	9.279
20	7.880	7.966	8.044	8.118	8.251	8.370	8.574	8.745	8.892	9.021	9.137
24	7.551	7.629	7.701	7.768	7.890	7.999	8.185	8.342	8.476	8.594	8.700
30	7.239	7.310	7.375	7.437	7.548	7.647	7.816	7.958	8.080	8.188	8.283
40	6.942	7.007	7.067	7.122	7.223	7.312	7.466	7.594	7.704	7.801	7.887
60	6.661	6.720	6.774	6.824	6.914	6.995	7.133	7.248	7.347	7.433	7.510
120	6.396	6.448	6.496	6.542	6.623	6.695	6.818	6.921	7.008	7.085	7.153
∞	6.144	6.191	6.234	6.274	6.347	6.411	6.520	6.611	6.689	6.756	6.816

Source: Adapted from H.L. Harter (1969). *Order Statistics and Their Use in Testing and Estimation, Volume* 1: *Tests Based on Range and Studentized Range of Samples from a Normal Population*. Washington, DC: U.S. Government Printing Office.

Index

Applied Probability and Statistics (Continued)

HOEL and JESSEN · Basic Statistics for Business and Economics, *Third Edition*
HOGG and KLUGMAN · Loss Distributions
HOLLANDER and WOLFE · Nonparametric Statistical Methods
HOSMER and LEMESHOW · Applied Logistic Regression
IMAN and CONOVER · Modern Business Statistics
JACKSON · A User's Guide to Principle Components
JESSEN · Statistical Survey Techniques
JOHN · Statistical Methods in Engineering and Quality Assurance
JOHNSON · Multivariate Statistical Simulation
JOHNSON and KOTZ · Distributions in Statistics
Discrete Distributions
Continuous Univariate Distributions—1
Continuous Univariate Distributions—2
Continuous Multivariate Distributions
JUDGE, GRIFFITHS, HILL, LÜTKEPOHL, and LEE · The Theory and Practice of Econometrics, *Second Edition*
JUDGE, HILL, GRIFFITHS, LÜTKEPOHL, and LEE · Introduction to the Theory and Practice of Econometrics, *Second Edition*
KALBFLEISCH and PRENTICE · The Statistical Analysis of Failure Time Data
KASPRZYK, DUNCAN, KALTON, and SINGH · Panel Surveys
KAUFMAN and ROUSSEEUW · Finding Groups in Data: An Introduction to Cluster Analysis
KEENEY and RAIFFA · Decisions with Multiple Objectives
KISH · Statistical Design for Research
KISH · Survey Sampling
KUH, NEESE, and HOLLINGER · Structural Sensitivity in Econometric Models
LAWLESS · Statistical Models and Methods for Lifetime Data
LEAMER · Specification Searches: Ad Hoc Inference with Nonexperimental Data
LEBART, MORINEAU, and WARWICK · Multivariate Descriptive Statistical Analysis: Correspondence Analysis and Related Techniques for Large Matrices
LEVY and LEMESHOW · Sampling of Populations: Methods and Applications
LINHART and ZUCCHINI · Model Selection
LITTLE and RUBIN · Statistical Analysis with Missing Data
McNEIL · Interactive Data Analysis
MAGNUS and NEUDECKER · Matrix Differential Calculus with Applications in Statistics and Econometrics
MAINDONALD · Statistical Computation
MALLOWS · Design, Data, and Analysis by Some Friends of Cuthbert Daniel
MANN, SCHAFER, and SINGPURWALLA · Methods for Statistical Analysis of Reliability and Life Data
MASON, GUNST, and HESS · Statistical Design and Analysis of Experiments with Applications to Engineering and Science
MILLER · Survival Analysis
MILLER, EFRON, BROWN, and MOSES · Biostatistics Casebook
MONTGOMERY and PECK · Introduction to Linear Regression Analysis, *Second Edition*
NELSON · Accelerated Testing, Statistical Models, Test Plans, and Data Analyses
NELSON · Applied Life Data Analysis
OCHI · Applied Probability and Stochastic Processes in Engineering and Physical Sciences
OSBORNE · Finite Algorithms in Optimization and Data Analysis
OTNES and ENOCHSON · Digital Time Series Analysis
PANKRATZ · Forecasting with Dynamic Regression Models
PANKRATZ · Forecasting with Univariate Box-Jenkins Models: Concepts and Cases
POLLOCK · The Algebra of Econometrics
RAO and MITRA · Generalized Inverse of Matrices and Its Applications
RÉNYI · A Diary on Information Theory
RIPLEY · Spatial Statistics
RIPLEY · Stochastic Simulation
ROSS · Introduction to Probability and Statistics for Engineers and Scientists
ROUSSEEUW and LEROY · Robust Regression and Outlier Detection
RUBIN · Multiple Imputation for Nonresponse in Surveys

*Now available in a lower priced paperback edition in the Wiley Classics Library.